SURROGATE TISSUE ANALYSIS

Genomic, Proteomic, and Metabolomic Approaches

SURROGATE TISSUE ANALYSIS

Genomic, Proteomic, and Metabolomic Approaches

Edited by

Michael E. Burczynski
John C. Rockett

Taylor & Francis
Taylor & Francis Group

Boca Raton London New York

A CRC title, part of the Taylor & Francis imprint, a member of the Taylor & Francis Group, the academic division of T&F Informa plc.

Published in 2006 by
CRC Press
Taylor & Francis Group
6000 Broken Sound Parkway NW, Suite 300
Boca Raton, FL 33487-2742

Printed in the United States of America on acid-free paper
10 9 8 7 6 5 4 3 2 1

International Standard Book Number-10: 0-8493-2840-3 (Hardcover)
International Standard Book Number-13: 978-0-8493-2840-4 (Hardcover)
Library of Congress Card Number 2005015679

Library of Congress Cataloging-in-Publication Data

Surrogate tissue analysis : genomic, proteomic and metabolomic approaches / edited by Michael E. Burczynski and John C. Rockett.
p. ; cm.
Includes bibliographical references and index.
ISBN-13: 978-0-8493-2840-4
ISBN-10: 0-8493-2840-3
1. Biochemical markers. 2. Genomics. 3. Proteomics.
[DNLM: 1. Biological Markers – Analysis. 2. Biological Markers – Blood. 3. Genetic Markers. 4. Genomics – Methods. 5. Metabolism. 6. Proteomics – Methods. QW 541 S962 2005] I. Burczynski, Michael E. II. Rockett, John C.

QH438.4.B55S87 2005
572.8—dc22 2005015679

informa
Taylor & Francis Group
is the Academic Division of Informa plc.

Visit the Taylor & Francis Web site at
http://www.taylorandfrancis.com

and the CRC Press Web site at
http://www.crcpress.com

To my lovely wife, Jennifer, and my son,
Michael William — M.E.B.

To those most dear to me — my loving wife, Gillian,
daughter Hannah Abigail,
and son Nathan David — J.C.R.

Preface

The "omic" revolution has spurred a variety of investigative techniques in a host of model systems. One of the many fields of biomedical inquiry that has benefited from the proliferation of high-throughput molecular screening methods has been the field of surrogate tissue analysis. The combination of "omic" technologies with surrogate tissue analysis has led to a rapid increase in the amount of data concerning levels of transcripts, proteins, metabolites, and other molecules present in surrogate tissues. Concomitant with this exponential increase in knowledge has been the simultaneous need to understand the relevance of these observations, and how they may be put to beneficial use.

Surrogate tissue analysis refers in general to the assessment of nontarget or off-target tissues in the body for biochemical, molecular, or cellular correlates or indicators. At its core, surrogate tissue analysis can lead to the identification of bona fide biomarkers with applications in drug discovery and development, toxicity and risk assessment, and even clinical patient management. The main attraction of surrogate tissue analysis lies in its obvious accessibility — the sampling of cerebral spinal fluid (CSF) to determine the effectiveness of a drug inhibiting neurodegeneration is eminently more feasible than the harvesting of a brain biopsy for the same purpose. Thus, it is in this manner that understanding molecular and cellular events in surrogate tissues in the context of disease, therapeutic intervention, and toxic exposure may ultimately provide the greatest benefit.

The present textbook, *Surrogate Tissue Analysis: Genomic, Proteomic, and Metabolomic Approaches*, represents a collection of chapters describing initial applications and considerations for "omic" technologies in the field of surrogate tissue analysis. The introductory chapter sets the stage for this field of inquiry and highlights some of the important issues to consider prior to conducting profiling studies in surrogate tissues. The next three sections of the textbook review specific advances in the field of genomic, proteomic, and metabolomic approaches in surrogate tissues.

In the first of these three sections, transcriptional profiling approaches in surrogate tissues are covered, and the preponderance of chapters focused on peripheral blood profiling provides hardy evidence that this field is rapidly spawning its own subspecialty — that of hemogenomics. Chapter 2 reviews the important considerations in peripheral blood profiling in great detail and summarizes results achieved when evaluations of various blood preparation platforms are used for the purpose of transcriptional profiling. Chapters 3 and 4 cover the relatively novel application of transcriptional profiling in neurological and oncological disease settings, respectively. Chapter 5 reviews the nature of surrogate tissue profiles of toxic exposure in preclinical studies where transcriptional effects in both target and surrogate tissues can be compared. Finally, Chapter 6 focuses on transcriptional profiling in a non-blood-based tissue, semen, which is utilized as a surrogate tissue for paternal exposure.

The next section focuses on proteomic and protein-based methods for identifying markers in surrogate tissues. Chapter 7 highlights mass spectrometry approaches for assessment of proteins in serum, with a focus on the obvious implications of protein-based biomarkers for detecting and monitoring early stages of cancer. Chapter 8 assesses the ability of circulating lymphocyte integrins to indicate endometrial receptivity, and Chapter 9 demonstrates how the surrogate tissue of nipple aspirate fluid can be used to detect and monitor breast cancer in afflicted patients.

The next section explores metabolomic approaches along with other novel molecular screens that can be applied in surrogate tissues for the purpose of finding biomarkers. Metabolomics is somewhat unique in that it is particularly suited to surrogate tissue analysis, since in contrast to most DNA, RNA, and intracellular proteins in the body, only metabolites (and secreted polypeptides) are freely found in surrogate tissues. Chapters 10 and 11 therefore review the field of metabolomics and how this technology is rapidly developing into a powerful technique for biomarker identification. Chapter 12 provides an excellent overview of a subfield of metabolomics, which focuses exclusively on the measurement of lipids and is appropriately termed lipidomics, and explores how the field of lipidomics can be used in surrogate tissues to provide an understanding of dynamic inflammatory responses in hosts. Chapter 13 reviews a PCR-based approach to detect and monitor metastatic cells in the circulation, and Chapter 14 covers a methylation profiling approach that can be used to accomplish a similar end.

The final section of the textbook attempts to look toward the horizon in more general terms, with chapters that focus on regulatory, economic, and pan-omic strategies, all of which will undoubtedly influence surrogate tissue analysis in the future. Chapter 15 summarizes generally applicable regulatory issues that will undoubtedly be important considerations for those biomarkers discovered in surrogate tissue profiling studies that support drug/co-diagnostic registration and require regulatory approval. Chapter 16 provides an esoteric and interesting evaluation of the value of profiling approaches to drug development in general; these sorts of economic analyses will prove of greater and greater value as the parameters affecting the "value" of biomarkers and profiling approaches become better understood. Chapter 17 reviews current concepts in pan-omic approaches during drug development where a compendium of data generated by multiple profiling approaches is assessed and evaluated—otherwise known, at least in part, as the holy grail of systems biology.

The last chapter provides a brief survey of findings in surrogate tissues that lie outside the covers of this textbook, summarizing important studies in this young field and looking to the future as well. It also discusses the burgeoning need for well-characterized and reproducible surrogate tissue analysis approaches as the requirement for biomarkers in the field of translational medicine is realized. One of the most exciting and simultaneously difficult characteristics of interpreting results from surrogate tissue profiling experiments today lies in the fact that there is often no precedent in the literature for the findings. Why do circulating peripheral blood mononuclear cells of renal cancer patients "look" different from those of healthy individuals at the transcriptional level? Are there clues to components of diseases that have been hitherto less explored — for instance, immunological responses of peripheral circulating cells to weakly immunogenic or nonimmunogenic solid tumors — and can this new knowledge be used to identify biomarkers of disease, but possibly to exploit mechanistically relevant pathways influencing disease progression by therapeutic intervention? These types of questions along with the constant efforts and the balance of innovative thinking with careful attention to details — both biological and technical — which are currently being exhibited by investigators in the field of surrogate tissue analysis would seem to ensure that this area of biomedical research will enjoy continued success in the years to come.

Editors

Dr. Michael E. Burczynski earned his Ph.D. in pharmacology from the University of Pennsylvania and is currently the head of Pharmacogenomics in the Biomarkers Laboratory at Wyeth Research in Collegeville, Pennsylvania. He is a member of the American Association of Cancer Research and Society of Toxicology and has authored more than 50 articles and abstracts, with some of his most recent articles appearing in *Cancer Research, Clinical Cancer Research,* and *Current Molecular Medicine*. He was the editor of *An Introduction to Toxicogenomics* published by CRC Press in 2003 and is also a published fiction author. He is currently working on his latest novel, tentatively entitled *The Orchard of Perdition*.

Dr. John C. Rockett earned his Ph.D. in biological sciences from the University of Warwick, England, and is currently a research fellow in the Preclinical Molecular Profiling group at Rosetta Inpharmatics (a wholly owned subsidiary of Merck & Co., Inc.) in Seattle, Washington. He is a past research fellow at the University of Surrey, England, and a research biologist with the U.S. Environmental Protection Agency in Triangle Park, North Carolina. He is a member of the Institute of Biology and the Society of Toxicology and has published more than 70 articles and abstracts in various scientific journals, most recently in *Biology of Reproduction, Environmental Health Perspectives*, *Genomics*, *Toxicological Sciences,* and *Toxicology and Applied Pharmacology*.

Contributors

Hikmat Al-Ahmadie
Department of Pathology
Memorial Sloan-Kettering Cancer Center
New York, New York

Satyajit Bhattacharya
Department of Pathology
Memorial Sloan-Kettering Cancer Center
New York, New York

Michael E. Burczynski
Molecular Profiling and Biomarker Discovery
Wyeth Research
Collegeville, Pennsylvania

Monica J. Cahilly
Green Mountain Quality Associates
Warren, Vermont

Katherine R. Calvo
FDA-NCI Clinical Proteomics Program
Laboratory of Pathology
National Cancer Institute
Bethesda, Maryland

Clary B. Clish
Beyond Genomics
Waltham, Massachusetts

Jennifer L. Colangelo
Pfizer Global Research and Development
Safety Sciences
Groton, Connecticut

Svenja Debey
Molecular Tumor Biology and Tumor Immunology
Clinic I for Internal Medicine
University of Cologne
Cologne, Germany

Stan N. Finkelstein
Program on the Phamaceutical Industry
Massachusetts Institute of Technology
Cambridge, Massachusetts

Ronald A. Ghossein
Department of Pathology
Memorial Sloan-Kettering Cancer Center
New York, New York

Donald L. Gilbert
Division of Neurology
Cincinnati Children's Hospital Medical Center
Cincinnati, Ohio

Tracy A. Glauser
Division of Neurology
Cincinnati Children's Hospital Medical Center
Cincinnati, Ohio

Julian L. Griffin
Department of Biochemistry
University of Cambridge
Cambridge, United Kingdom

S.M. Gupta
Immunology Division
National Institute for Research in Reproductive Health
Indian Council of Medical Research
Parel, Mumbai, India

Andrew D. Hershey
Division of Neurology
Cincinnati Children's Hospital Medical Center
Cincinnati, Ohio

Stephen A. Krawetz
Department of Obstetrics and Gynecology
Center for Molecular Medicine and Genetics
Institute for Scientific Computing
Wayne State University
Detroit, Michigan

Michael P. Lawton
Pfizer Global Research and Development
Safety Sciences
Groton, Connecticut

Lance A. Liotta
FDA-NCI Clinical Proteomics Program
Office of Cell Therapy and Gene Therapy
Food and Drug Administration
Bethesda, Maryland

Aigang Lu
MIND Institute
University of California at Davis
Sacramento, California

P. K. Meherji
Immunology Division
National Institute for Research in Reproductive Health
Indian Council of Medical Research
Parel, Mumbai, India

Deborah P. Mounts
Bioinformatics Systems Development
Wyeth Research
Cambridge, Massachusetts

Judith L. Oestreicher
Molecular Profiling and Biomarker Discovery
Wyeth Research
Cambridge, Massachusetts

G. Charles Ostermeier
Department of Obstetrics and Gynecology
Center for Molecular Medicine and Genetics
Wayne State University
Detroit, Michigan

William D. Pennie
Pfizer Global Research and Development
Safety Sciences
Groton, Connecticut

Emanuel F. Petricoin III
FDA-NCI Clinical Proteomics Program
Office of Cell Therapy and Gene Therapy
Food and Drug Administration
Bethesda, Maryland

Raji Pillai
Genomics Collaborations
Affymetrix, Inc.
Santa Clara, California

Ruiqiong Ran
MIND Institute
University of California at Davis
Sacramento, California

K.V.R. Reddy
Immunology Division
National Institute for Research in Reproductive Health
Indian Council of Medical Research
Parel, Mumbai, India

Shawn Ritchie
Phenomenome Discoveries
Saskatoon, Saskatchewan, Canada

John C. Rockett
Preclinical Molecular Profiling
Rosetta Inpharmatics
Seattle, Washington

Edward Sauter
University of Missouri
Columbia, Missouri

Joachim L. Schultze
Molecular Tumor Biology and Tumor Immunology
Clinic I for Internal Medicine
University of Cologne
Cologne, Germany

Charles N. Serhan
Center for Experimental Therapeutics and Reperfusion Injury
Department of Anesthesiology, Perioperative and Pain Medicine
Brigham and Women's Hospital and Harvard Medical School
Boston, Massachusetts

Frank R. Sharp
MIND Institute
University of California at Davis
Sacramento, California

Anthony J. Sinskey
Department of Biology, Health Sciences and Technology and Program on the Pharmaceutical Industry
Massachusetts Institute of Technology
Cambridge, Massachusetts

Lisa A. Speicher
Translational Research
Wyeth Research
Collegeville, Pennsylvania

Sarah C. Stallings
Massachusetts Institute of Technology
Cambridge, Massachusetts

Yang Tang
MIND Institute
University of California at Davis
Sacramento, California

William L. Trepicchio
Clinical Pharmacogenomics
Millennium Pharmaceuticals
Cambridge, Massachusetts

Nigel J. Waters
DMPK Research
AstraZeneca Research and Development
Charnwood, Loughborough, United Kingdom

Maryann Z. Whitley
Expression Profiling Informatics
Wyeth Research
Cambridge, Massachusetts

Ivy H.N. Wong
Department of Obstetrics and Gynaecology
The Chinese University of Hong Kong
Hong Kong, China

Julia Wulfkuhle
FDA-NCI Clinical Proteomics Program
Laboratory of Pathology
National Cancer Institute
Bethesda, Maryland

Huichun Xu
MIND Institute
University of California at Davis
Sacramento, California

Thomas Zander
Molecular Tumor Biology and Tumor Immunology
Clinic I for Internal Medicine
University of Cologne
Cologne, Germany

Acknowledgments

The editors specifically thank the authors of the individual chapters in this textbook for their excellence in science and their dedication to the project.

I first thank Dr. Rockett for his initial suggestion of a book centered on this exciting topic to which we've both grown attached. I thank the members of my own laboratory, past and present, who have contributed to the ongoing field of surrogate tissue analysis — specifically Judy Oestreicher, Jennifer Stover, Natalie Twine, Krystyna Zuberek, and Christine Reilly — all of whom have helped make tremendous strides in understanding the power of surrogate tissue profiling in human disease. I also thank my many collaborators and colleagues at Wyeth Research in the departments of Biological Technologies, Translational Research, Translational Development, and Clinical Research and Development who are too numerous to mention but have made tremendous strides in this field possible. Specifically, I must thank Dr. Andy Dorner, Dr. Ron Salerno, and Dr. John Ryan for their unwavering commitment to pharmacogenomics and their guidance and support of Wyeth's endeavors in the field of surrogate tissue analysis. Most importantly I thank the patients, whose selfless contributions of samples to increase our understanding of disease make the clinically oriented investigations into the field of surrogate tissue analysis possible in the first place. — M.E.B.

I first of all thank Dr. Burczynski for initiating this project and showing me the ropes on my first foray into book editing. I also owe a debt of gratitude to the many dedicated, knowledgeable, and able colleagues, past and present, who have contributed practically and mentally to my scientific development and experience; in particular I thank David Dix, Sally Darney, and Bob Kavlock, whose support and encouragement were instrumental in initiating and advancing my interest and research in surrogate tissue analysis. — J.C.R.

Contents

Section I

Introduction to Surrogate Tissue Analysis

CHAPTER 1

Introduction to Surrogate Tissue Analysis

John C. Rockett and Michael E. Burczynski

CONTENTS

1.1 INTRODUCTION

Postgenomic technologies, including those used to analyze genomic, transcriptomic, proteomic, metabonomic, and other "omic" targets, have made it possible to define molecular physiology in exquisite detail, when tissues are accessible for sampling. However, many target tissues are not accessible for human experimental or epidemiological studies, or clinical evaluations, creating the need for surrogates that afford insight into exposures and effects in such tissues. A "surrogate" can be defined simply as "one that takes the place of another." In surrogate tissue analysis (STA), one tissue takes the place of another. More specifically, an accessible tissue takes the place of an inaccessible target tissue. For example, one might examine a patient's peripheral blood lymphocytes (PBLs) to determine whether that person has suboptimal endometrial receptivity (Chapter 8), has suffered from neurological damage (Chapter 3), has developed a nonlymphatic neoplasm (Chapter 4), or has been exposed to a toxicant (Chapter 5). An alternative STA paradigm is to measure or analyze parts or products of a target tissue that originate from the target, but are collected or measured distal to it, in the surrogate tissue. For example, it is possible to isolate and analyze sperm from semen and use the data to help understand molecular events occurring in the testis (Chapter 6). In a similar manner, peripheral blood can be a source of circulating tumor cells that have detached or have been shed from their parent neoplasm. These can be isolated and used as a source of information about the original neoplasm (Chapters 13 and 14). In other cases soluble proteins, metabolites, or lipids are secreted or excreted from target tissues, and these can be detected and measured in fluids such as blood (Chapters 7, 11, and 12), cerebrospinal fluid (Chapter 11), nipple aspirate (Chapter 9), seminal fluid, milk, saliva, and urine. Drugs, drug metabolites, and toxicants can also be detected in such fluids (Chapters 10 and 11).

Surrogate "tissue" is a convenient, though perhaps misleading term. Where "tissue" is specified, the term is in fact used broadly to refer to any biologically derived material (biospecimen) used to report on events in a specific target tissue. Indeed, the majority of samples that offer potential application in STA are usually not considered tissues according to traditional definitions. The majority of surrogate tissues (Table 1.1) consist of either body fluids (e.g., urine, milk, tears, saliva, blood, and semen), or populations of cells extracted from body fluids (e.g., epithelial cells from urine, milk, or tears; lymphocytes from blood; sperm from semen), while some (e.g., hair follicle, hair, nail) are neither tissue nor free cells.

1.2 AREAS THAT COULD BENEFIT FROM SURROGATE TISSUE ANALYSIS

STA is not a new concept. Indeed, evidence that accessible tissues can be used to monitor events in an inaccessible tissue has been around for many years. For example, Nesnow et al. (1993) showed that the DNA adduct formation, a potential method of measuring exposure to environmental genotoxicants, exhibited a similar pattern in rat PBLs, lung, and liver following exposure to polycyclic hydrocarbons, and that this was detectable at least 56 days after treatment.

The development of "omic" technologies has led many researchers to look again, or more closely, or anew at the utility and application of STA, since such technologies have broadened both the range of tissues that can be examined and the number of targets that can be analyzed in a single experiment. In particular, there is widespread interest in how STA might be developed into a new paradigm for monitoring human health. The potential benefits include:

Table 1.1 Accessible "Tissues" That Can Potentially Be Used as Surrogate Tissues

Surrogate Tissue	Targets for Analysis	Potential Source
Blood	Cells, DNA, RNA, protein, drug metabolites, heavy metals	All
Breath condensate	Proteins, metabolites	All
Bronchial lavage	Cells, DNA, RNA, protein	All
Buccal cells	Cells, DNA, RNA, protein	All
Cord blood	Cells, DNA, RNA, protein	Postpartum females
Colostrum	DNA, RNA, protein	Postpartum females
Cerebrospinal fluid	Protein	All
Cerumen (earwax)	Protein	All
Hair shaft	DNA, protein, heavy metals, drug metabolites	All
Hair follicle	Cells, DNA, RNA, protein	All
Meconium	DNA, RNA, protein	Newborn infants
Milk	Cells, DNA, RNA, protein	Postpartum females
Nail	DNA, protein, heavy metals, drug metabolites	All
Nasal lavage	DNA, RNA, protein	All
Nipple aspirate	Cells, DNA, RNA, protein	All
Placenta	Cells, DNA, RNA, protein	Postpartum females
Saliva	DNA, RNA, protein	All
Semen	Cells, DNA, RNA, protein	Adult males
Skin	Cells, DNA, RNA, protein	All
Sputum	Cells, DNA, RNA, protein	All
Stool	DNA, RNA, protein	All
Sweat	Protein	All
Tear duct secretions	DNA, RNA, protein	All
Endocervical epithelium	DNA, RNA, protein	Adult females
Vaginal epithelium	DNA, RNA, protein	Adult females
Urine	DNA, RNA, protein, drug metabolites, heavy metals	All

Source: Adapted from Rockett, 2002.

1. The ability to monitor for and measure toxicant exposure without foreknowledge of the type of exposure
2. The ability to monitor clinically healthy internal organs at the molecular level without directly sampling those organs
3. The ability to identify possible pathological events at the preclinical stage and therefore administer preventative action
4. If disease is already apparent, an ability to identify the specific type and stage without invasive biopsy
5. The ability to determine which drug regimens offer the best chance of success in treating a specific disease
6. The ability to determine if a drug is working according to its proposed mechanism of action

These benefits fall into three broad areas: monitoring toxicant exposure and effect, monitoring disease development and progression, and drug efficacy testing.

1.2.1 Monitoring Toxicant Exposure and Effect

Toxicogenomics is a postgenomic approach to toxicology that uses primarily genomic techniques to elucidate mechanisms of toxicant action by studying the genome-wide effects of xenobiotics. One of the primary tenets of toxicogenomics is that the effects of toxicants on cellular functions are mediated through gene expression changes, or at least cause gene changes to occur as secondary effects. In most cases these gene changes occur prior to clinical manifestation of toxicity, which provides a window of opportunity for preclinical diagnosis of possible toxic end points that may arise as a result of the exposure. Such a diagnosis would employ the use of gene expression profiling (GEP), either on a global or restricted scale. GEP offers the potential to classify toxicant exposures (Burczynski et al., 2000; Bartosiewicz et al., 2001; Thomas et al., 2001; Hamadeh et al., 2002a, 2002b), predict clinical outcome of such exposures (Waring et al., 2001a; Hamadeh et al., 2002c), and provide mechanistic data useful for risk assessments (Waring et al., 2001b). Recent studies have also demonstrated that early gene expression changes can predict a pathological outcome days in advance of its occurrence (Kier et al., 2004). Consequently, GEP may eventually provide a vehicle for developing exposure, diagnostic, and prognostic tests for at-risk populations or individuals.

However, using GEP to monitor for toxicant exposure and/or effect in an inaccessible tissue is a difficult prospect, since direct biopsy of such tissue is not feasible unless strong medical reason (usually indicated by clinical symptoms) dictates otherwise. A less invasive method must therefore be developed if monitoring programs are to be developed based on this toxicogenomic approach. One possible solution is the use of STA. It has been proposed that gene expression changes in accessible (surrogate) tissues (e.g., nucleated blood cells) often reflect those in inaccessible (target) tissues, thus offering a convenient biomonitoring method to provide insight into the effects of environmental toxicants on target tissues (Rockett, 2002). This subject is discussed in more detail in Chapter 5.

1.2.2 Monitoring Disease Development and Progression

One of the most intriguing concepts to have recently evolved in the field of clinical pharmacogenomics is the possibility that surrogate tissues (often the circulating cells of the peripheral blood) may contain transcriptional profiles that correlate with disease, disease status, or other clinical measures of outcome in human patients. Currently in the field of oncology it is unknown whether, in the context of solid tumor burden, such "analogous" transcriptional profiles in surrogate tissues exist. While alterations in transcriptional profiles of PBMCs of patients with cancer may not share identity with those observed in the primary tumor, such patterns would nonetheless be of tremendous physiological relevance and bear obvious diagnostic value in the assessment of this disease.

1.2.3 Drug Efficacy Testing

STA has also been used in clinical pharmacology, whereby pharmacodynamic assays are being developed for the measurement of drug action in tumor and surrogate tissue. The need to demonstrate that a drug is working according to its proposed mechanism is of paramount importance. Researchers at places such as the CRC in London (http://www.icr.ac.uk) are trying to determine whether such studies may be able to utilize PBLs as surrogate tissue by comparing gene expression changes in PBLs with those in cancer biopsies following administration of test drugs (http://www.icr.ac.uk/cctherap/clinical.htm). Gene expression profiling of blood has also been used to differentiate patients who respond to a drug treatment from those who do not, thus providing a mechanism for the early determination of drug efficacy. In this way, should a certain disease prove refractory to a prescribed drug, the lack of efficacy of that drug can be determined at an earlier stage than would otherwise be the case. This increases the chance of patient survival since an alternative drug regimen or treatment method can be given at an earlier stage. Examples of these and related uses of surrogate tissues in clinical pharmacology are found throughout the present text.

1.3 CHALLENGES TO THE USE OF SURROGATE TISSUES

Although there have been some promising studies in the area of STA, like all new methods and approaches there are likely to be a number of challenges to overcome before it can be determined where and when STA is both applicable and appropriate. Some challenges that have been identified so far include specimen collection, specimen availability, specimen contamination, specimen homogeneity, specimen suitability, specimen specificity, and data interpretation.

1.3.1 Specimen Collection

The biological specimens that might be used in human STA are listed in Table 1.1. With such a varied selection of samples available, one of the first challenges is to develop appropriate methods for collection, storage, and transportation of tissues at and between sites of collection and analysis. "Appropriate" means that:

1. Sufficient specimen must be collected to enable extraction of reasonable amounts of good quality target material.
2. The collection, transportation, and storage procedures must not permit degradation of the target biomolecules. For example, RNA (used for gene expression analysis) is notoriously quick to degrade in *ex vivo* samples and must be protected in such a way as to inhibit the activity of RNAses. Chapter 2 discusses this issue in depth from the perspective of blood collection for genomic analysis.
3. To obtain an accurate profile from a subject or experimental animal at the time of specimen collection, the population of RNAs (the "transcriptome") or proteins (the "proteome") or other "ome" under investigation in a specimen must not change between collection of the specimen and extraction of the target biomolecules (RNA, protein, etc.) from the specimen in the laboratory.

Actual measurement of the level of individual biomolecules, be they members of the transcriptome, proteome, metabonome, lipidome, or other "ome," can be achieved in a number of ways. However, many of the newer techniques are not yet fully validated. For example, where the use of DNA arrays is concerned, many accessible tissues provide only small amounts of sample, yielding only small amounts of RNA. To overcome this, protocols have been developed that incorporate RNA amplification steps prior to labeling and hybridization of the sample. Fink et al. (2002) used this approach successfully in carrying out microarray analysis of RNA extracted from laser capture microdissection samples. However, the reliability of array data from amplified RNA samples has yet to be fully determined. In addition, the accuracy and reliability of much of the published microarray data are still a matter of open debate, and the methods for assuring data quality are not well established (Chipping Forecast II, 2002).

1.3.2 Specimen Availability

In some cases, a potential surrogate tissue may be useful only at certain times. For example, human hair follicles exist in several different growing states, with the majority (80%) in anaphase (actively growing). These are the best for RNA extraction. In cataphase, the hair follicles are moribund, and are consequently much smaller and yield correspondingly small quantities of RNA. In other cases a potential surrogate tissue may only be available from certain populations (e.g., sperm from adult males) or at certain times (e.g., placental tissue and cord blood from postpartum females, and milk from lactating females). Another factor that might occasionally limit availability of samples is cultural, religious, or personal beliefs that prohibit the provision of certain biospecimens, most notably blood or semen.

1.3.3 Specimen Contamination

The issue of contamination must also be addressed where many surrogate tissues are concerned. This arises from the fact that since many of them are externally accessible, they may be contaminated with nonhuman biological material, including bacteria, viruses, and fungi. Stool, nail, and saliva are perhaps the best example of this.

1.3.4 Specimen Homogeneity

Many surrogate tissues are homogeneous, in that they are composed of a number of different components, including fluid (e.g., serum in blood and seminal fluid in semen) and different populations of cells (e.g., leukocytes and erythrocytes in blood and leukocytes, epithelial cells, and spermatozoa in semen). It may be necessary (depending on the cell population being sought after or the "omic" technique being used) to selectively remove or separate specific cell populations from the surrogate tissue specimen. This can be done using magnetic beads or fluorescence-activated cell sorting (FACS) if appropriate antibodies are available to cell-specific antigens, by using separation gradients, e.g., Ficoll and Percoll (Amersham Biosciences), or by using selective lysis. In the isolation of sperm from semen, for example, a wash step is included, which lyses somatic cells (epithelial and inflammatory), leaving the highly resistant sperm cells intact (see Chapter 6).

1.3.5 Specimen Suitability

Surrogate tissues vary in the types of analysis that can be carried out on them. For example, DNA can be obtained from nail and hair (Tanigawara et al., 2001), but these tissues do not yield RNA. Hair follicles, on the other hand, are a good source of RNA, and work published by Mitsui et al. (1997) indicates that as much as 900 ng of total RNA can be extracted from a single human hair follicle. Buccal cells yield both DNA and RNA. Unfortunately, since these particular cells, which are obtained by swabbing the inside cheek, are typically moribund, the RNA obtained from them is not of sufficiently good quality to use on arrays, although it has been used for reverse transcriptase-polymerase chain reaction (RT-PCR) (Smith et al., 1996).

1.3.6 Specimen Specificity

Another issue is that in some cases toxicant action can be very specific, and there may be no appropriate surrogate tissue. In other cases, certain surrogate tissues may be more useful than others depending on the target tissue being studied. For example, sperm is likely to be the best surrogate tissue for monitoring events occurring in the testis (Ostermeier et al., 2002), whereas intuitively one could reasonably hypothesize that PBLs are probably most useful as surrogates for thymus, spleen, tonsils, bone marrow, or glandular tissues. Indeed, when Ember et al. (2000) compared Ha-ras and p53 expression in PBLs with several target tissues (lung, liver, lymph nodes, kidneys, spleen) following exposure to a carcinogenic agent, similar expression patterns were found only in PBLs and spleen. Thus, some appropriate matching of targets and surrogate tissues is called for. Of course, there may have been many other genes that did correlate in these studies but were not analyzed. Therefore, the ability to monitor expression of many thousands of genes or proteins in one experiment, as permitted by DNA or protein arrays, makes the application of such technology to STA highly desirable.

1.3.7 Data Interpretation

Perhaps the greatest challenge of all will be the interpretation and appropriate utilization of all the "omic" and other data obtained from target and surrogate tissues. Validating the relationship between gene expression or protein profiles and toxicant exposure or disease state has already begun. If and when these relationships have been fully verified in target tissues, then the relationship between gene or protein expression in target and surrogate tissues must be established. In doing so, it will be necessary to determine whether genetic or proteomic biomarkers of toxicity or disease in target tissues are reflected in the surrogate tissue across a range of doses, time points, and disease states. Alternatively, omic biomarkers in surrogate tissues may be of high clinical value but fail to share identity with markers in the primary tissue. For example, one such scenario might involve the transcriptional response of circulating peripheral blood leukocytes due to tumor regression induced by successful chemotherapy, in which the transcriptional responses of PBMCs accurately "predict" beneficial tumor response.

One of the best hopes for successful utilization of STA lies in identifying unique biomarkers (e.g., changes in expression of a single gene/protein or a small number of such genes/proteins) of exposure and effect that show concordant modulation in surrogate and target tissues following toxicant exposure. What is ideally needed to utilize such biomarkers is one or more large relational databases through which newly generated data can be compared against previously documented toxicant exposures and effects. This would facilitate diagnosis of the type and likely outcome of any particular exposure. Of course, gene and/or protein expression levels alone may be insufficient to make an accurate diagnosis or prognosis. Other factors, such as the presence of polymorphisms in drug metabolizing and detoxifying enzymes, may need to be incorporated to improve the reliability and accuracy of this approach. Until such a time, perhaps a decade or more away, when such databases are available, it will in most cases be an enormous challenge to interpret the biological meaning and significance of the data.

1.4 SUMMARY

Surrogate tissue analysis is currently a relatively small but rapidly growing area of research. The ability to investigate biological mechanisms and obtain diagnostic and prognostic information about an inaccessible target tissue by using accessible surrogate tissues and fluids has significant and far reaching implications for health care as well as basic and clinical research. Initial proof-of-principal experiments in humans and animal models, many of which are described in this text, have provided encouraging results that suggest that STA can be applied in a large number of different scenarios. Whether STA becomes an integral component of future human health monitoring programs, a tool of limited situation-specific use, or a dead end idea, will be determined only after further studies have been conducted. However, the future of STA appears to be linked quite closely with the advancement of omic technologies, and given the large and widespread investment in these, further advances in the utility and application of STA seem quite likely.

REFERENCES

Bartosiewicz, M., Penn, S., and Buckpitt, A., 2001. Applications of gene arrays in environmental toxicology: fingerprints of gene regulation associated with cadmium chloride, benzo(*a*)pyrene, and trichloroethylene. *Environ. Health Perspect.* 109, 71–74.

Burczynski, M.E., McMillian, M., Ciervo, J., Li, L., Parker, J.B., Dunn, R.T., II, Hicken, S., Farr, S., and Johnson, M.D., 2000. Toxicogenomics-based discrimination of toxic mechanism in HepG2 human hepatoma cells. *Toxicol. Sci.* 58(2), 399–415.

Chipping Forecast II, 2002. *Nat. Gen. Suppl.,* Vol. 32, December.

Ember, I., Kiss, I., Gyongyi, Z., and Varga, C.S., 2000. Comparison of early onco/suppressor gene expressions in peripheral leukocytes and potential target organs of rats exposed to the carcinogen 1-nitropyrene. *Eur. J. Cancer Prev.* 9, 439–442.

Fink, L., Kohlhoff, S., Stein, M.M., Hanze, J., Weissmann, N., Rose, F., Akkayagil, E., Manz, D., Grimminger, F., Seeger, W., and Bohle, R.M., 2002. cDNA array hybridization after laser-assisted microdissection from nonneoplastic tissue. *Am. J. Pathol.* 160, 81–90.

Hamadeh, H.K., Bushel, P.R., Jayadev, S., DiSorbo, O., Bennett, L., Li, L., Tennant, R., Stoll, R., Barrett, J.C., Paules, R.S., Blanchard, K., and Afshari, C.A., 2002a. Prediction of compound signature using high density gene expression profiling. *Toxicol. Sci.* 67(2), 232–240.

Hamadeh, H.K., Bushel, P.R., Jayadev, S., Martin, K., DiSorbo, O., Sieber, S., Bennett, L., Tennant, R., Stoll, R., Barrett, J.C., Blanchard, K., Paules, R.S., and Afshari, C.A., 2002b. Gene expression analysis reveals chemical-specific profiles. *Toxicol. Sci.* 67(2), 219–231.

Hamadeh, H.K., Knight, B.L., Haugen, A.C., Sieber, S., Amin, R.P., Bushel, P.R., Stoll, R., Blanchard, K., Jayadev, S., Tennant, R.W., Cunningham, M.L., Afshari, C.A., and Paules, R.S., 2002c. Methapyrilene toxicity: anchorage of pathologic observations to gene expression alterations. *Toxicol. Pathol.* 30(4), 470–482.

Kier, L.D., Neft, R., Tang, L., Suizu, R., Cook, T., Onsurez, K., Tiegler, K., Sakai, Y., Ortiz, M., Nolan, T., Sankar, U., and Li, A.P., 2004. Applications of microarrays with toxicologically relevant genes (tox genes) for the evaluation of chemical toxicants in Sprague Dawley rats *in vivo* and human hepatocytes *in vitro. Mutat. Res.* 549, 101–113.

Mitsui, S., Ohuchi, A., Hotta, M., Tsuboi, R., and Ogawa, H., 1997. Genes for a range of growth factors and cyclin-dependent kinase inhibitors are expressed by isolated human hair follicles. *Br. J. Dermatol.* 137(5), 693–698.

Nesnow, S., Ross, J., Nelson, G., Holden, K., Erexson, G., Kligerman, A., and Gupta, R.C., 1993. Quantitative and temporal relationships between DNA adduct formation in target and surrogate tissues: implications for biomonitoring. *Environ. Health Perspect.* 101(Suppl. 3), 37–42.

Ostermeier, G.C., Dix, D.J., Miller, D., Khatri, P., and Krawetz, S.A., 2002. Spermatozoal RNA profiles of normal fertile men. *Lancet* 360, 772–777.

Rockett, J.C., Surrogate tissue analysis for monitoring the degree and impact of exposures in agricultural workers. *AgBiotechnet* 4, 1–7.

Smith, J.K., Chi, D.S., Krishnaswamy, G., Srikanth, S., Reynolds, S., and Berk, S.L., 1996. Effect of interferon alpha on HLA-DR expression by human buccal epithelial cells. *Arch. Immunol. Ther. Exp.* (Warsz) 44, 83–88.

Tanigawara, Y., Kita, T., Hirono, M., Sakaeda, T., Komada, F., and Okumura, K., 2001. Identification of *N*-acetyltransferase 2 and CYP2C19 genotypes for hair, buccal cell swabs, or fingernails compared with blood. *Ther. Drug Monitoring* 23, 341–346.

Thomas, R.S., Rank, D.R., Penn, S.G., Zastrow, G.M., Hayes, K.R., Pande, K., Glover, E., Silander, T., Craven, M.W., Reddy, J.K., Jovanovich, S.B., and Bradfield, C.A., 2001. Identification of toxicologically predictive gene sets using cDNA microarrays. *Mol. Pharmacol.* 60(6), 1189–1194.

Waring, J.F., Jolly, R.A., Ciurlionis, R., Lum , P.Y., Praestgaard, J.T., Morfitt, D.C., Buratto, B., Roberts, C., Schadt, E., and Ulrich, R.G., 2001a. Clustering of hepatotoxins based on mechanism of toxicity using gene expression profiles. *Toxicol. Appl. Pharmacol.* 175(1), 28–42.

Waring, J.F., Ciurlionis, R., Jolly, R.A., Heindel, M., and Ulrich, R.G., 2001b. Microarray analysis of hepatotoxins *in vitro* reveals a correlation between gene expression profiles and mechanisms of toxicity. *Toxicol. Lett.* 120(1–3), 359–368.

Section II

Genomic Approaches

CHAPTER 2

Impact of Sample Handling and Preparation on Gene Signatures as Exemplified for Transcriptome Analysis of Peripheral Blood

Joachim L. Schultze, Svenja Debey, Raji Pillai, and Thomas Zander

CONTENTS

2.1 INTRODUCTION

Over the last decade, genomic technologies have revolutionized the way we think about disease, diagnostics, and prognosis — and this is only the beginning. The power of many of the landmark studies applying genomic technologies to clinical and health care questions has been breathtaking. However, as with almost every technological development, initial euphoria needs to be followed by vigorous implementation of clinically applicable standards, allowing those new technologies to become part of a medical routine. Although such standards have been appreciated for nearly every classical test in medicine, the critical importance of standardization for genomic technologies assessing hundreds to thousands of genes simultaneously is unprecedented to date. Every aspect of the complex procedures involved in genomic-based assays needs to be carefully assessed concerning standardization, including sample processing, isolation of RNA or DNA, microarray hybridization, and bioinformatic analysis. National and international consortia have been assembled to standardize experimental procedures and bioinformatics, but only recently have researchers started to systematically assess the impact of sample source and processing on the final results in surrogate tissue–based studies. In this chapter, this important issue is discussed for peripheral blood, which will probably constitute the most important tissue source for diagnostic and prognostic assessment of drug effects and disease.

It is apparent that genomic technologies will revolutionize and dramatically change modern medicine. Massive parallel analysis of gene expression has already significantly improved our understanding of complex diseases like cancer. Transcriptome and proteome analyses have been applied to many aspects of human biology, e.g., identification of signaling cascades[1–4] and regulated expression of cell cycle associated genes,[5,6] or description of stress responses of human cells.[7–9] In clinical and translational research studies, gene signatures have been applied to better define biological processes associated with disease and therapeutic responses or severe adverse events due to therapeutic intervention. As exemplified in clinical cancer research, gene signatures have been used to understand the basic mechanism of cancer biology[3,6,10,11] or metastasis,[12,13] to describe diagnostically relevant gene clusters serving as future biomarkers for disease,[14] to develop a comprehensive molecular nomenclature of cancer diseases,[15–17] and to stratify patients' therapy based on the identification of distinguished gene signatures associated with good or bad prognosis.[18–21] While most gene expression profiling studies have been conducted on tissue samples, it has been recently appreciated that peripheral blood is a highly accessible biospecimen that might be used to address important questions concerning diagnosis and prognosis of disease, therapeutic efficacy or identification of patients at risk for severe adverse events derived from therapy. Indeed, gene signatures of peripheral blood mononuclear cells (PBMCs) have already been used to determine variation of expression in healthy individuals,[22] to assess differences between patients with cancer and healthy individuals,[23] to determine underlying mechanisms of diseases, e.g., systemic lupus erythematosus (SLE),[24,25] and to investigate the influence of bacteria on gene expression patterns.[26] While it was clearly demonstrated that interindividual differences in gene expression exist in healthy individuals,[22] the differences observed for PBMC derived from healthy individuals and patients with autoimmune diseases,[24,25] patients suffering from bacterial infection,[26] or renal cell cancer[23] were clearly more pronounced. These findings strongly support the further evaluation of peripheral blood for diagnosis of systemic diseases and/or for monitoring drug effects.[27] In addition to genetic or metabolic disorders, diseases associated with dysregulated immunity (such as cancer and autoimmune diseases) are characterized by changes in the cellular compartment of peripheral blood. It therefore comes as no surprise that such changes are also reflected on a molecular level as determined by transcriptome or proteome analysis. Similar to current biochemical tests such as the assessment of liver enzymes in peripheral blood, we and others have proposed that disease processes outside the bloodstream still can be assessed by gene signatures within the peripheral blood. As with solid tissue sample processing, it is critical to develop appropriate standard operating procedures for the use with blood specimens to move from an exploratory research phase to one of clinical applicability.[28]

2.2 APPLICATION OF STANDARDS TO GENOMIC TECHNOLOGIES

Based on the exciting preliminary discoveries in the field of surrogate tissue profiling, it is appropriate to start addressing more translational questions such as clinical applicability, robustness, specificity, and sensitivity of the new techniques. Although initial studies in several areas have demonstrated the superior diagnostic value of gene signatures over classical approaches of combined clinical, biochemical, and imaging diagnostics,[29] the true value of genomics will be appreciated only if it can be applied easily and inexpensively to day-by-day medical routine. Several international consortia such as the Tumor Analysis Best Practices Working Group[30] and the Lymphoma/Leukemia Molecular Profiling Project[16] have now begun to develop standard procedures for clinical use. Standardization will be required for each level throughout the process, including sample handling, cell and RNA processing, cRNA preparation, scanning, data acquisition and storage, as well as the sophisticated data analysis required when analyzing thousands of genes in parallel. With the introduction of guidelines for the reporting and annotation of microarray data from the Microarray Gene Expression Data (MGED) Society[31] — including the so-called Minimum Information about a Microarray Experiment (MIAME) standard[32] and the MAGE-ML mark-up language[33] — a first step toward standardization of data reporting has been made in an international academic-industry partnership. Thus, while many experimental procedures subsequent to RNA isolation have been standardized for the major transcriptome analysis platforms, surprisingly little agreement has been reached concerning the procurement, transport, and processing of blood and other surrogate tissues.

There is widespread acceptance of the necessity of standardized sample preparation and rapid RNA isolation. The importance of standardization of sample biopsy procedures, procurement, transport, storage, and cell isolation is discussed here in more detail using peripheral blood–based transcriptome analysis as an example. It is important to note that similar considerations have to be made for any other tissue sample prior to genomics analysis. While additional aspects need to be dealt with in proteome analysis (see Chapters 10 and 11), the principal approach to developing clinically suitable tools, methodology, and applications is similar.

To date, the impact of cell and RNA isolation procedures as well as physical influences on clinical specimens have not been systemically addressed until recently, particularly for peripheral blood.[34] The impact of such influences, as shown here, was surprisingly pronounced.[34] Because large-scale clinical trials utilize multiple centers, simple and standardized sample preparation procedures are required for large-scale clinical investigation and practice.[28] It is important to understand the limits and caveats associated with the variety of approaches available for surrogate tissue and blood handling. The recent adaptation of RNA stabilization techniques (see next section) is just one example of the type of innovation that may ultimately be suitable for microarray-based analyses in peripheral blood.[35]

2.3 DIFFERENT CELL AND RNA PREPARATION METHODS FROM WHOLE BLOOD: AN INTRODUCTION

Many different techniques are available to process blood samples, to separate subsets of specific cells from whole blood, or to prepare RNA. Here we give an introduction to the principles of commonly used cell and RNA preparation methods including the PAXgene™ Blood RNA Isolation System (PreAnalytiX, GmbH, Hombrechtikon, Switzerland), QIAamp® RNA Blood Mini Kit (QIAGEN, Hilden, Germany), classical Ficoll-Hypaque protocols, and the VACUTAINER® CPT™ Cell Preparation Tube (BD-CPT; Becton Dickinson, Heidelberg, Germany).

2.3.1 Isolation of RNA from Whole Blood by the PAXgene Method

Accurate quantification of mRNA in whole blood is one of the biggest challenges in gene expression analysis, due to unintended *ex vivo* gene induction and simultaneous degradation of mRNA transcripts caused by sample collection, handling, and storage.[36–38] To overcome this challenge, PreAnalytiX developed the PAXgene system, which enables the collection, stabilization, and storage of whole blood samples and provides a rapid standard protocol for subsequent RNA isolation. Blood samples are drawn into evacuated blood collection tubes, which contain a blend of additives that lyse the cells and provide stabilization of the cellular RNA profile.[35,39] The isolation of RNA (using the PAXgene Blood RNA Kit) is performed after at least a 2-h incubation at room temperature, and begins with sedimentation of nucleic acids and subsequent removal of proteins by Proteinase K digestion. Silica-gel membrane technology is then used to isolate the cellular RNA. In all, 2.5 ml of blood can be processed per PAXgene collection tube. Because of the variable RNA yield, which is highly donor dependent, it is recommended to process replicates for every blood donor to obtain sufficient RNA for downstream gene expression profiling experiments. The main benefit of this method is the immediate stabilization of the RNA of the samples. It also allows transportation and storage of the samples for several days without introducing *ex vivo* changes. Wide application of this methodology might be hampered by the relatively high costs and the inability to isolate specific cell types from the sample at a later time point. Furthermore, we have demonstrated some major challenges for the application of this method to microarray technology, which is described later in this chapter.

2.3.2 Isolation of RNA from Whole Blood with the QIAamp Method

The QIAamp RNA Blood Mini Kit provides an alternative of isolating RNA from whole-blood samples. In contrast to the PAXgene System the QIAamp method provides no stabilization of the cellular RNA profile. The RNA obtained is mainly from leukocytes since erythrocytes are selectively lysed by a hypotonic buffer and leukocytes are recovered by centrifugation. RNA from leukocytes (lymphocytes, monocytes, and granulocytes) is then isolated by silica-gel membrane column technology. On a per-column basis, approximately 1.5 ml whole blood or 1×10^7 leukocytes can be processed. As with the PAXgene method, larger amounts of whole blood per donor should be processed to obtain sufficient RNA amounts for downstream gene expression profiling experiments. The main advantage of the QIAamp method is the relatively fast and simple protocol, although the number of samples processed in parallel can be limited and often a small extent of erythrocyte contamination occurs.

2.3.3 Ficoll-Hypaque Type Isolation of Mononuclear Peripheral Blood Cells

2.3.3.1 Ficoll-Hypaque Method

So far, most gene expression analyses of peripheral blood have been carried out on PBMCs isolated from anticoagulated phlebotomy samples,[25,29,40,41] because PBMCs are the most transcriptionally active cells in blood along with granulocytes. During phlebotomy, whole blood is usually drawn in standard blood collection tubes, but different anticoagulants, e.g., EDTA, sodium citrate, and heparin, are used. The subsequent separation of PBMC from other blood cells, such as erythrocytes, can be performed via a density-gradient medium (Ficoll-Hypaque).[42,43] In this method, the anticoagulated blood/buffy coat sample is layered on the Ficoll solution and centrifuged for a short time period. Differential sedimentation in the density gradient medium during centrifugation results in separation of the different cell types. The duration of the whole procedure is about 2 h, depending on the number of samples processed in parallel. The isolated PBMC are then lysed and subjected to RNA isolation, e.g., with TRIzol® Reagent (Molecular Research Center, Inc., Cincinnati, OH, U.S.), a phenol/chloroform guanidine thiocyanate–based isolation method according to Chomczynski.[44,45] The main advantage of this approach is that it constitutes a relatively inexpensive method for the isolation of lymphocytes and monocytes from blood samples. However, in comparison to the QIAamp method, isolation of PBMC by Ficoll-based technique is time-consuming, requires skilled processing, and multiple variants of the procedure exist.

2.3.3.2 BD-CPT Method

Because the isolation of PBMCs by Ficoll-based technique is laborious and time-consuming, Becton Dickinson developed the VACUTAINER® CPT™ Cell Preparation Tube (BD-CPT), which combines an evacuated blood collection tube containing an anticoagulant (sodium citrate or sodium heparin) with a Ficoll-Hypaque density fluid and a polyester gel, which separates the two liquids. Whole blood is collected, centrifuged, and processed entirely within these tubes. During centrifugation, the PBMC move from the plasma in the density gradient, while the polyester gel forms a stable barrier isolating them from erythrocytes and granulocytes. In comparison to the classical Ficoll-Hypaque method, the isolation of PBMC with the BD-CPT method has some appreciable qualities. The method is faster (preparation time ~70 to 90 min) and simple and therefore more applicable in clinical studies. The cost is higher than Ficoll, and PBMC pellets isolated with BD-CPT tubes often show slight erythrocyte contamination.

2.4 COMPARISON OF DIFFERENT PREPARATION TECHNIQUES OF WHOLE BLOOD SAMPLES

In a recent study we compared the impact of the aforementioned isolation techniques and classified important variables influencing gene expression profiles performed by oligonucleotide microarrays with respect to sensitivity and variability in array performance, as well as the identification of genes that are sensitive to *ex vivo* changes prior to RNA isolation and microarray analysis.

Using the different isolation techniques for PBMC by Ficoll-Hypaque centrifugation, we established that even when using a particular methodology of cell isolation, the use of different devices or prehandling of cells prior to isolation has an impact on the expression of certain genes. Aspects such as blood isolation platform (whole blood vs. PBMC), temperature (room temperature or 8°C) and the two different PBMC isolation procedures (BD-CPT versus classical Ficoll method) were assessed. Table 2.1 shows the different experimental groups, which showed comparable detection sensitivity (mean percentage present call rates) with the exception of Paxgene. There were differences in terms of variability. In fact, the degree of variability for each gene that was associated with the different isolation techniques and conditions was lowest for PBMC isolated by Ficoll at 8°C (FI-8°C), followed by PBMC isolated with BD-CPT (BD), Ficoll at room temperature (FI-RT), and from buffy coat (Buffy), respectively (Figure 2.1).

Extensive statistical and descriptive analysis revealed few biologically relevant differences between cells derived from venous blood and cells from buffy coats, as well as between the two different isolation approaches or the different isolation temperatures, respectively. The genes that were prone to *ex vivo* changes in our sample set when comparing different Ficoll isolation temperatures could mainly be related to immediate-early genes, transcription factors, translation-initiation factors, enzymes, and cytokines, which are induced in response to stress and which exhibited upregulated expression in samples isolated at room temperature.[34]

Table 2.1 Percentage of Expressed Genes According to Microarray Analysis

Method	Ficoll RT	Ficoll 8°C	Ficoll ON	BD-CPT	Buffy	QIAamp	GRP	PAX
Mean % present call	56.7	58.0	51.4	57.9	55.7	56.3	55.7	45.4
SD % present call	2.8	0.8	2.4	1.5	1.6	2.2	2.4	3.6

Note: The percentage of expressed genes was assessed with dChip software. Given is the mean ± SD.

Abbreviations: Ficoll RT: PBMC isolated with Ficoll-Hypaque at room temperature; Ficoll 8°C: PBMC isolated with Ficoll-Hypaque at 8°C; Ficoll ON: PBMC isolated with Ficoll-Hypaque after 24-h incubation at room temperature; BD-CPT: PBMC isolated with BD-CPT tubes at room temperature; Buffy: PBMC isolated from buffy coats with Ficoll-Hypaque at room temperature; QIAamp: RNA isolated from whole blood samples with the QIAamp method; PAX: RNA isolated from whole blood samples with the PAXgene technique; GRP: RNA isolated with the PAXgene technique followed by globin reduction protocol.

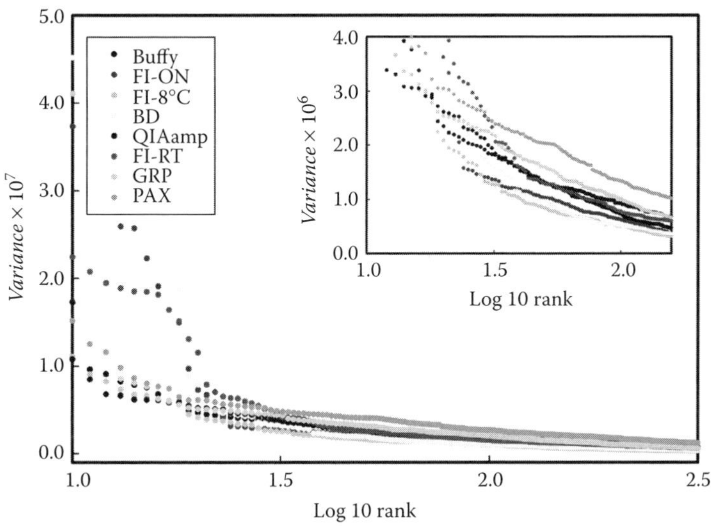

Figure 2.1 (Color figure follows p. 138.) Overall variance for each probe set and technique. The variance for all genes was calculated within the respective groups, ordered by rank and plotted against the decade logarithm of the rank. Highlighted are the ranks with the highest variability. Abbreviations according to those described in Table 2.1.

Further analysis of the different blood cell and RNA isolating methods using unsupervised hierarchical cluster analysis revealed additional aspects (Figure 2.2). In this analysis, samples prepared by other techniques (PAX or QIAamp) were also included. It is important to note that all PBMC samples are grouped into one subcluster regardless of isolation method, isolation temperatures, and time of storage, or sampling technique. This might suggest that the differences are minor and might be neglected. However, in most cases, samples derived from a single donor prepared by different isolation techniques did not cluster together (data not shown), indicating

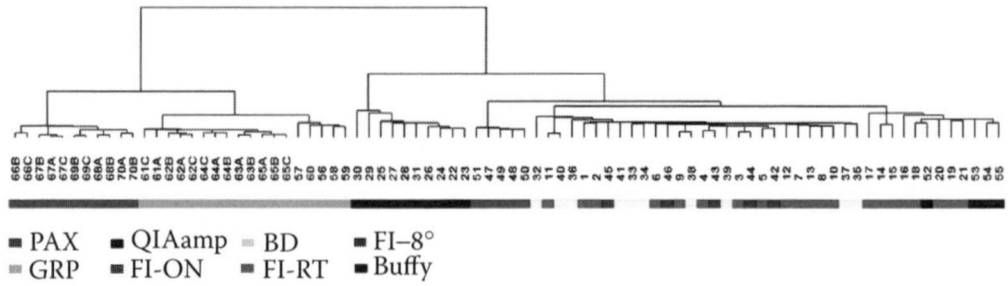

Figure 2.2 **(Color figure follows p. 138.)** Hierarchical clustering of peripheral blood samples. Gene clusters associated with RNA isolation methods from whole blood (PAX, GRP, QIAamp) and isolation methods/conditions of PBMCs (FI-RT, FI-8°, BD, FI-ON, Buffy). Hierarchical clustering of samples was performed with Pearson's correlation algorithm and precalculated distances using dChip software. Prior clustering analysis genes were filtered with a statistical filter (0.5 < SD/mean < 10). Replicates are indicated by alphabetical suffixes.

that changes introduced by the different methods have an important impact on the final profile. These data clearly underline the necessity of using a single protocol consistently throughout a study to obtain meaningful results and to minimize variability introduced by different isolation techniques.

2.5 DISTINCT GENE EXPRESSION PATTERNS IN PERIPHERAL BLOOD AFTER DELAYED PREPARATION

The use of genomic technologies in a clinical setting will require initial single-center trials, followed by multicenter trials and subsequent introduction into routine use. While highly skilled laboratories at major medical centers around the world have established the value of gene signatures for many aspects of medicine, mostly in single-center studies, such analysis has not really entered later stage clinical development scenarios. Before entering multicenter trials numerous aspects of logistics need to be taken into account. While many parameters assessed in peripheral blood might be stable even after prolonged time of transportation to a central laboratory, this may not be the case for genomic and proteomic analysis. We have addressed these issues in a recent study.[34] Peripheral blood was obtained from healthy individuals and either directly processed to obtain RNA from PBMC (FI-RT) or processing was delayed by 24 h (FI-ON) to mimic a typical time for overnight shipment conditions. The statistical analysis revealed a high impact of time delay in PBMC preparation with a large number of genes that were significantly differentially expressed. The adverse impact of delayed PBMC preparation was also reflected by a significantly lower number of mean percentage present call rates (Table 2.1). Furthermore, all samples with delayed PBMC preparation were clearly separated as a subgroup from all other PBMC samples when performing unsupervised hierarchical cluster analysis (Figure 2.2) indicating again a high impact of this parameter on changes in gene expression profiles. Gene ontology analyses revealed that changes introduced by delayed blood processing occur not by chance, but are characterized by distinct signatures. One of the most obvious events is an induction of a hypoxia signature, caused by the changes of oxygen homeostasis in the samples after blood withdrawal. Therefore, many stress-associated and hypoxia-induced genes showed elevated expression in delayed PBMC samples, e.g., the transcription factor HIF1α and several downstream target genes like VEGF,[46,47] ADM,[48] PFKFB3,[48–50] and transferrin receptor.[51] On the other hand, several genes associated with important physiological functions like cell cycle, proliferation, transcription, and metabolism showed decreased expression in delayed PBMC samples. A large number of genes associated with immune function (e.g., chemokines, cytokine receptors, and cell surface receptors) also revealed reduced expression and genes associated with apoptosis showed downregulated expression of both proapoptotic and antiapoptotic factors. These findings indicate the initiation of a complex regulatory machinery to compensate for the potentially lethal microenvironment during delayed sample handling. Overall, this study leads to the conclusion that delay in handling and processing biopsy material can introduce significant *ex vivo* effects that will clearly reduce the informative value of any data set obtained in multicenter clinical trials, for which each sample will have its individual timeframe from blood draw to cell and RNA isolation.

In our experience the changes introduced by delayed sample handling are significantly higher compared to the changes observed between healthy individuals and patients suffering from end-stage cancer diseases (Debey, Zander, Schultze, unpublished results). We thus conclude that immediate sample preparation is optimal for multicenter trials as well as routine clinical use. However, this is currently impractical for large-scale clinical trials since many smaller clinical sites lack the resources to perform immediate sample preparation. Where this is the case, blood samples should be handled identically within a given study.

2.6 COMPARISON OF THE QIAAMP METHOD TO PBMC

The QIAamp method provides direct isolation of RNA from white blood cells without laborious cell isolation procedures. In comparison to PBMC isolated by the Ficoll method, the QIAamp technique exhibits similar detection sensitivity (Table 2.1) as well as a similar degree of variability in levels of gene expression (Figure 2.1). However, statistical analysis revealed huge differences in the expression profiles between these two methods, most likely due to the different cellular subtypes that were analyzed. Most of the transcripts varying between the two methods showed an increased expression in the QIAamp samples, presumably due to the lack of granulocytes in the PBMC samples, since many transcripts upregulated in QIAamp samples were granulocyte-specific genes.[22,34] The differences in the gene signatures of samples prepared with QIAamp or Ficoll, respectively, are also reflected in the hierarchical cluster analysis (Figure 2.2). Here, QIAamp samples were clearly separated by forming a subgroup next to the PBMC samples. This simple comparison again clearly illustrates that it seems impossible to compare clinical studies that use different cell isolation and RNA preparation procedures. These data not only highlight the impact of cell and RNA isolation technologies, but also emphasize the impact of differing cellular compositions within samples. This will be particularly important when analyzing complex tissue compositions from tissues other than blood.

2.7 GENE EXPRESSION PROFILING OF WHOLE BLOOD

A major obstacle for applying expression profiling in clinical studies is the high variability of sample handling, particularly prior to arrival of the sample in the laboratory, which in turn can jeopardize the acquisition of high-quality data as we have exemplified for PBMCs. As mentioned previously, in multicenter trials it is nearly impossible to (1) guarantee a standardized sample handling if the procedures themselves are not very simple and (2) ensure that all clinical sites will have the capability of immediate sample processing.

To deal with this issue, PreAnalytiX developed the PAXgene system, which allows the storage of the primary material at room temperature without RNA degradation. The quality of this technique was assessed by quantitative polymerase chain reaction (PCR) for several genes, and a high degree of stability was clearly demonstrated.[22,34,35,39] Such results could lead to the wide acceptance of this technique for sample storage, especially for future analysis of patients treated with new drugs where studies on single gene's level may be performed.[52] This technique also seems very promising for application in whole genome expression profiling, although questions remain. In one of our recent studies we examined expression profiles from RNA samples isolated with the PAXgene system and compared them to several different sample preparation techniques, including QIAamp, Ficoll, and BD-CPT.[34] Several findings were quite different between those samples prepared by the PAXgene method and the other techniques. A much smaller number of genes were called present as assessed by microarray analysis software (mean 45.4 ± 3.6 PAX vs. mean 56.9 ± 1.0 all others). This reduction in present call rate was due to the high amount of RNA mainly derived from globin genes present in reticulocytes and early erythrocytes, masking genes present in other cell types. As a result of the high number of red blood cells, globin RNA may account for up to 70% of the RNA isolated from whole blood. In contrast to all other techniques, RNA present in red blood cells is isolated only by the PAXgene system. This masking by reticulocyte derived genes also explains the wide variety of genes highly affected by the PAXgene method, such as genes associated with protein biosynthesis, regulation of translation, mRNA processing, regulation of transcription, nucleic acid metabolism, growth arrest, apoptosis, and mitochondrial electron transport. Another difference specific to PAXgene is the high degree of variability found between the different samples. As demonstrated in Figure 2.1, samples prepared by the PAXgene method clearly exhibit more variability than all other samples. This constitutes a major drawback for applying the original PAXgene technique for whole genome expression profiling, as this high degree of variability may mask biological differences.

To overcome the masking of expressed genes in leukocytes by the high amounts of RNA derived from red blood cells, the globin reduction protocol (GRP) was established (http://www.affymetrix.com/support/technical/technotes/blood2_technote.pdf). By the specific binding of oligonucleotides to globin RNA and subsequent digestion of DNA/RNA dimers globin transcripts are withdrawn from the sample. This reduction of globin transcripts is performed after RNA isolation. Globin RNA reduction clearly increased the present call rate to a level seen in all other techniques (mean 55.7 ± 2.4% vs. mean 56.9 ± 1.0% of all other sample groups excluding PAX and FI-ON). Most interestingly, the high degree of variability was also dramatically reduced (Figure 2.1).

An additional approach to test the quality of expression profiles obtained by different isolation techniques is to perform technical replicates and then analyze the similarity between the replicates. When performing unsupervised hierarchical clustering on technical replicates, the PAXgene samples did not cluster next to each other in contrast to those samples subjected to the globin RNA reduction protocol (see Figure 2.2). This was true using different gene filters as exemplified here for the most variable genes. In preliminary studies we have also addressed whether predictors or classifiers developed within PBMC sample sets associated with specific biological characteristics can be applied to PAXgene or PAX-GRP samples (Zander, Debey, Schultze, unpublished data). This is an important aspect since many predictors for peripheral blood have been and will be established in PBMC under single-center conditions, while multicenter settings will more likely require methodologies such as PAX-GRP to be utilized. Our initial results suggest that predictors or classifiers established in a PBMC sample set can indeed be transferred to samples sets prepared by the PAXgene method, provided that the globin RNA reduction protocol is applied (Zander, Debey, Schultze, unpublished results). It should be noted, however, that other laboratories have identified increases in 3′/5′ ratios for actin and GAPDH control genes, suggesting that the nuclease-dependent procedure for globin reduction is not entirely specific to the globin DNA/RNA hybrids (Dr. M.E. Burczynski, Wyeth Research, personal communication). For this reason, additional non-nuclease-dependent globin reduction methods are being developed, which may enable a more robust globin depletion method for analysis of PAXgene stabilized whole blood.

2.8 REQUISITES FOR FUTURE CLINICAL TRANSCRIPTOME STUDIES OF PERIPHERAL BLOOD

Careful sample handling and a high degree of standardization are crucial when comprehensive expression profiling of peripheral blood is to be performed. Delay in processing leads to major changes in the expression profile reflecting the physiological response of the cells to this stress. In many studies PBMCs might be the most interesting cells and classical density gradient techniques for isolation of PBMCs can be reliably used in well-controlled settings. There might also be situations where granulocytes are of major interest for the expression profiling project. In this case the QIAamp method may be chosen in a well-controllable setting. A high reproducibility of the data can be obtained, but sample transportation and storage prior to RNA isolation still seem to be the most crucial aspects to be considered within study designs. The PAXgene method followed by a globin RNA reduction protocol might be suitable to overcome this pitfall, although this needs further confirmation and characterization of additional aspects, e.g., the comparability of classifiers defined within PBMC samples and PAX-GRP samples. Certainly, the final goal should be the combination of high reproducibility and low variability while at the same time allowing prolonged sample storage prior to further sample processing. Additional studies within this field of translational research are required to develop a gold standard for widely applicable genomic analysis as part of routine medical diagnostics in the future.

2.9 CONCLUSIONS AND FUTURE DIRECTIONS

Transcriptome and proteome analyses will become important tools for diagnostics, disease prognosis, pharmacogenomics, and pharmacoprediction in the future. The successful installation of these techniques will require a new level of interdisciplinary interaction within medical centers. Those centers achieving the establishment of interdisciplinary research teams quickly will possess an advantage over competitors. As for many other medical tests in the past, the early phase of excitement regarding genomic assays has now been replaced by a period of introducing robust standard operation procedures necessary for these new technologies to become clinical tools. While tissue specimens will remain a biopsy source for transcriptome and proteome analysis, it is very likely, as for many medical tests in the past, that peripheral blood will become an important clinical specimen for genomic and proteomic analysis. For gene expression profiling of peripheral blood to become a routine tool, recent studies have clearly demonstrated that a high level of standardization of sample procurement, transportation, or storage is required to ensure high-quality data.[28,34,61,62] While techniques downstream of RNA isolation are very reliable[28,61,63] and even data analysis with the emergence of ever more sophisticated software becomes less of a challenge,[64–66] our own work has demonstrated that standardized isolation techniques for cells and RNA need to be introduced for gene expression analysis within large-scale clinical investigation.[34,61,62] Clinical use of genomic technology will need robust and preferably simple methodology. So far, none of the methods we and others have tested has been established as a gold standard. While some methods are less variable and probably more informative in their gene expression profiles, others might have advantages in handling under clinical conditions. The method to be used depends on (1) the practicality of the situation and (2) the cellular component that is most informative with changes in gene expression due to disease or therapy. Studies assessing the impact of sample handling as presented here and in our previous work will help to optimize gene expression profiling for large, multicenter trials and subsequent routine use in the clinical arena. In the best case scenario, blood samples and RNA should be processed immediately after isolation to avoid interference of the *in vivo* gene expression signature with *ex vivo* stress responses, but the practical problems encountered with implementing this strategy in large-scale multicenter trials are very real. Further studies will be necessary to finally define a gold standard for routine clinical use that overcomes the limitations of each of the methods currently available. Such a gold standard should be an inexpensive and simple method that allows for prolonged transportation of samples without introducing *ex vivo* responses, provides sufficient sensitivity for the detection of relevant informative transcripts, and keeps variability to a minimum.

ACKNOWLEDGMENTS

This work was supported in part by a Sofja Kovalevskaja award from the Alexander von Humboldt-Foundation (J.L.S.) and a fellowship by the Frauke-Weiskam Foundation (T.Z.).

REFERENCES

1. Fambrough, D., McClure, K., Kazlauskas, A., and Lander, E.S. Diverse signaling pathways activated by growth factor receptors induce broadly overlapping, rather than independent, sets of genes. *Cell* 97, 727–741, 1999.
2. Shaffer, A.L. et al. BCL-6 represses genes that function in lymphocyte differentiation, inflammation, and cell cycle control. *Immunity* 13, 199–212, 2000.
3. Coller, H.A. et al. Expression analysis with oligonucleotide microarrays reveals that MYC regulates genes involved in growth, cell cycle, signaling, and adhesion. *Proc. Natl. Acad. Sci. U.S.A.* 97, 3260–3265, 2000.
4. Diehn, M. et al. Genomic expression programs and the integration of the CD28 costimulatory signal in T cell activation. *Proc. Natl. Acad. Sci. U.S.A.* 99, 11796–11801, 2002.
5. Whitfield, M.L. et al. Identification of genes periodically expressed in the human cell cycle and their expression in tumors. *Mol. Biol. Cell* 13, 1977–2000, 2002.
6. Mirza, A. et al. Global transcriptional program of p53 target genes during the process of apoptosis and cell cycle progression. *Oncogene* 22, 3645–3654, 2003.
7. Iyer, V.R. et al. The transcriptional program in the response of human fibroblasts to serum. *Science* 283, 83–87, 1999.
8. Belcher, C.E. et al. The transcriptional responses of respiratory epithelial cells to *Bordetella pertussis* reveal host defensive and pathogen counter-defensive strategies. *Proc. Natl. Acad. Sci. U.S.A.* 97, 13847–13852, 2000.
9. Guillemin, K., Salama, N.R., Tompkins, L.S., and Falkow, S. Cag pathogenicity island-specific responses of gastric epithelial cells to *Helicobacter pylori* infection. *Proc. Natl. Acad. Sci. U.S.A.* 99, 15136–15141, 2002.
10. Lessnick, S.L., Dacwag, C.S., and Golub, T.R. The Ewing's sarcoma oncoprotein EWS/FLI induces a p53-dependent growth arrest in primary human fibroblasts. *Cancer Cell* 1, 393–401, 2002.
11. Chang, B.D. et al. Molecular determinants of terminal growth arrest induced in tumor cells by a chemotherapeutic agent. *Proc. Natl. Acad. Sci. U.S.A.* 99, 389–394, 2002.
12. Clark, E.A., Golub, T.R., Lander, E.S., and Hynes, R.O. Genomic analysis of metastasis reveals an essential role for RhoC. *Nature* 406, 532–535, 2000.
13. Ramaswamy, S., Ross, K.N., Lander, E.S., and Golub, T.R. A molecular signature of metastasis in primary solid tumors. *Nat. Genet.* 33, 49–54, 2003.
14. Welsh, J.B. et al. Analysis of gene expression profiles in normal and neoplastic ovarian tissue samples identifies candidate molecular markers of epithelial ovarian cancer. *Proc. Natl. Acad. Sci. U.S.A.* 98, 1176–1181, 2001.

15. Golub, T.R. et al. Molecular classification of cancer: class discovery and class prediction by gene expression monitoring. *Science* 286, 531–537, 1999.
16. Alizadeh, A.A. et al. Distinct types of diffuse large B-cell lymphoma identified by gene expression profiling. *Nature* 403, 503–511, 2000.
17. Bhattacharjee, A. et al. Classification of human lung carcinomas by mRNA expression profiling reveals distinct adenocarcinoma subclasses. *Proc. Natl. Acad. Sci. U.S.A.* 98, 13790–13795, 2001.
18. Shipp, M.A. et al. Diffuse large B-cell lymphoma outcome prediction by gene-expression profiling and supervised machine learning. *Nat. Med.* 8, 68–74, 2002.
19. van't Veer, L.J. et al. Gene expression profiling predicts clinical outcome of breast cancer. *Nature* 415, 530–536, 2002.
20. van de Vijver, M.J. et al. A gene-expression signature as a predictor of survival in breast cancer. *N. Engl. J. Med.* 347, 1999–2009, 2002.
21. Rosenwald, A. et al. The use of molecular profiling to predict survival after chemotherapy for diffuse large-B-cell lymphoma. *N. Engl. J. Med.* 346, 1937–1947, 2002.
22. Whitney, A.R. et al. Individuality and variation in gene expression patterns in human blood. *Proc. Natl. Acad. Sci. U.S.A.* 100, 1896–1901, 2003.
23. Twine, N.C. et al. Disease-associated expression profiles in peripheral blood mononuclear cells from patients with advanced renal cell carcinoma. *Cancer Res.* 63, 6069–6075, 2003.
24. Bennett, L. et al. Interferon and granulopoiesis signatures in systemic lupus erythematosus blood. *J. Exp. Med.* 197, 711–723, 2003.
25. Baechler, E.C. et al. Interferon-inducible gene expression signature in peripheral blood cells of patients with severe lupus. *Proc. Natl. Acad. Sci. U.S.A.* 100, 2610–2615, 2003.
26. Boldrick, J.C. et al. Stereotyped and specific gene expression programs in human innate immune responses to bacteria. *Proc. Natl. Acad. Sci. U.S.A.* 99, 972–977, 2002.
27. Brown, P.O. and Hartwell, L. Genomics and human disease — variations on variation. *Nat. Genet.* 18, 91–93, 1998.
28. Frank, R. and Hargreaves, R. Clinical biomarkers in drug discovery and development. *Nat. Rev. Drug Discov.* 2, 566–580, 2003.
29. Valk, P.J. et al. Prognostically useful gene-expression profiles in acute myeloid leukemia. *N. Engl. J. Med.* 350, 1617–1628, 2004.
30. Expression profiling--best practices for data generation and interpretation in clinical trials. *Nat. Rev. Genet.* 5, 229–237, 2004.
31. Ball, C.A. et al. Standards for microarray data. *Science* 298, 539, 2002.
32. Brazma, A. et al. Minimum information about a microarray experiment (MIAME) — toward standards for microarray data. *Nat. Genet.* 29, 365–371, 2001.
33. Spellman, P.T. et al. Design and implementation of microarray gene expression markup language (MAGE-ML). *Genome Biol.* 3, research 0046.1–0046.9, 2002.
34. Debey, S. et al. Comparison of different isolation techniques prior gene expression profiling of blood derived cells: impact on physiological responses, on overall expression and the role of different cell types. *Pharmacogenomics J.* 4, 193–207, 2004.
35. Rainen, L. et al. Stabilization of mRNA expression in whole blood samples. *Clin. Chem.* 48, 1883–1890, 2002.
36. Tanner, M.A. et al. Substantial changes in gene expression level due to the storage temperature and storage duration of human whole blood. *Clin. Lab. Haematol.* 24, 337–341, 2002.
37. Pahl, A. and Brune, K. Gene expression changes in blood after phlebotomy: implications for gene expression profiling. *Blood* 100, 1094–1095, 2002.

38. Hartel, C., Bein, G., Muller-Steinhardt, M., and Kluter, H. *Ex vivo* induction of cytokine mRNA expression in human blood samples. *J. Immunol. Methods* 249, 63–71, 2001.
39. Thach, D.C. et al. Assessment of two methods for handling blood in collection tubes with RNA stabilizing agent for surveillance of gene expression profiles with high density microarrays. *J. Immunol. Methods* 283, 269–279, 2003.
40. DePrimo, S.E. et al. Expression profiling of blood samples from an SU5416 Phase III metastatic colorectal cancer clinical trial: a novel strategy for biomarker identification. *BMC Cancer* 3, 3, 2003.
41. Bomprezzi, R. et al. Gene expression profile in multiple sclerosis patients and healthy controls: identifying pathways relevant to disease. *Hum. Mol. Genet.* 12, 2191–2199, 2003.
42. Boyum, A. Separation of white blood cells. *Nature* 204, 793–794, 1964.
43. Ting, A. and Morris, P.J. A technique for lymphocyte preparation from stored heparinized blood. *Vox Sang* 20, 561–563, 1971.
44. Chomczynski, P. A reagent for the single-step simultaneous isolation of RNA, DNA and proteins from cell and tissue samples. *Biotechniques* 15, 532–534, 536–537, 1993.
45. Chomczynski, P. and Sacchi, N. Single-step method of RNA isolation by acid guanidinium thiocyanate-phenol-chloroform extraction. *Anal. Biochem.* 162, 156–159, 1987.
46. Levy, A.P., Levy, N.S., Wegner, S., and Goldberg, M.A. Transcriptional regulation of the rat vascular endothelial growth factor gene by hypoxia. *J. Biol. Chem.* 270, 13333–13340, 1995.
47. Forsythe, J.A. et al. Activation of vascular endothelial growth factor gene transcription by hypoxia-inducible factor 1. *Mol. Cell. Biol.* 16, 4604–4613, 1996.
48. Garayoa, M. et al. Hypoxia-inducible factor-1 (HIF-1) up-regulates adrenomedullin expression in human tumor cell lines during oxygen deprivation: a possible promotion mechanism of carcinogenesis. *Mol. Endocrinol.* 14, 848–862, 2000.
49. Marsin, A.S., Bouzin, C., Bertrand, L., and Hue, L. The stimulation of glycolysis by hypoxia in activated monocytes is mediated by AMP-activated protein kinase and inducible 6-phosphofructo-2-kinase. *J. Biol. Chem.* 277, 30778–30783, 2002.
50. Minchenko, A. et al. Hypoxia-inducible factor-1-mediated expression of the 6-phosphofructo-2-kinase/fructose-2,6-bisphosphatase-3 (PFKFB3) gene. Its possible role in the Warburg effect. *J. Biol. Chem.* 277, 6183–6187, 2002.
51. Tacchini, L., Bianchi, L., Bernelli-Zazzera, A., and Cairo, G. Transferrin receptor induction by hypoxia. HIF-1-mediated transcriptional activation and cell-specific post-transcriptional regulation. *J. Biol. Chem.* 274, 24142–24146, 1999.
52. Muller, M.C. et al. Improvement of molecular monitoring of residual disease in leukemias by bedside RNA stabilization. *Leukemia* 16, 2395–2399, 2002.
53. Ichikawa, W. et al. Thymidylate synthase and dihydropyrimidine dehydrogenase gene expression in relation to differentiation of gastric cancer. *Int. J. Cancer*, 112, 967–973, 2004.
54. Kim, J.O. et al. Differential gene expression analysis using paraffin-embedded tissues after laser microdissection. *J. Cell. Biochem.* 90, 998–1006, 2003.
55. Korbler, T., Grskovic, M., Dominis, M., and Antica, M. A simple method for RNA isolation from formalin-fixed and paraffin-embedded lymphatic tissues. *Exp. Mol. Pathol.* 74, 336–340, 2003.
56. Specht, K. et al. Identification of cyclin D1 mRNA overexpression in B-cell neoplasias by real-time reverse transcription-PCR of microdissected paraffin sections. *Clin. Cancer Res.* 8, 2902–2911, 2002.

57. Grotzer, M.A. et al. Biological stability of RNA isolated from RNAlater-treated brain tumor and neuroblastoma xenografts. *Med. Pediatr. Oncol.* 34, 438–442, 2000.
58. Ellis, M. et al. Development and validation of a method for using breast core needle biopsies for gene expression microarray analyses. *Clin. Cancer Res.* 8, 1155–1166, 2002.
59. Florell, S.R. et al. Preservation of RNA for functional genomic studies: a multidisciplinary tumor bank protocol. *Mod. Pathol.* 14, 116–128, 2001.
60. Dreskin, S.C. et al. Measurement of changes in mRNA for IL-5 in noninvasive scrapings of nasal epithelium taken from patients undergoing nasal allergen challenge. *J. Immunol. Methods* 268, 189–195, 2002.
61. Xiang, Z., Yang, Y., Ma, X., and Ding, W. Microarray expression profiling: analysis and applications. *Curr. Opin. Drug Discov. Dev.* 6, 384–395, 2003.
62. Leiva, I.M., Emmert-Buck, M.R., and Gillespie, J.W. Handling of clinical tissue specimens for molecular profiling studies. *Curr. Issues Mol. Biol.* 5, 27–35, 2003.
63. Gerhold, D.L., Jensen, R.V., and Gullans, S.R. Better therapeutics through microarrays. *Nat. Genet.* 32(Suppl.), 547–551, 2002.
64. Miller, L.D. et al. Optimal gene expression analysis by microarrays. *Cancer Cell* 2, 353–361, 2002.
65. Zhou, Y. and Abagyan, R. Algorithms for high-density oligonucleotide array. *Curr. Opin. Drug Discov. Dev.* 6, 339–345, 2003.
66. Slonim, D.K. From patterns to pathways: gene expression data analysis comes of age. *Nat. Genet.* 32(Suppl.), 502–508, 2002.

CHAPTER 3

Blood Genomic Fingerprints of Brain Diseases

Yang Tang, Donald L. Gilbert, Tracy A. Glauser, Andrew D. Hershey, Aigang Lu, Ruiqiong Ran, Huichun Xu, and Frank R. Sharp

CONTENTS

3.1 INTRODUCTION

Global gene expression profiling with DNA microarrays is one of the most powerful tools in genomics research. Microarrays work by hybridization of RNA or DNA molecules from biological samples to DNA sequences immobilized on an array surface. The hybridization of a sample to an array is similar to the classical

Northern or Southern blotting analyses, at least in principle. However, combined with complete sequence information, this technology provides a platform to examine the transcriptional activity of tens of thousands of genes in a highly parallel fashion.

The application of microarray technology has shown great potential in understanding and managing human diseases. Among other uses, the expression profile, which consists of many individual measurements, can serve as fingerprints for disease diagnosis, classification, and prognosis. In particular, cancer expression profiling studies have demonstrated the resolving power to distinguish tumor subtypes, to evaluate the sensitivity to chemotherapy, and to predict clinical outcomes.[1–3] This technology has also been applied to examining the brain genomic changes of many neurological and psychiatric diseases including Alzheimer's diseases,[4] multiple sclerosis,[5] schizophrenia,[6] autism,[7] and others. Although these studies provide important insights into the pathogenesis of these diseases, the results from brain genomic profiling cannot be readily used to guide clinical practice in neurology due to the difficulty of obtaining routine brain biopsy samples.

This made us consider whether genomic profiling of peripheral blood could provide meaningful surrogate markers for brain diseases. As a proof of concept, we examined the gene expression profile in blood of rats subjected to a variety of acute neurological insults, including ischemic stroke, hemorrhagic stroke, sham surgeries, kainate-induced seizures, hypoxia, and insulin-induced hypoglycemia.[8] We found that at 24 h following each of these insults, specific gene expression changes in blood can be identified by microarrays. In addition, we found a common blood gene expression pattern in rats correlated with the occurrence of neuronal cell death regardless of the causes[9] (Figure 3.1). These data from the animal models support the hypothesis that the expression profiles of peripheral blood cells may be used to detect acute pathological changes in brain. Since then, we have also conducted a series of human studies on several chronic neurological conditions.

In this chapter, we briefly review our data on blood gene expression patterns for neurofibromatosis type 1 (NF1)[10,44] and anticonvulsant drugs in pediatric epilepsy,[11] and Tourette syndrome (TS).[45] The possibilities and limitations of this novel approach are also discussed.

3.2 METHODS

After an informed consent was obtained, a 10- to 15-ml blood sample was drawn from the cubital vein and mixed with Trizol LS reagent (Invitrogen, Carlsbad, CA) within 15 min. Total RNA was isolated according to the protocol provided by the manufacturer and was further purified using RNeasy mini kit (Qiagen, Chatsworth, CA). Sample labeling, hybridization to U95A arrays, and image scanning were carried out as described in the Affymetrix Expression Analysis Technical Manual. The arrays were normalized with "Invariant Set Normalization" method and the expression values were calculated with "PM-only model-based expression index" with dchip software v. 1.2.[12]

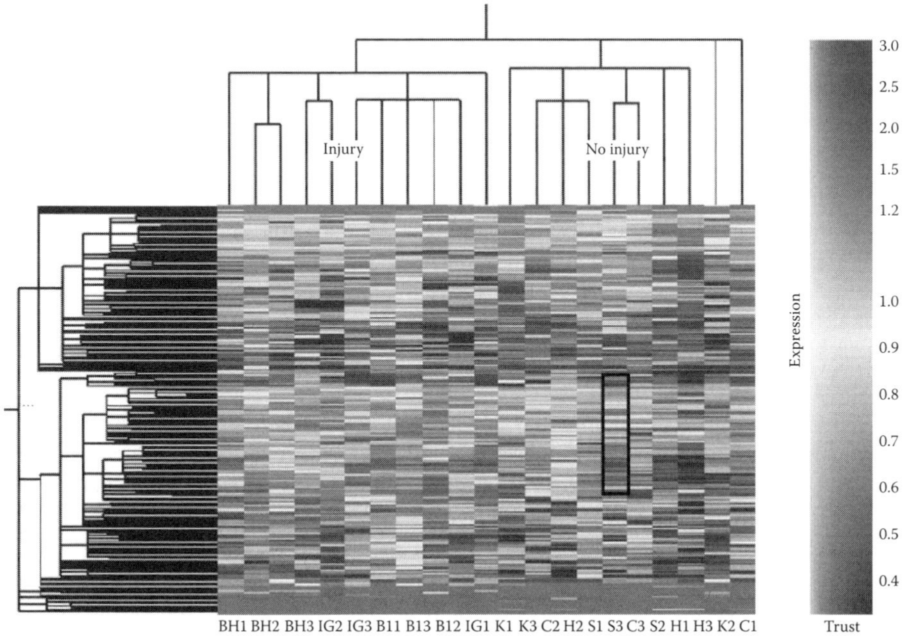

Figure 3.1 Genes upregulated (dark gray) and downregulated (dark gray) in blood monocytes of animals with brain injury (left side: BH1, BH2, BH3, IG2, IG3, BI1, BI3, BI2, IG1, K1, K3) compared to animals without brain injury (right side: C2, H2, S1, S3, C3, S2, H1, H3, "K2," C1). The plot shows the hierarchical clustering of 197 regulated genes (*y*-axis) from blood monocytes of 21 different rat samples (*x*-axis). At 24 h after brain hemorrhage (BH), insulin-induced hypoglycemia (IG), brain ischemia (BI), kainate-induced seizures (K), 8% hypoxia for 6 h (H), sham surgery (S), or being assigned as untouched controls (C), adult rats were sacrificed and *mononuclear cells* were separated. Total RNA was isolated and gene expression assessed with Affymetrix U34A microarrays (3 arrays/group). An "Injury" group of animals that included brain hemorrhage (BH), brain ischemia (BI), kainate (K), and insulin-glucose (IG) subjects were compared to a "No injury" group that included untouched (C), sham-operated (S), and hypoxia (H) subjects. A nonparametric Wilcoxon–Mann–Whitney test was used to screen genes that are differentially expressed between "Injury" samples and "No injury" samples. A Benjamini and Hochberg false discovery rate of < 0.3 was used as a significance threshold. For each of 197 genes that met the threshold, the raw expression data were normalized to the median value of 21 measurements if the median value was greater than 100. Hierarchical clustering was performed with Genespring (Silicon Genetics, Redwood City, CA). The color bar indicates the normalized expression level. For genes with low expression values at the bottom of the figure, because the median of the 21 measurements was less than 100 and the expression data were normalized to 100, most appear as a dark gray color. Genes in the black box are upregulated genes, some of which are shared between one "No injury" sample (C3) and "Injury" samples. (From Tang, Y. et al., *J. Cereb. Blood Flow Metab.* 23(3), 2003. With permission.)

3.3 RESULTS

3.3.1 Variation of Blood Gene Expression in Healthy Subjects and Patients

To test the ability of gene arrays to identify the blood gene expression patterns caused by diseases and drug treatments, we decided first to characterize the inter- and intrasubject variation between samples. The variation of gene expression in human blood has been related to many factors, which would tend to create noise, or possible spurious results (type I error) when comparing disease cases to controls. Non-disease-related factors, including relative proportions of the different blood cell types, gender, age, and time of blood draw, have all been shown to affect gene expression patterns in blood.[13]

To explore these variations in our data set, we analyzed 14 samples taken on two separate days from seven different healthy donors. A total of 266 genes were selected whose expression varied by a minimum of twofold from the median in at least 2 of the 14 samples, and subjected to unsupervised cluster analysis[14] with Genespring 6.0 (Silicon Genetics, Redwood City, CA). It was anticipated that if the gene expression variation due to temporal or technical factors exceeded inter-individual variation, the cluster analysis would not cluster samples together from the same individual. This would indicate that comparisons between individuals with different diseases might yield meaningless differences. However, if variations in gene expression patterns within individuals at two time points were smaller than inter-individual variations, then the cluster analysis should cluster each individual's two samples together. This would indicate that blood gene expression patterns may reflect the genetic makeup and/or environmental factors unique to each individual and are sufficiently stable to be used for disease-control comparisons.

With unsupervised cluster analysis based on the 266 genes with the highest variation, all the duplicate samples from the same individual clustered side by side (Figure 3.2). It is noteworthy that the distinct blood gene expression pattern among different individuals is very robust and sensitive to the gene selection criteria. This indicates that the variation in blood gene expression related to the temporal and technical reasons is smaller than the pattern caused by the intrinsic difference between different individuals, which may reflect the genetic makeup and/or environmental factors unique to each individual.

Among the genes with greatest variation are those from red blood cells and interferon-related genes. Also, genes from the Y chromosome have a higher expression in male blood (Figure 3.3), whereas genes from lymphocytes, especially a group of immunoglobins, have a higher expression in children than in adults (Figure 3.4).[10]

3.3.2 Blood Gene Expression and Chronic Neurological Disease

The primary hypotheses tested were that blood gene expression patterns in patients with neurologic diseases of interest were different from those of healthy and diseased controls. It was anticipated that some differences in gene expression levels would be found by chance since over 10,000 genes were surveyed simulta-

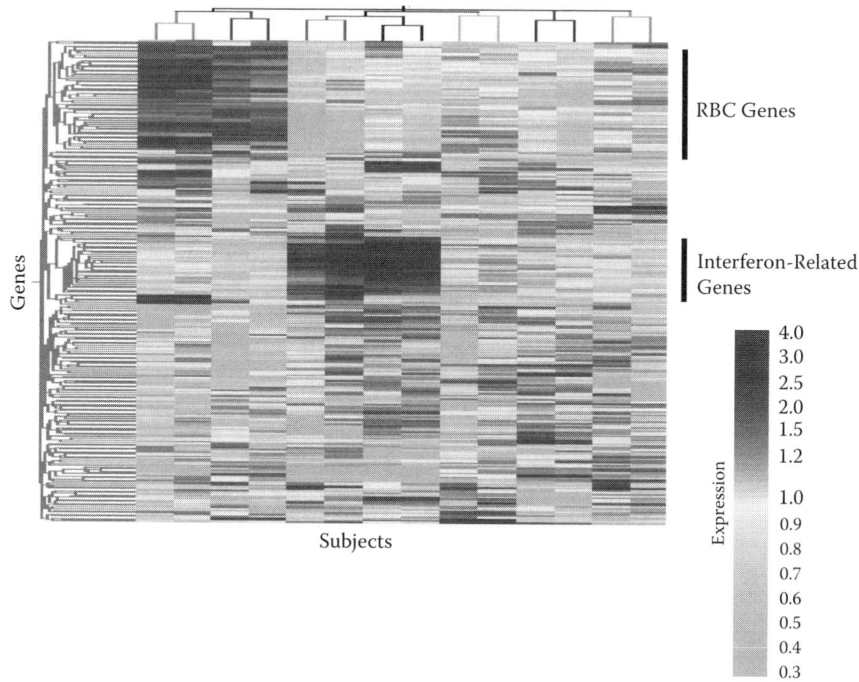

Figure 3.2 Hierarchical cluster analysis of inter- and intra-subject variation. The 14 blood samples were taken on two separate occasions from seven healthy donors. A hierarchical cluster analysis was performed on 266 selected genes whose expressions vary by a minimum of twofold from the median in at least 2 of the 14 samples. Genes with similar expression profiles were grouped in rows whereas the samples with similar impacts on the overall expression were clustered in columns. The branches of the dendrogram are presented in grayscale to indicate different donors.

neously. To determine whether the differences in gene expression exceeded what would occur by chance, permutation analysis (BRB array tools) was used to compare the predefined classes: NF1 vs. age- and gender-matched controls; TS vs. age- and gender-matched controls; and children with epilepsy treated with anticonvulsants vs. drug-free children with epilepsy.

Permutation analysis first performs a parametric *t* test for each gene and determines the number of genes that are differentially expressed at an appropriate significance level. The analysis then performs random permutations of the class labels (i.e., which samples correspond to which classes) and computes the proportion of the random permutations that gave as many genes significant at the same significance level used for the predefined classes. This proportion provides a global test of whether the expression profiles in the predefined classes were significantly different from noise. Using this approach, the significance of the global gene expression patterns for NF1, TS, and pediatric epilepsy were assessed. A *p* value of less than 0.05 is sufficient to establish that class-associated differences in gene expression exceed what would be expected by chance. However, this does not allow one to

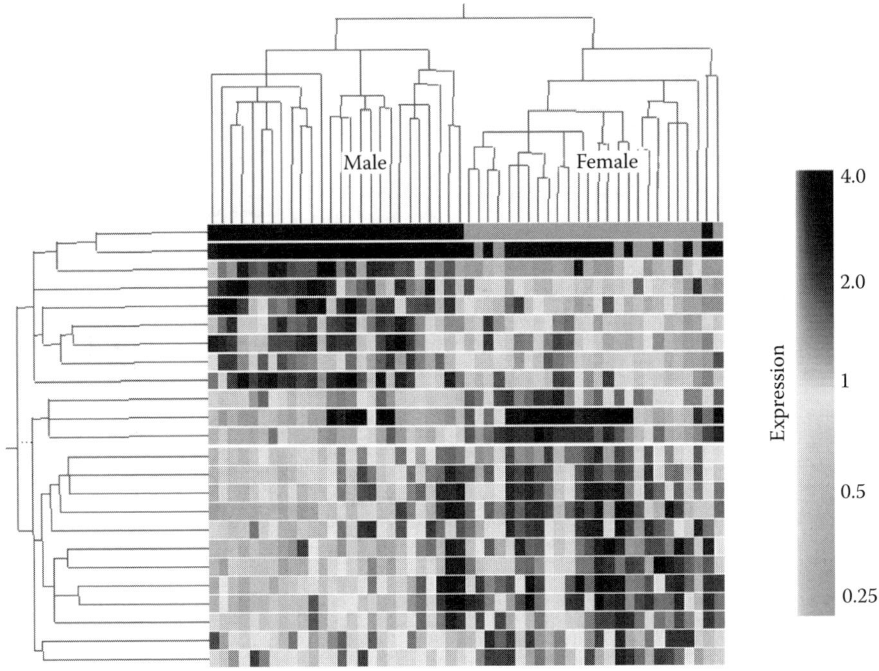

Figure 3.3 Genes differentially expressed in the blood of males compared to females. Hierarchical cluster analysis of 24 genes demonstrates differential expression between 26 male and 26 female blood samples. A parametric *t*-test (BRB-Array Tools 2.0) was performed on 4528 genes that were highly expressed in blood to derive a group of 24 genes that were significantly regulated in males vs. females ($P < 0.001$). These genes were subjected to a hierarchical cluster analysis using Genespring software. Each gene was normalized to the median of 52 measurements so that its relative expression in each sample was indicated by the fold change relative to the median as represented by the density of the squares. (From Tang, Y. et al., *Mol. Brain Res.,* 132, 155–167, 2004. With permission.)

distinguish whether class differences in expression levels of individual genes are real. Determining whether specific genes are expressed differently in diseased vs. control patients would require a larger study and independent confirmation with other expression assays.

3.3.3 Blood Genomic Expression Pattern of NF1

Since genetic factors play important roles in the pathogenesis of many neurological and psychiatric diseases, we wished to determine whether blood gene expression profiling can provide molecular markers for these genetic factors and help understand the genotype–phenotype correlation. It was postulated that gene/chromosome abnormalities passed through the germ line should be present in blood cells and should produce downstream transcriptional changes even in the absence of obvious blood phenotypes. This hypothesis was tested using NF1, an autosomal

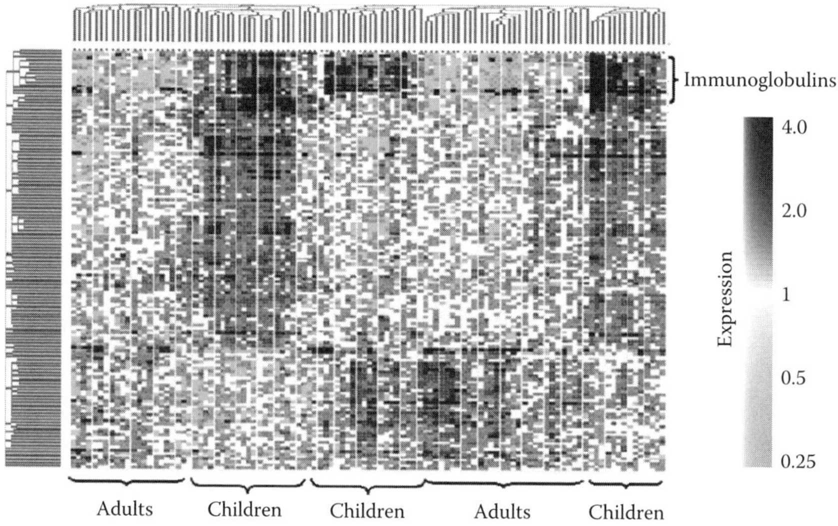

Figure 3.4 **(Color figure follows p. 138.)** Age affects blood genomic expression. Hierarchical cluster analysis of 144 genes regulated between different age groups. Children and adults can be roughly separated although there are some misclassifications. The cluster of genes that correlates best with age relates to the immunoglobulins. (From Tang, Y. et al., *Mol. Brain Res.*, 2004. With permission.)

dominant genetic disease caused by mutations of the *NF1* gene on chromosome 17q11.2.

In comparing NF1 to the three control groups, the p values for permutation analyses were 0.023, 0.02, and 0.007, indicating a specific gene expression pattern of NF1 in blood. NF1 samples clustered separately from each set of controls by hierarchical cluster analysis (Figure 3.5). It was of interest that many genes dysregulated in NF1 blood are related to tissue remodeling, bone development, and tumorigenesis, which may provide important insights into the role of *NF1* gene in target organs.[10,44]

3.3.4 Valproic Acid Blood Genomic Expression Patterns in Children with Epilepsy

In addition, we wished to determine whether gene expression patterns in blood could be used to search for markers and possibly mechanisms of medication responses in neurological diseases. Great variation exists in the way people respond to medications in terms of efficacy and toxicity. Identifying individuals prior to or early in therapy with low therapeutic response or high risk for toxicity would represent a significant advance in pharmacotherapy. This presents considerable challenges, however. Drug responses are polygenic traits. Virtually all the genes involved in pharmacokinetics and pharmcodynamics may harbor genetic polymorphisms/mutations that contribute to a combined inherited basis of drug response.[15,16]

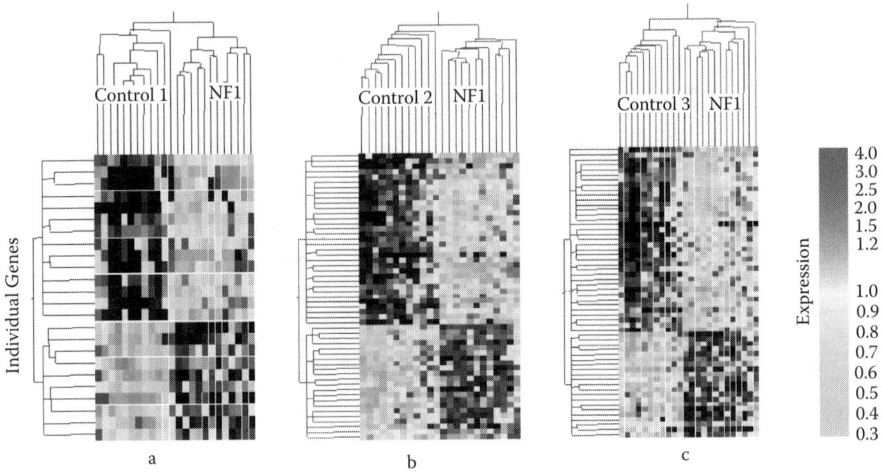

Figure 3.5 Blood gene expression profile of patients with NF1. The 12 NF1 blood samples were compared to three independent sets of controls, with each set of controls consisting of 12 blood samples that were age and gender matched with 12 patients with NF1. Genes that were significant for each comparison were used for the hierarchical cluster analysis. Each gene was normalized to the median of all measurements so that its relative expression in each sample was indicated by the fold change relative to the median as represented by the density of the squares. (From Tang, Y. et al., *Mol. Brain Res.*, 132, 155–167, 2004. With permission.)

We reasoned that an unbiased, hypothesis-free, high-throughput approach using gene expression patterns might yield important insights into treatment responses.

The global permutation-based test showed that the expression patterns in blood caused by both drugs were significantly different from controls (p = 0.005 for 11 VPA samples vs. 7 drug-free samples, and p = 0.02 for 6 carbemazepine [CBZ] samples vs. 7 drug-free samples). A hierarchical cluster analysis automatically segregated the VPA samples into two subclusters with different expression profiles. Interestingly, one subcluster included all three VPA-resistant patients while the other included all eight VPA-responsive patients, suggesting that part of the inter-patient variation of VPA blood genomic pattern might be associated with its efficacy (Figure 3.6).

In addition, it was found that many mitochondrial genes, especially those related to electron transport and oxidative phosphorylation, are overexpressed in VPA responders, which points to the possible involvement of mitochondria in the determination of VPA efficacy.

3.3.5 Blood Gene Expression Profiling Discloses T Lymphocyte Activation in a Subgroup of Patients with Tourette Syndrome

To determine whether blood gene expression patterns distinguish heritable neurologic diseases for which no causative gene has been identified, we measured blood gene expression profiles in patients with familial Tourette syndrome (TS). TS is a chronic, childhood-onset disorder characterized by motor and vocal tics, which are

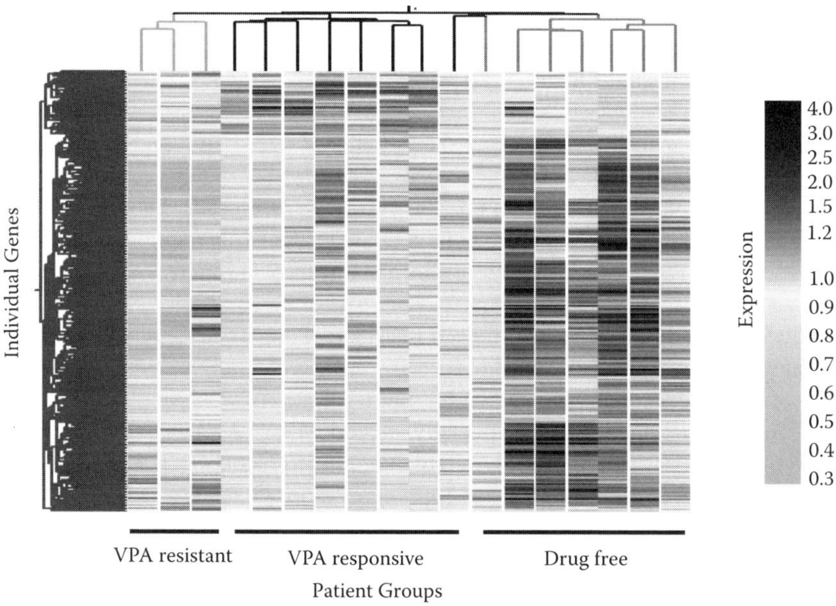

Figure 3.6 Hierarchical cluster analysis of 461 genes regulated by chronic VPA monotherapy. A parametric *t*-test (BRB-Array Tools 2.0) was performed on 5053 genes that were highly expressed in blood to derive a group of 461 genes that were significantly regulated in the VPA group ($n = 11$) compared to the drug-free group ($n = 7$) (FDR < 0.1). Each gene was normalized to the median of 18 measurements so that its relative expression in each sample was indicated by the fold change relative to the median as represented by the density of the squares. The cluster analysis yielded three distinct clusters that correlate with whether the patients were drug free, seizure free while on VPA (VPA responsive), or continued having seizures while on VPA (VPA resistant). (From Tang, Y. et al., *Acta Neurol. Scand.* 109(3), 2004. With permission.)

often accompanied by obsessive compulsive disorder (OCD) and attention-deficit hyperactivity disorder (ADHD).[17,18] Multigenerational family, twin, and adoption studies show evidence of autosomal dominant inheritance with varying penetrance and a more severe phenotype in cases of bilineal transmission.[19–21] Despite the identification of large kindreds, the search for genes and linkage has been inconclusive to date.[22]

A variety of nongenetic factors are also associated with the onset or increased severity of tics and co-morbid OCD in patients with TS.[23–28] In addition, it has been suggested that a subgroup of patients suffers from an autoimmune form of this disorder, termed pediatric autoimmune neuropsychiatric disorders associated with streptococcal infection (PANDAS).[29,30] Modeled on the paradigm of Sydenham's chorea, the PANDAS phenotype is characterized clinically by the temporal association of symptom onset or exacerbations with Group A beta hemolytic streptococcal (GABHS) infections.[31] Thus, it appears possible that a number of environmental factors could modulate gene(s) that influence the vulnerability to or otherwise affect the TS phenotype. It is also possible that the TS phenotype has multiple genotypes, some of which confer susceptibility to contributing environmental factors.

In this study, we tried to use blood genomic profiling to detect the effects of possible genetic and environmental factors in TS. We reasoned that a high-throughput genomic approach might identify TS or TS subtypes that could then be subjected to further genetic analysis.

Permutation analysis showed that the blood gene expression pattern associated with TS has a p value = 0.2. In other words, 20% of random permutations of class label generated the same number of up- and downregulated genes. Thus, no evidence was found that the clinical diagnosis of TS is associated with a single, unique gene expression profile in whole blood that is significantly different from normal or diseased controls.

Subgroup analysis, however, showed that there were six upregulated genes and one downregulated gene in TS ($p < 0.05$ for each of 8 comparisons). Expression levels of these genes in TS are shown in Figure 3.4. These were all genes known to be expressed by lymphocytes, especially natural killer (NK) cells. *Granzyme B* (tested and confirmed by real-time polymerase chain reaction (RT-PCR) on 16 TS and 16 age-matched controls with 1.8-fold increase and Student's t test $p = 0.09$) is involved in the target killing process of cytotoxic T cells (CTL) or Natural Killer cells (NK) (Lord, 2003). *NKG2E* encodes a lectin-like receptor, which plays a role in the recognition of the MHC molecules by NK cells and some CTL cells. *CD94* is also preferentially expressed by NK cells and forms heterodimers with NKG subunits.[33] NK-p46 participates in NK-cell-mediated lysis of cells infected with intracellular bacteria.[34]

The one downregulated gene, *IMPA2*, is also of interest. *IMPA2* plays a crucial role in the phosphatidylinositol signaling pathway. In the brain, its expression is substantially higher in subcortical regions, most prominently in the caudate, a region shown in many neuroimaging studies to be involved in TS and OCD.[35,36] It is also considered to be a strong candidate gene for bipolar disorder.[37,38]

Although these genes are significantly regulated in TS compared to other groups, they all have a large variance within the TS group, which raised the question whether these genes might serve as markers to identify a subgroup of patients with TS. Using k-means cluster analysis, TS and control samples were stratified into two clusters with samples in cluster A that are low expressers and samples in cluster B that are high expressers of these 6 CTL/NK genes (Figure 3.7 and Figure 3.8). Although there are a few higher expressers (cluster B) in each control group, the proportion of TS subjects in the higher expression group was significantly greater than the proportions in the control groups (chi square, $p < 0.05$). Permutation-based analysis showed that less than 4% of random permutations generated the same number of differentially expressed genes as those in clusters A and cluster B. This analysis suggests that the differences in expression profiles between the two TS clusters are not due to chance ($p < 0.05$).

3.4 DISCUSSION

Global gene expression profiling holds great potential for classifying diseases and predicting clinical outcomes based on the molecular features. As one of the

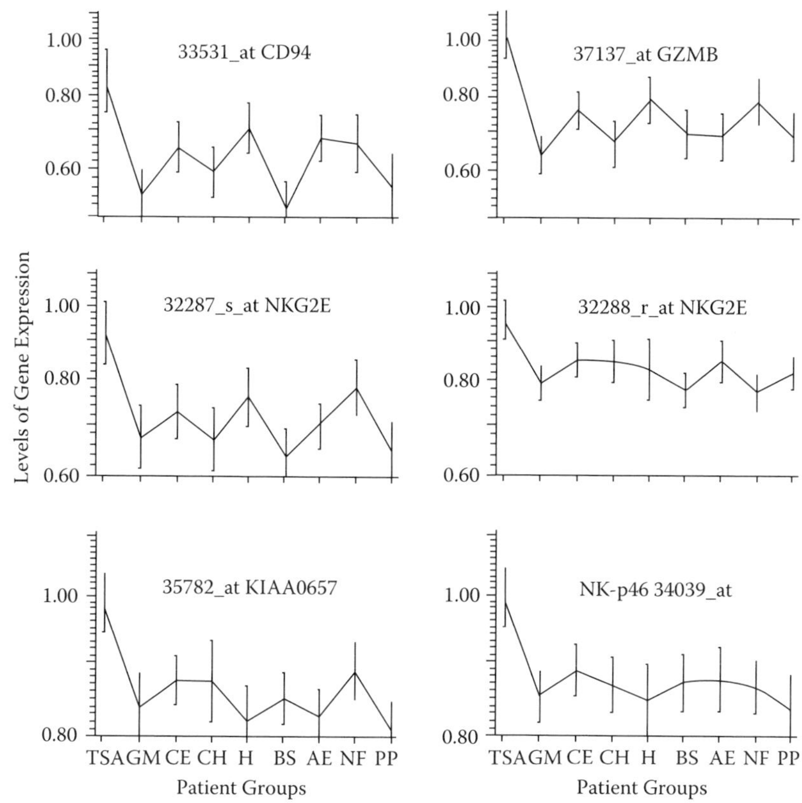

Figure 3.7 The expression profiles of six genes that are specifically regulated in patients with TS. The *x*-axis indicates the subject groups: TS = Tourette syndrome; AGM = age- and gender-matched controls; CE = children with epilepsy; CH = children with headache; H = healthy controls; BS = bipolar disorder and schizophrenia; AE = adult epilepsy; NF = neurofibromatosis type 1; PP = Parkinson's disease and progressive supranuclear palsy. The *y*-axis indicates the relative expression value for each gene. All the values were normalized to the average of the patients with TS and expressed as mean ± SEM. (From Tang, Y. et al., *Arch. Neurol.,* 62, 210–215, 2005. With permission.)

most accessible tissues, blood gene expression profiling has been used to explore hematological malignancies,[39] autoimmune disorders,[40,41] and infectious disorders.[42,43] Our preliminary studies suggest this approach can be extended to neurological and psychiatric diseases.

Brain tissue is not accessible *in vivo* for the vast majority of neurologic and psychiatric diseases. Although the majority of brain diseases do not have obvious phenotypes in blood, there are two reasons measuring blood gene expression is a plausible approach. First, genetic factors play a crucial role in the development of many chronic brain diseases and the determination of responses to therapeutic interventions. Peripheral blood cells inherit the same genetic information as brain cells, so blood genomic profiling may mirror the changes of gene regulatory networks of the brain and other target organs. Second, peripheral blood cells are equipped

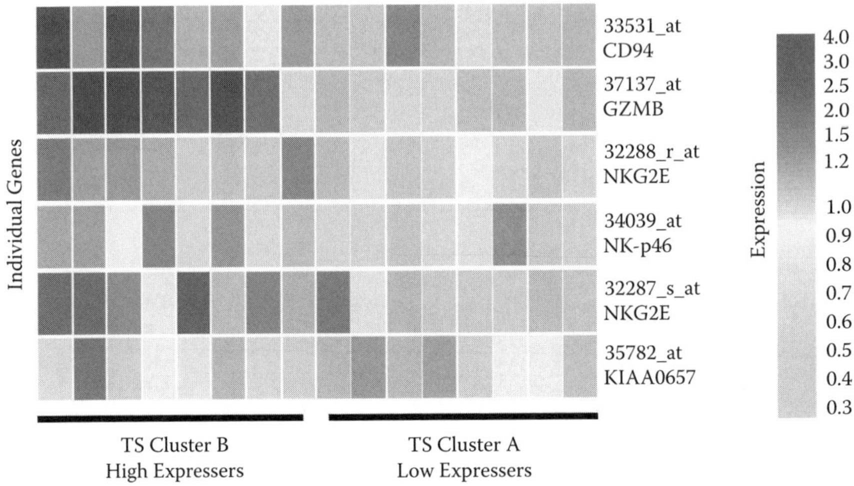

Figure 3.8 The expression of 6 CTL/NK genes in 16 TS samples. The 16 TS samples (*x*-axis) were aligned from left to right in the sequence determined by *k*-means cluster analysis. (From Tang, Y. et al., *Arch. Neurol.,* 62, 210–215, 2005. With permission.)

with abundant receptors and signaling pathways that likely respond to pathological changes in the brain.

Our findings in NF1 support the notion that monogenetic neurologic or multiorgan disorders can be identified by distinct gene expression profiles. The extent to which this approach can be potentially more usefully generalized to many complex traits/disorders is suggested by the results of our small studies of pediatric epilepsy and TS. It has been suggested that a functional genomics approach could define a common pathway of functional abnormalities that could narrow the search for responsible genes.[44,45] In this regard, blood genomic profiling might provide an accessible platform to categorize a polygenic condition into molecular subtypes. As exemplified in the anticonvulsant study, part of the inter-individual variation in VPA efficacy may be related to an identifiable drug effect at the blood transcriptional level. It is possible that blood gene expression markers, either prior to or early in drug therapy, may be able to distinguish those individuals who will respond to medication from those who will not before the clinical end points are reached.

We did not identify a unique, significant blood expression pattern in familial TS. However, our findings could also be consistent with the hypothesis that TS is a heterogeneous disorder including multiple subgroups identifiable by gene arrays. Our finding that a group of genes involved in the functioning of T lymphocytes and NK cells is upregulated in the blood of a subgroup of patients with TS is particularly intriguing, given studies suggesting that autoimmune mechanisms triggered by infection with GABHS are involved in the pathogenesis of some but not all patients with TS.[29] Compared to the traditional single marker approach, using the pattern of gene expression provided by microarrays may prove more informative for subgrouping patients with TS with different causes and/or mechanisms of disease.

With that said, the results of these and future blood gene expression studies in neurologic diseases must be interpreted cautiously. First, most genes regulated in blood have low-fold changes and high variability. The low-fold change is not unexpected considering the absence of obvious blood phenotypes. The variation in gene expression patterns in peripheral blood come from multiple sources such as age, gender, the relative proportions of the different blood cell types, time of blood draw, allelic polymorphisms, and other factors.[13] In addition, the different experimental protocols and time delay in RNA isolation can drastically affect the blood gene expression patterns, further confounding case-control comparisons or obscuring disease associated patterns (see Chapter 2).

Gene expression patterns have to be confirmed and refined with larger studies in which the concomitant factors can be fully characterized and the association of genomic pattern and various clinical features can be probed. In the case of TS, future studies should involve a larger, representative sample of patients with TS with comorbid OCD and ADHD. Special emphasis should be given to determining whether patients with TS who meet some or all of the clinical criteria for PANDAS, as well as atypical or apparently non-PANDAS TS patients, have distinct genomic profiles. The possible role of GABHS or other environmental triggers could be assessed using this methodology in longitudinal studies.

Finally, as exemplified in cancer genomic studies, the diagnostic classifiers must be cross validated. The predictive value of any set of dysregulated genes in an initial patient group should be validated in a second, independent patient sample. Also, it appears that different neurological diseases may affect different subsets of blood cells. Therefore, the gene expression pattern of blood cell subsets should be explored and may in some cases provide better resolution than whole blood.

ACKNOWLEDGMENTS

These studies were supported by NS 41920 (D.L.G.); NS28167, AG19561, NS38084, NS42774, NS43252, and an American Heart Association Bugher Award (F.R.S.); NS040261, NS044956 (T.A.G.); and NS045752 (A.D.H.).

REFERENCES

1. Golub, T.R., Slonim, D.K., Tamayo, P., Huard, C., Gaasenbeek, M., Mesirov, J.P., Coller, H., Loh, M.L., Downing, J.R., Caligiuri, M.A., Bloomfield, C.D., and Lander, E.S., Molecular classification of cancer: class discovery and class prediction by gene expression monitoring. *Science* 286(5439), 531–537, 1999.
2. Yeoh, E.J., Ross, M.E., Shurtleff, S.A., Williams, W.K., Patel, D., Mahfouz, R., Behm, F.G., Raimondi, S.C., Relling, M. V., Patel, A., Cheng, C., Campana, D., Wilkins, D., Zhou, X., Li, J., Liu, H., Pui, C.H., Evans, W.E., Naeve, C., Wong, L., and Downing, J.R., Classification, subtype discovery, and prediction of outcome in pediatric acute lymphoblastic leukemia by gene expression profiling. *Cancer Cell* 1(2), 133–143, 2002.

3. Staunton, J.E., Slonim, D.K., Coller, H.A., Tamayo, P., Angelo, M.J., Park, J., Scherf, U., Lee, J.K., Reinhold, W.O., Weinstein, J.N., Mesirov, J.P., Lander, E.S., and Golub, T.R., Chemosensitivity prediction by transcriptional profiling. *Proc. Natl. Acad. Sci. U.S.A.* 98(19), 10787–10792, 2001.
4. Blalock, E.M., Geddes, J.W., Chen, K.C., Porter, N.M., Markesbery, W.R., and Landfield, P.W., Incipient Alzheimer's disease: microarray correlation analyses reveal major transcriptional and tumor suppressor responses. *Proc. Natl. Acad. Sci. U.S.A.* 101(7), 2173–2178, 2004.
5. Steinman, L. and Zamvil, S., Transcriptional analysis of targets in multiple sclerosis. *Nat. Rev. Immunol.* 3(6), 483–492, 2003.
6. Mirnics, K., Middleton, F.A., Marquez, A., Lewis, D.A., and Levitt, P., Molecular characterization of schizophrenia viewed by microarray analysis of gene expression in prefrontal cortex. *Neuron* 28(1), 53–67, 2000.
7. Purcell, A.E., Jeon, O.H., Zimmerman, A.W., Blue, M.E., and Pevsner, J., Postmortem brain abnormalities of the glutamate neurotransmitter system in autism. *Neurology* 57(9), 1618–1628, 2001.
8. Tang, Y., Lu, A., Aronow, B.J., and Sharp, F.R., Blood genomic responses differ after stroke, seizures, hypoglycemia, and hypoxia: blood genomic fingerprints of disease. *Ann. Neurol.* 50(6), 699–707, 2001.
9. Tang, Y., Nee, A.C., Lu, A., Ran, R., and Sharp, F.R., Blood genomic expression profile for neuronal injury. *J. Cereb. Blood Flow Metab.* 23(3), 310–319, 2003.
10. Tang, Y, Lu, A., Ran, R., Aronow, B.J., Schorry, E.K., Hopkin, R.J., Gilbert, D.L., Glauser, T.A., Hershey, A.D., Richtand, N.W., Privitera, M.D., Dalvi, A., Sahay, A., Szaflarski, J.P., Ficker, D.M., Ratner, N., and Sharp, F.R., Human blood genomics: distinct profiles for gender, age and neurofibromatosis type 1. *Mol. Brain Res.* 132(2), 155–167, 2004.
11. Tang, Y., Glauser, T.A., Gilbert, D.L., Hershey, A.D., Privitera, M.D., Ficker, D.M., Szaflarski, J.P., and Sharp, F.R., Valproic acid blood genomic expression patterns in children with epilepsy — a pilot study. *Acta Neurol. Scand.* 109(3), 159–168, 2004.
12. Li, C. and Wong, W.H., Model-based analysis of oligonucleotide arrays: expression index computation and outlier detection. *Proc. Natl. Acad. Sci. U.S.A.* 98(1), 31–36, 2001.
13. Whitney, A.R., Diehn, M., Popper, S.J., Alizadeh, A.A., Boldrick, J.C., Relman, D.A., and Brown, P.O., Individuality and variation in gene expression patterns in human blood. *Proc. Natl. Acad. Sci. U.S.A.* 100(4), 1896–1901, 2003.
14. Eisen, M.B., Spellman, P.T., Brown, P.O., and Botstein, D., Cluster analysis and display of genome-wide expression patterns. *Proc. Natl. Acad. Sci. U.S.A.* 95(25), 14863–14868, 1998.
15. Evans, W.E. and Relling, M.V., Pharmacogenomics: translating functional genomics into rational therapeutics. *Science* 286(5439), 487–491, 1999.
16. Goldstein, D.B., Pharmacogenetics in the laboratory and the clinic. *N. Engl. J. Med.* 348(6), 553–556, 2003.
17. Jankovic, J., Tourette's syndrome. *N. Engl. J. Med.* 345(16), 1184–1192, 2001.
18. Singer, H.S., Current issues in Tourette syndrome. *Movement Disord.* 15(6), 1051–1063, 2000.
19. Kurlan, R., Eapen, V., Stern, J., McDermott, M.P., and Robertson, M.M., Bilineal transmission in Tourette's syndrome families. *Neurology* 44(12), 2336–2342, 1994.
20. McMahon, W.M., van de Wetering, B.J., Filloux, F., Betit, K., Coon, H., and Leppert, M., Bilineal transmission and phenotypic variation of Tourette's disorder in a large pedigree. *J. Am. Acad. Child Adolesc. Psychiatr.* 35(5), 672–680, 1996.

21. Hanna, P.A., Janjua, F.N., Contant, C.F., and Jankovic, J., Bilineal transmission in Tourette syndrome. *Neurology* 53(4), 813–818, 1999.
22. Barr, C.L., Wigg, K.G., Pakstis, A.J., Kurlan, R., Pauls, D., Kidd, K.K., Tsui, L.C., and Sandor, P., Genome scan for linkage to Gilles de la Tourette syndrome. *Am. J. Med. Genet.* 88(4), 437–445, 1999.
23. Hyde, T.M., Aaronson, B.A., Randolph, C., Rickler, K.C., and Weinberger, D.R., Relationship of birth weight to the phenotypic expression of Gilles de la Tourette's syndrome in monozygotic twins. *Neurology* 42(3 Pt. 1), 652–658, 1992.
24. Leckman, J.F., Dolnansky, E.S., Hardin, M.T., Clubb, M., Walkup, J.T., Stevenson, J., and Pauls, D.L., Perinatal factors in the expression of Tourette's syndrome: an exploratory study. *J. Am. Acad. Child Adolesc. Psychiatr.* 29(2), 220–226, 1990.
25. Santangelo, S.L., Pauls, D.L., Goldstein, J.M., Faraone, S.V., Tsuang, M.T., and Leckman, J.F., Tourette's syndrome: what are the influences of gender and comorbid obsessive-compulsive disorder? *J. Am. Acad. Child Adolesc. Psychiatr.* 33(6), 795–804, 1994.
26. Silva, R.R., Munoz, D.M., Barickman, J., and Friedhoff, A.J., Environmental factors and related fluctuation of symptoms in children and adolescents with Tourette's disorder. *J. Child Psychol. Psychiatr.* 36 (2), 305–312, 1995.
27. Chouinard, S. and Ford, B., Adult onset tic disorders. *J. Neurol. Neurosurg. Psychiatr.* 68 (6), 738–743, 2000.
28. Krauss, J.K. and Jankovic, J., Severe motor tics causing cervical myelopathy in Tourette's syndrome. *Movement Disord.* 11(5), 563–566, 1996.
29. Swedo, S.E., Leonard, H.L., Garvey, M., Mittleman, B., Allen, A.J., Perlmutter, S., Lougee, L., Dow, S., Zamkoff, J., and Dubbert, B.K., Pediatric autoimmune neuropsychiatric disorders associated with streptococcal infections: clinical description of the first 50 cases. *Am. J. Psychiatr.* 155(2), 264–271, 1998.
30. Kurlan, R., Tourette's syndrome and "PANDAS": will the relation bear out? Pediatric autoimmune neuropsychiatric disorders associated with streptococcal infection. *Neurology* 50(6), 1530–1534, 1998.
31. Swedo, S.E., Leonard, H.L., Mittleman, B.B., Allen, A.J., Rapoport, J.L., Dow, S.P., Kanter, M.E., Chapman, F., and Zabriskie, J., Identification of children with pediatric autoimmune neuropsychiatric disorders associated with streptococcal infections by a marker associated with rheumatic fever. *Am. J. Psychiatr.* 154(1), 110–112, 1997.
32. Lord, S.J., Rajotte, R.V., Korbutt, G.S., and Bleackley, R.C., Granzyme B: a natural born killer. *Immunol. Rev.* 193(1), 31–38, 2003.
33. Chang, C., Rodriguez, A., Carretero, M., Lopez-Botet, M., Phillips, J.H., and Lanier, L.L., Molecular characterization of human CD94: a type II membrane glycoprotein related to the C-type lectin superfamily. *Eur. J. Immunol.* 25(9), 2433–2437, 1995.
34. Vankayalapati, R., Wizel, B., Weis, S.E., Safi, H., Lakey, D.L., Mandelboim, O., Samten, B., Porgador, A., and Barnes, P.F., The NKp46 receptor contributes to NK cell lysis of mononuclear phagocytes infected with an intracellular bacterium. *J. Immunol.* 168(7), 3451–3457, 2002.
35. Peterson, B.S., Skudlarski, P., Anderson, A.W., Zhang, H., Gatenby, J.C., Lacadie, C.M., Leckman, J.F., and Gore, J.C., A functional magnetic resonance imaging study of tic suppression in Tourette syndrome. *Arch. Gen. Psychiatr.* 55(4), 326–333, 1998.
36. Albin, R.L., Koeppe, R.A., Bohnen, N.I., Nichols, T.E., Meyer, P., Wernette, K., Minoshima, S., Kilbourn, M.R., and Frey, K.A., Increased ventral striatal monoaminergic innervation in Tourette syndrome. *Neurology* 61(3), 310–315, 2003.

37. Yoshikawa, T., Turner, G., Esterling, L.E., Sanders, A.R., and Detera-Wadleigh, S.D., A novel human myo-inositol monophosphatase gene, IMP.18p, maps to a susceptibility region for bipolar disorder. *Mol. Psychiatr.* 2(5), 393–397, 1997.
38. Yoshikawa, T., Padigaru, M., Karkera, J.D., Sharma, M., Berrettini, W.H., Esterling, L.E., and Detera-Wadleigh, S.D., Genomic structure and novel variants of myo-inositol monophosphatase 2 (IMPA2). *Mol. Psychiatr.* 5(2), 165–171, 2000.
39. Alizadeh, A.A., Eisen, M.B., Davis, R.E., Ma, C., Lossos, I.S., Rosenwald, A., Boldrick, J.C., Sabet, H., Tran, T., Yu, X., Powell, J.I., Yang, L., Marti, G.E., Moore, T., Hudson, J., Jr., Lu, L., Lewis, D.B., Tibshirani, R., Sherlock, G., Chan, W.C., Greiner, T.C., Weisenburger, D.D., Armitage, J.O., Warnke, R., Staudt, L.M., et al., Distinct types of diffuse large B-cell lymphoma identified by gene expression profiling [see comments]. *Nature* 403(6769), 503–511, 2000.
40. Bennett, L., Palucka, A.K., Arce, E., Cantrell, V., Borvak, J., Banchereau, J., and Pascual, V., Interferon and granulopoiesis signatures in systemic lupus erythematosus blood. *J. Exp. Med.* 197(6), 711–723, 2003.
41. Baechler, E.C., Batliwalla, F.M., Karypis, G., Gaffney, P.M., Ortmann, W.A., Espe, K.J., Shark, K.B., Grande, W.J., Hughes, K.M., Kapur, V., Gregersen, P.K., and Behrens, T.W., Interferon-inducible gene expression signature in peripheral blood cells of patients with severe lupus. *Proc. Natl. Acad. Sci. U.S.A.* 100 (5), 2610–2615, 2003.
42. Baldwin, D.N., Vanchinathan, V., Brown, P.O., Theriot, J.A., Boldrick, J.C., Alizadeh, A.A., Diehn, M., Dudoit, S., Liu, C. L., Belcher, C.E., Botstein, D., Staudt, L.M., and Relman, D.A., A gene-expression program reflecting the innate immune response of cultured intestinal epithelial cells to infection by *Listeria monocytogenes. Genome Biol.* 4(1), R2, 2003.
43. Boldrick, J.C., Alizadeh, A.A., Diehn, M., Dudoit, S., Liu, C.L., Belcher, C.E., Botstein, D., Staudt, L.M., Brown, P.O., and Relman, D.A., Stereotyped and specific gene expression programs in human innate immune responses to bacteria. *Proc. Natl. Acad. Sci. U.S.A.* 99(2), 972–977, 2002.
44. Tang, Y., Schapito, M.B., Franz, D.N., Patterson, B.J., Hickey, F.J., Schorry, E.K., Hopkin, R.J., Wylie, M., Narayan, T., Glauser, T.A., Gilbert, D.N., Hershey, A.D., and Sharp, F.R., Blood expression profiles for tuberous sclerosis complex Z, Neurofibromatosis typel, and Down's syndrome. *Annals of Neurology*, 56(6), 808–814, 2004.
45. Tang, Y., Gilbert, D., Glauser, TA., Hershey, A., and Sharp, F.R., Blood gene expression profiling of neurological diseases — a pilot microarray study. *Arch. Neurology* 62(2), 210–215, 2005.

CHAPTER 4

Transcriptional Profiling of Peripheral Blood in Oncology

Michael E. Burczynski

CONTENTS

4.1 INTRODUCTION

Since the introduction of microarrays more than a decade ago, many studies describing global transcriptional profiles in human tissues and model systems have been published. The field of oncology has experienced a specific boom in expression profiling research sufficient to coin its own subspecialty of oncogenomics. Initial studies cataloged transcriptional alterations in primary tumors that were significantly distinct from normal tissues[1,2] or defined molecular subclasses of tumors.[3,4] In more recent years, the focus has turned to the identification of transcriptional patterns in tumors that appear to correlate with patient outcomes in general[5–8] or even predict response to certain therapies or therapeutic classes of compounds.[9,10]

These findings have fueled great interest in the application of transcriptional profiling to samples available from real-time clinical trials, and clinical pharmacogenomic objectives utilizing transcriptional profiling strategies are becoming increasingly incorporated into clinical trial study designs when tumor tissue is available. Despite the great promise afforded by this technology, the ultimate benefit of applying transcriptional profiling in prospective clinical trials has yet to be realized because a number of practical impediments to this process exist. There are a number of circumstances under which primary tumor tissue biopsies may not be available, or appropriate, for expression profiling studies in clinical oncology trials. The most common instance is often encountered in early phase trials of novel oncology therapeutics in patients with advanced cancer who have already failed surgery and one or more courses of radio-, immuno-, or chemotherapies that comprise the typical standards of care for the disease. The oncology patients enrolled in early phase trials have often already undergone tumor resection, and present at the time of enrollment as patients afflicted with advanced metastatic disease. Metastatic tumor biopsies may or may not be available, depending on the trial protocol. Other circumstances that can preclude availability of tumor tissues comprise certain diseases of the central nervous system, or other diseases where surgical access to the tumor tissue is not a safe or feasible option.

The unavailability of primary tumor tissues, however, does not necessarily obviate the use of expression profiling strategies in oncology studies. An area of active research in clinical pharmacogenomics is the investigation of various surrogate tissues as alternative sources of expression profiles that may be informative in the treatment of certain diseases. This approach involves the expression profiling of so-called "surrogate" tissues — peripheral blood, serum, cerebral spinal fluid (CSF), skin biopsies, etc. — in an attempt to identify profiles that may be associated with disease, drug efficacy, or drug toxicity (for a review, see Reference 11). As discussed throughout this book, the main theories behind surrogate tissue profiling are that cells or molecules in the surrogate tissue (1) will reflect some aspect of the disease state, (2) respond differentially following therapeutic intervention, and (3) possibly even be predictive of eventual patient outcomes. In oncology diseases the alterations in surrogate tissue profiles may be directly due to the presence of the tumor (for instance, the detection of a tumor specific antigen in the serum of oncology patients) or secondary responses of the surrogate tissue to the tumor (e.g., transcriptional responses of circulating mononuclear cells to the presence of the tumor). Regardless of the ultimate source of transcriptional alterations in surrogate tissues, surrogate tissue profiling represents a potential method that can provide an alternative global approach for the identification of useful biomarkers in certain oncology indications. The rest of this chapter summarizes (1) strategies for surrogate tissue profiling in the context of translational medicine; (2) analytical approaches in surrogate tissue profiling studies; (3) early evidence in breast cancer profiling studies that surrogate tissues (e.g., infiltrating lymphocytes) might harbor informative transcriptomes; (4) recent efforts dedicated to the application of surrogate tissue profiling in the field of renal cancer; and (5) additional recent examples of surrogate tissue profiling in the field of oncology.

4.2 SURROGATE TISSUE PROFILING IN TRANSLATIONAL MEDICINE AND ONCOLOGY DRUG DEVELOPMENT

Chemotherapeutics are undergoing a revolution in the field of cancer drug development. Whereas previous chemotherapeutic strategies typically included the identification of highly toxic agents designed to kill tumor cells in a nonspecific manner, more and more oncology drug development programs are focusing on agents that target specific molecular features of specific types of tumors. The recent development of Trastuzumab (Herceptin, Genentech, San Francisco, CA) and Imatinib (Gleevec, Novartis, Basel, Switzerland) provide excellent examples of translational strategies implementing predictive biomarkers that identify patients who will most likely respond to specific therapies (for a more in-depth review, see Reference 12). Trastuzumab is a recombinant antibody developed against HER2 based on the role of HER2 in cellular proliferation[13,14] and its overexpression in breast cancer and association with poor prognosis in this population.[15–17] HER2 assessment by an immunohistochemical assay was co-developed with Trastuzumab on the basis of several preclinical and clinical observations that constituted an overall successful translational strategy for this drug (reviewed in Reference 18). Imatinib is a small molecule inhibitor of the ABL tyrosine kinase and was optimized for its ability to inhibit the BCR-ABL tyrosine kinase transforming oncogene that is found in more than 95% of the cases of chronic myeloid leukemia.[19,20] Reverse transcription polymerase chain reaction (RT-PCR)-based screening for the presence of this translocation marker prior to and during therapy offers an opportunity to identify appropriate patients for this therapy and to monitor the course of their response during treatment.[21]

As the number of targeted molecular therapies developed for oncology indications increases, these types of therapeutics will be accompanied by increasingly complex translational activities associated with the identification of relevant biomarkers that enable assessment of the efficacy of the drug candidate in humans. The identification of assays that characterize biomarkers that predict patient responses, and efficacy biomarkers that suitably measure downstream molecular indicators of drug efficacy in human beings, comprises the burgeoning field of translational medicine.

One of the considerations in drug development translational strategies in the future will be whether tumor tissues will be available for biomarker analyses. Early knowledge of the drug candidate's primary indication and proof of mechanism/proof of concept strategy will thus inform decision making concerning the suitability of surrogate tissue profiling.

At the outset it is important to distinguish between pharmacodynamic biomarkers (molecular cellular indicators of the activity of a drug following administration): efficacy biomarkers (molecular/cellular indicators of a beneficial effect of a drug following administration) and predictive biomarkers (molecular/cellular indicators predictive of drug efficacy *prior* to drug administration). The best translational medicine approaches are those similar to Imatinib, in which both types of biomarkers (markers predictive of patient response and markers of drug efficacy) are identified and available for assessment. However, even if an oncology therapeutic is accompanied by a robust translational strategy with biomarkers of pharmacodynamic effect and drug efficacy, and even biomarkers that would be expected to predict clinical benefit, there is still an opportunity for predictive biomarkers to be identified after a drug enters human studies. In scenarios where drugs either do not have predictive biomarker strategies or rationally selected predictive biomarkers ultimately do not predict patient response, clinical pharmacogenomic strategies implemented during early clinical studies can add tremendous value to a drug development program. In such studies retrospective analysis can reveal transcriptional signatures that are correlated with patient responsiveness and development of adverse events within a preselected subpopulation at greater resolution than that provided by the rationally selected predictive biomarker alone. For instance, for an antibody-conjugated cytotoxic therapy the presence of a cell-specific surface marker on the tumor may indicate that a patient is an appropriate candidate for that targeted therapy. Thus, the presence of the cell-specific surface marker will likely comprise the predictive biomarker strategy for that compound as it leaves the drug discovery phase and enters clinical development. However, one can imagine a scenario in which the compound is found to only be effective in 30% of the patients, despite the fact that all of the prestratified patients possess tumors bearing the cell specific surface marker. Clinical pharmacogenomic analyses executed during these clinical studies can provide the opportunity to identify those additional molecular determinants that ultimately define the responsiveness of patients. Perhaps a subset of proteases that can deactivate the therapeutic are overexpressed in approximately 70% of the population — clinical pharmacogenomic studies increase the chances of identifying these additionally predictive biomarkers and enhancing the predictive strength of a composite diagnostic that could accompany the therapeutic in later development. It is in these types of situations that expression profiling of surrogate tissues may afford an opportunity to identify additional molecular determinants that either reflect or directly influence, and therefore predict, the responsiveness of patients to a therapeutic intervention.

4.3 CLASS DISCOVERY AND CLASS DISTINCTION IN SURROGATE TISSUE PROFILING STUDIES

General strategies involved in applying unsupervised and supervised strategies in surrogate tissue profiling studies are presented in Figure 4.1. Samples of RNA from peripheral blood mononuclear cells (PBMCs) or other surrogate tissues are isolated and hybridized to cDNA- or oligonucleotide-based microarrays, and expression profiles are generated for each sample. At this point, sample expression profiles

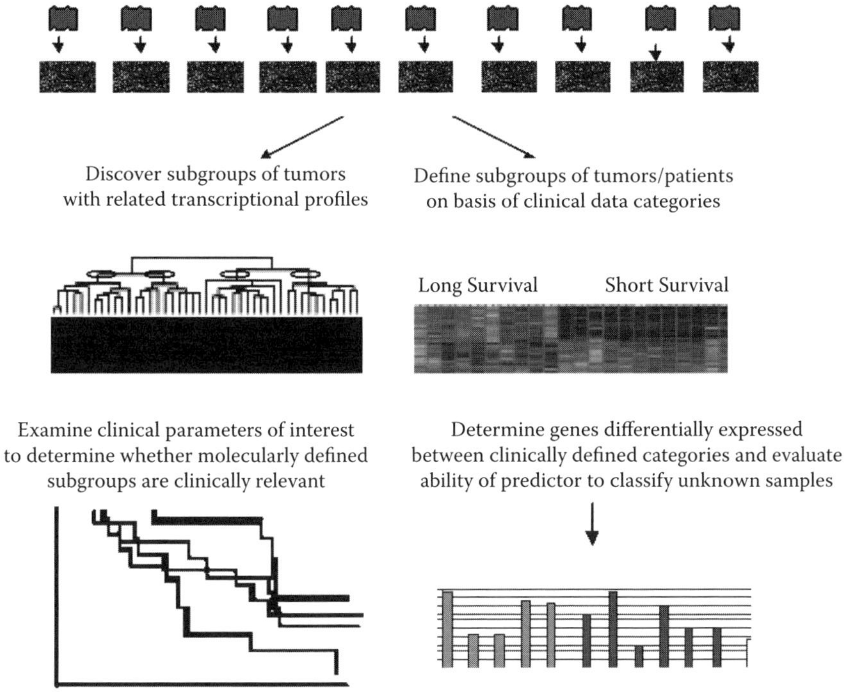

Figure 4.1 Unsupervised and supervised strategies in transcriptional profiling studies. After tumor profiles are hybridized to microarrays and expression profiles are generated, two approaches can be used for the purposes of class discovery or class prediction. In a class discovery approach, unsupervised analysis can be performed (left side) in which the relationships of expression patterns are organized by any number of unsupervised methods (hierarchical clustering is provided as an example here). Once sample relationships have been discovered on the basis of transcriptional profiles, the discovered subgroups are evaluated with respect to their clinical characteristics (Kaplan–Meier analysis of overall survival is provided as an example here). Alternatively, in a class prediction approach expression profiles can be organized according to their clinical characteristics in supervised fashion (right side) and gene selection is performed to identify transcriptional differences between profiles from patients in clinically relevant subgroups of interest (a signal-to-noise ratio metric discovered by a *k*-nearest-neighbors algorithm is presented here). The predictive value of the gene classifier can then be evaluated on an independent set of samples to determine the predictive utility of the discovered classifier (confidence scores for a weighted voting algorithm are depicted here). (Reprinted with permission from Burczynski et al., *Curr. Mol. Med.*, 5, 83–102, 2005.)

can be assessed using either unsupervised strategies to discover novel classes of samples or supervised strategies to distinguish between known classes of samples.

Unsupervised approaches can include hierarchical clustering, principal component analysis, *k*-means clustering, and other methods that discover transcriptional relationships between PBMCs from different patients and thus define molecular subclasses of PBMC profiles on the basis of their transcriptional signatures. The relevance of these molecular subclasses may or may not be related to eventual patient outcome, but that hypothesis can be tested by examining whether clinical parameters are significantly distinct between the discovered classes. Transcriptionally related sets of PBMC profiles can be assessed for similarities or differences in clinical characteristics such as time to disease progression, overall survival, or any other relevant clinical parameter that was measured.

Supervised approaches include nearest-neighbors algorithms, support vector machines, and other class prediction methods that divide profiles into subclasses based on known clinical characteristics and then identify transcriptional differences that can be exploited to predict patient outcomes in future/independent sets of samples. Samples in the clinical classes of interest are typically compared to identify transcriptional differences in PBMCs; this phase of class prediction is also called gene selection, and the samples used in this analysis comprise the training set of samples. Cross-validation approaches can be used on these samples to estimate the predictive value of the discovered gene sets by removing a subset of the samples from the training set, rebuilding the gene classifier, and classifying the removed samples based on their gene expression patterns. However, cross-validation approaches by themselves do not prove predictiveness of the gene classifiers.[22,23] The only approach that can begin to establish the predictive value of a gene classifier is an independent, prospective evaluation of the discovered gene classifier's ability to correctly assign an independent set of samples (a test set).

In the next section we provide a description of an early indication that surrogate tissue profiling in the compartment of peripheral blood may constitute a relevant endeavor in certain oncology situations. Subsequently we discuss in more detail the practical considerations involved in conducting pharmacogenomic studies in real-time oncology clinical trials for the purpose of identifying suitable patient populations that will respond to specific therapies. In the final section we review preliminary results in surrogate tissue profiling during the evaluation of an investigational agent in renal cell carcinoma, and suggest a pathway forward for implementing pharmacogenomic study designs in early phase clinical trials that can facilitate the identification and validation of gene expression-based classifiers in surrogate tissues that enhance the safety and efficacy of therapeutics in patients in certain oncology indications.

4.4 RELEVANCE OF PERIPHERAL BLOOD IN ASSESSMENT OF PATIENTS WITH SOLID TUMORS

Of the major tumors afflicting the worldwide population, more progress has been made in applying transcriptional profiling to breast cancer than any other tumor. An initial expression profiling study in breast cancer characterized the differences between breast carcinoma tissue and human mammary epithelial cells,[1] while another examined transcriptional profiles in laser dissected normal, malignant, and metastatic breast cancer cell populations from the same patient.[2] The latter study identified many differences between normal and malignant profiles, confirming that transcriptional patterns of breast tumors would be distinct from transcriptional patterns in nonmalignant tissue. Soon thereafter multiple laboratories extended these results by investigating larger sets of breast tumor profiles and demonstrating correlations between tumor expression patterns and additional observed characteristics in the tumors. Several groups demonstrated early on that estrogen receptor (ER) status is an important determinant of breast tumor transcriptional patterns.[3,4]

Perou et al. showed that unsupervised analysis of breast tumor profiles using 496 informative genes exhibiting large intervariability (high variation across different individuals' tumors) and small intravariability (low variation within replicate samples of each individual's tumor) readily distinguished ER-positive tumors from ER-negative tumors.[4] In that same study they also identified many functionally related clusters of transcripts exhibiting similar patterns of expression in transcriptionally defined subtypes of breast cancers. Interestingly, these included genes in an endothelial cell gene expression cluster, a stromal/fibroblast cluster, a breast basal epithelial cluster, a B-cell cluster, an adipose-enriched/normal breast cluster, a macrophage cluster, a T-cell cluster, and a breast luminal epithelial cell cluster. The identification of blood cell-specific expression patterns in the various tumors suggested the presence of inflammatory infiltrates contributed in part to the grossly dissected tumor profiles. Thus, this early study not only described a wealth of information for breast cancer, but also provided indirect evidence that surrogate tissues (PBMCs, etc.) might possess informative transcriptional profiles in the context of solid tumor diseases.

One of the next major advances in breast cancer profiling came as a result of research studies directed solely toward the identification of tumor profiles predictive of the clinical outcome of survival.[7,8] These confirmed initial indications that profiles in breast tumors appeared correlated with patient outcome.[5,6] van't Veer and colleagues evaluated 98 primary breast tumor specimens (only samples with greater than 50% tumor cells were analyzed) from patients that were accompanied by a larger pool of associated tumor and clinical outcome annotation — histological grade, *BRCA 1* status, *ER* status, tumor typing, an index of angioinvasion, degree of lymphocytic infiltration, the onset of eventual metastases and overall survival.[8] Unsupervised analysis of these breast tumor profiles using ~5000 genes significantly regulated across the samples identified two main subgroups characterized by differences in expression of *ER*-alpha and *ER*-alpha coregulated genes, and differences in the levels of genes expressed in B and T cells. The authors thus arrived at the same conclusions as those previously described by Perou et al.,[4] namely, that unsupervised clustering detects two main subgroups of breast cancer that appear strongly related to *ER* status and lymphocytic infiltration.

The sum total of data from these and other studies has provided evidence that transcriptional patterns in breast tumors appear to be (1) reflective of genetic/epigenetic alterations that have contributed to the development of the cancer; (2) reflective of the degree of immunological infiltration in the tumor tissue; and (3) prognostic in the context of metastasis-free survival and overall survival. Of relevance to surrogate tissue profiling, a gene signature characteristic of lymphocyte infiltration was identified as a recognizable subcluster of gene expression in breast tumors,[4,8] and lymphocyte infiltration is a hallmark of many other tumors.[24–26] It is not clear whether lymphocytes or other cells destined to infiltrate/interact with tumors will contribute to expression profiles in the periphery that are specific to disease, but this hypothesis can be explored. The data from these studies alone by no means provide proof that infiltrating lymphocytes will exhibit transcriptional patterns in the periphery reflective of tumor aggressiveness/tumor status, but the identification of B- and T-cell transcriptional signatures that appear to define subclasses of tumors with respect to prognosis seems to justify the exploration of surrogate tissuc profiles in the field of oncology. Such assessments should help determine whether the circulating profiles of lymphocytes and monocytes may provide biomarkers of relevance in solid tumor diseases.

4.5 PHARMACOGENOMIC ANALYSIS OF PBMCS IN RENAL CELL CARCINOMA: A CASE STUDY

4.5.1 Disease-Associated Transcripts in PBMCs of Patients with Renal Cancer

Renal cell carcinoma (RCC) comprises the majority of all cases of kidney cancer and is one of the 10 most common cancers in industrialized countries.[27] The 5-year survival rate for advanced RCC is less than 5%.[28] RCC is usually detected by imaging methods, and 30% of apparently nonmetastatic patients undergo relapse after surgery and eventually succumb to disease.[29] Recent expression profiling studies have demonstrated that the transcriptional profiles of primary malignancies are radically altered from the transcriptional profiles of the corresponding normal tissue (for a review, see Reference 30). Specific microarray studies examining RCC tumor transcriptional profiles in detail[31] have identified many classes of genes altered between normal kidney tissue and primary RCC tumors.

Despite the progress in expression profiling of primary malignant tissues, until very recently it was unknown whether in the context of RCC or any other active solid tumor burden there would exist correspondingly distinct markers of gene expression in the PBMCs of affected individuals. In a study conducted by our laboratory, global expression profiles of PBMCs from patients with RCC were compared with PBMC profiles from normal volunteers using oligonucleotide arrays for the purpose of identifying surrogate transcriptional markers of disease in the blood of patients with RCC.[32]

Gene expression patterns were analyzed in 20 disease-free individuals in parallel with the 45 baseline PBMC samples from patients with RCC consented for pharmacogenomic analysis. Expression profiling analysis of the 20 normal PBMC RNA samples and 45 RCC PBMC RNA samples revealed that of the 12,626 genes on the HgU95A chip, 5249 genes met the initial criteria for further analysis (at least one present call, at least one frequency > 10 ppm). On average, 4023 transcripts were detected as "present" in the 45 RCC PBMCs, while 4254 expressed transcripts were detected as "present" in the 20 normal PBMCs. An initial unsupervised cluster analysis approach grouped the majority of RCC PBMCs (42/45) into a single RCC-specific cluster, while expression patterns of normal PBMCs and a small subset of RCC PBMCs (3/45) formed a separate cluster (Figure 4.2). A fold change analysis and Student's *t*-test (two-tailed distribution; two-sample unequal variance) identified 132 transcripts that differed on average by twofold or greater between normal and

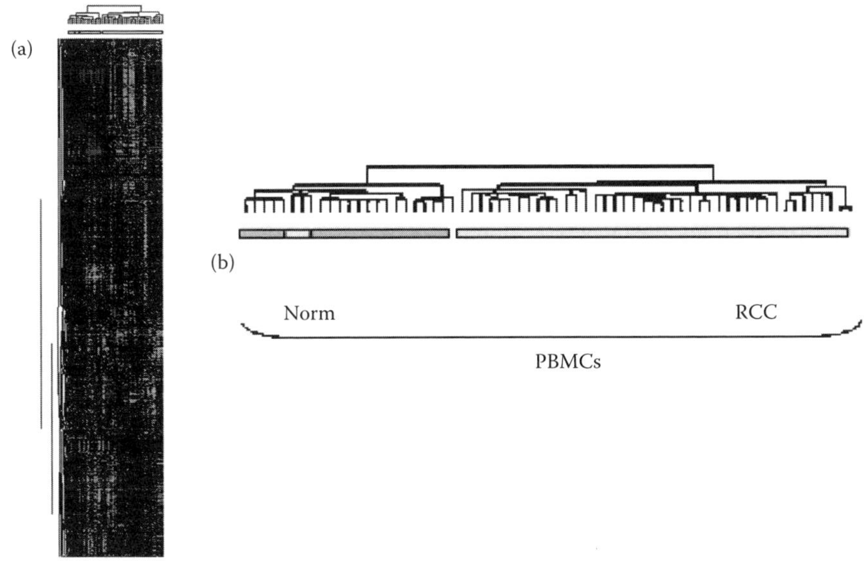

Figure 4.2 **(Color figure follows p. 138.)** Global expression analysis of PBMCs from patients with RCC and normal volunteers. Total RNA obtained from PBMCs of 45 patients with RCC and PBMCs from 20 normal subjects were analyzed on oligonucleotide arrays containing more than 12,000 full-length human genes. In total, 65 samples were analyzed on individual arrays. In no case were samples pooled. (A) Unsupervised hierarchical cluster analysis of normal and RCC PBMCs using all expressed genes present in at least one sample and possessing a frequency of greater than 10 ppm in at least one sample (5249 genes total). Red indicates genes that are elevated relative to the average expression value across all experiments and green indicates genes that are decreased relative to the average expression value. (B) A dendrogram of sample relatedness using all expressed gene expression values is shown. RCC patient PBMC expression profiles are denoted by yellow bars and normal volunteer PBMC expression profiles are indicated by gray bars. (With permission from Twine et al., *Cancer Res.*, 63, 6069–6075, 2003.)

RCC PBMCs with an unadjusted p value below 0.001 and were expressed in at least 15% of the PBMC samples analyzed. These results are reminiscent of a recent publication that identified profiles in circulating T cells of patients with melanoma that are distinct from those of healthy individuals,[33] demonstrating that the transcriptional profiles of circulating CD8+ T cells also appear distinct in the context of melanoma. It will be interesting to determine whether less immunogenic solid tumors bear similar disease-associated signatures.

It is theoretically possible that transcriptional alterations in the blood of patients with RCC were due to the presence of metastatic cells in the circulation contributing to the transcriptional profiles that were measured (see Chapters 13 and 14 for further discussion on the analysis of circulating tumor cells from peripheral blood). However, investigation of the disease-associated genes in RCC PBMCs failed to reveal commonality with the tumor-associated genes following evaluation of the transcripts most strongly upregulated in RCC tumors (n = 47) relative to normal kidney tissue

(n = 60) using profiles downloaded from the Bioexpress Database (GeneLogic, Gaithersburg, MD). This lack of overlap suggested that shed RCC tumor cells did not contribute significantly to the disease-associated transcripts identified in PBMCs isolated from patients with RCC.

An *ex vivo* T-cell activation model was also used to determine whether any of the transcripts observed as elevated or repressed in RCC PBMCs were common to those elevated or repressed following T-cell activation *ex vivo*. In this approach we identified 14 of the 132 transcripts that were differentially expressed in RCC PBMCs and differentially expressed between unstimulated CD4+ T cells (n = 3 normal donors) and CD4+ T cells (n = 3 normal donors) stimulated with anti-CD3 and anti-CD28 in culture.

Also investigated was the question of whether the differentially expressed transcriptional patterns in PBMCs of patients with RCC were similar to PBMC transcriptional responses observed in non-RCC end-stage renal failure. The differentially expressed genes in RCC PBMCs were compared with genes differentially expressed between PBMCs from patients with non-RCC end-stage renal failure (n = 9 individuals) and PBMCs from normal volunteers (n = 4 individuals). Of these, 9 transcripts differentially expressed in PBMCs from patients with renal failure were also disease-associated transcripts in RCC PBMCs. Thus, the marker gene list from PBMCs of patients with RCC contains a smaller subset of markers commonly involved in immune responses measured *ex vivo* (activated CD4+ T-cell profiles) and in responses of circulating leukocytes to renal dysfunction (PBMCs from patients with non-RCC renal failure) observed *in vivo*.

The potential practical utility of these results was demonstrated by determining the ability of minimal gene set(s) to classify RCC vs. RCC disease-free status using expression patterns in the peripheral blood. To initially build and subsequently train the classifiers, 70% of the RCC PBMCs (n = 31) and 70% of the normal PBMCs (n = 14) were randomly selected and used as the training set. Genecluster's default correlation metric[34] was used to identify genes with expression levels most highly correlated with the classification vector characteristic of the training set. All 5249 genes meeting the main filter criteria were screened using this approach. Prediction was also performed in Genecluster using the weighted voting method. In this method, the expression level of each gene in the classifier set contributes to an overall vote on the classification of the sample.[35] The prediction strength is a combined variable that indicates the support for one class or the other, and can vary between 0 (narrow margin of victory) and 1 (wide margin of victory) in favor of the predicted class. An 8-gene classifier set containing the four top genes upregulated in RCC and the four top genes downregulated in RCC was found to yield the highest cross-validation prediction accuracy on the training set. Classification of the remaining test set of samples using the 8-gene classifier showed that the predicted class matched the true class in all cases, though for one of the 19 test samples the prediction strength was negligible. These studies therefore demonstrated the feasibility of predicting advanced RCC vs. non-RCC status using expression patterns found in a limited number of gene transcripts in mononuclear cells from peripheral blood. However, since the patients in this study were patients with advanced cases of renal cancer, the potential relevance of these transcriptional signatures for early detection of renal disease is unknown.

4.5.2 Outcome-Correlated Patterns in Pretreatment PBMCs of Patients with RCC

Our laboratory has also recently tested the hypothesis that several other types of transcriptional biomarkers may be characteristic of the circulating PBMCs of human patients. In the best-case scenario, expression profiling of surrogate tissues *prior* to drug therapy would reveal transcriptional biomarkers or patterns that are predictive of whether a patient will ultimately respond to a given therapeutic regimen *before* drug therapy is ever initiated. Such transcriptional signatures may be generally prognostic, representing a patient's general disease status, or they may be more therapeutically predictive, indicating the responsiveness of a patient to a given therapy. We have recently identified transcriptional patterns in PBMCs of patients with RCC that appear associated with times to disease progression and overall survival, by both Cox proportional hazard regression and supervised classification algorithms.[36] An unsupervised analysis of the RCC PBMC profiles suggested that subsets of patients with distinct expression patterns in PBMCs exhibited differences in survival (Figure 4.3). Supervised analyses identified outcome-associated patterns in PBMCs in a training set of sample within the Phase II study, which were subsequently evaluated on an independently withheld test set of samples from the same Phase II study. The clinically defined classes of patient samples with poor and favorable outcomes that were used to develop the gene classifiers in the training set were not confounded by any of the recorded clinical and technical parameters, including patient demographics, technical gene chip parameters, and cellular composition of the isolated PBMCs. Thus, the results of this study using multiple analytical approaches appear to imply that PBMC profiles can provide an early indicator of eventual patient outcomes, and the predictive models discovered in this Phase II study are currently undergoing evaluation in an ongoing Phase III trial in renal cancer.

The implications for these latter types of surrogate biomarkers of response in clinically accessible tissues are enormous and could ultimately influence clinical trial design and label indications for therapeutics in the future. It is therefore critical to validate the results of these smaller proof-of-concept studies in larger trials, and to understand the mechanistic relevance (if any) of transcriptional profiles in surrogate tissues that are ultimately correlated with clinical courses. Additionally, in the future it will be important to establish rigorous standards that will allow the utilization of assays, platforms, and reference standards that can accurately determine transcriptional profiles in surrogate tissues of human patients.

4.6 OTHER SURROGATE TISSUE PROFILING STUDIES IN ONCOLOGY

Pharmacogenomic studies conducted on PBMCs of patients with melanoma receiving interleukin-2 (IL-2) therapy[37] demonstrated that treatment with this cytokine results in large gene expression changes in circulating PBMCs. By carefully comparing expression patterns in the circulating mononuclear cells with expression changes identified in the tumor microenvironment, the authors did not find substantial

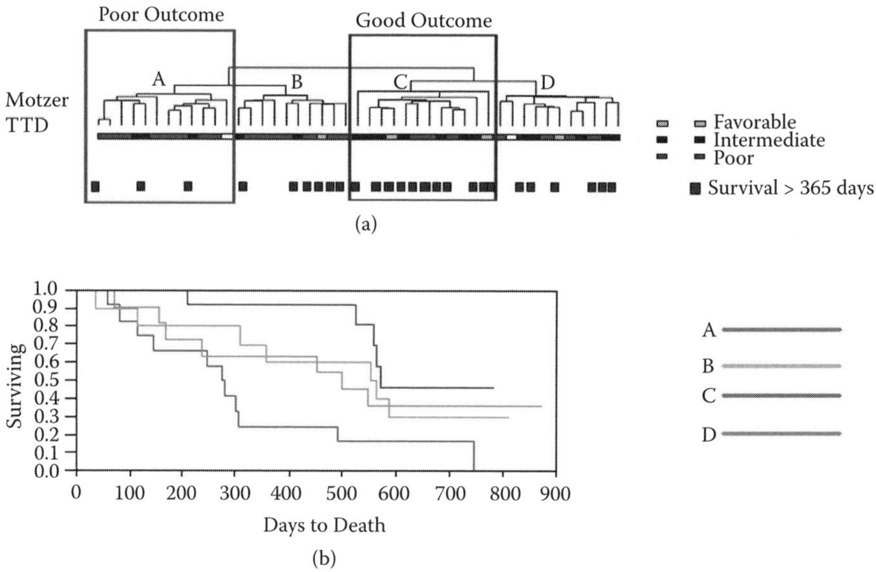

Figure 4.3 **(Color figure follows p. 138.)** Unsupervised hierarchical clustering of RCC patient PBMC profiles and correlation with overall survival. (a) The dendrogram of sample relatedness using all 5424 genes' expression values is shown. Four distinct nodes were identified (nodes A, B, C, and D). Of the 12 patients with PBMC profiles in cluster A, 9 exhibited survival of less than 1 year (red outline), while 10 of the 12 patients with PBMC profiles in cluster C exhibited survival greater than 1 year (blue outline). The associated Motzer risk classifications (green = favorable, black = intermediate, red = poor, yellow = unassigned) are presented underneath the dendrogram. Year-long survival (blue squares indicate > 1 year survival) is also presented. (b) Kaplan–Meier survival curves for patients in the unsupervised analysis. Patients in cluster A possessed significantly shorter survival (median survival = 281 days) relative to patients in clusters B (median survival = 566 days), C (median survival = 573), and D (median survival = 502 days). (With permission from Burczynski et al., *Clin. Cancer Res.,* 11, 1181–1189, 2005.)

overlap, but rather evidence for an IL-2-based activation of antigen-presenting monocytes, release of chemoattractants, and the activation of lytic systems in monocytes and natural killer (NK) cells. On the basis of the results the authors postulated that the main effect of IL-2 administered *in vivo* may be the facilitation of T-cell effector function rather than sustaining their proliferation and noted that if this hypothesis proves true then adoptive transfer of effector T cells should follow, rather than precede, administration of IL-2.

In addition to the above analysis of a cytokine, pharmacogenomic analyses of small molecule inhibitors have also been conducted in PBMCs. DePrimo et al.[38] evaluated the effects of the kinase inhibitor SU5416 in PBMCs of patients with colorectal cancer and identified four transcripts that could accurately reflect control and treatment arms. Since SU5416 is an antagonist of the vascular endothelial growth factor (VEGF) receptor, the authors reasoned that PBMC profiling may reflect SU5416 exposure through direct effects of VEGF receptor antagonism on VEGF receptor-expressing monocytes, or through indirect effects of therapy-induced perturbations in circulating cells. Although transcripts that appeared specific to SU5416 were identified, no transcripts appeared to correlate with responses measured in the study.

4.7 ISSUES AND CAVEATS WITH PBMC PROFILING IN ONCOLOGY STUDIES

PBMC profiling represents a difficult task, both in terms of logistics and interpretation. Our own internal studies and more recent reports have documented significant alterations in select subsets of transcripts in PBMCs following overnight incubation of whole blood at room temperature under conditions that mimic those encountered in clinical trials, where blood samples are drawn at clinical sites and then shipped overnight to a central processing laboratory.[39,40] Debey et al.[39] reevaluated the transcripts our laboratory identified as disease associated in patients with RCC[32] and identified only 12 of the 132 disease-associated transcripts as belonging to the group of transcripts subject to significant fluctuation following overnight incubation. This was an expected result, since our original analysis involved PBMCs from disease-free individuals and patients with RCC that were handled under identical conditions of overnight storage of whole blood prior to processing to PBMCs. However, the authors correctly opine that, while the majority of disease-associated transcripts in RCC PBMCs do not appear to be subject to significant alteration during *ex vivo* incubation, it is possible that our laboratory missed transcripts that might have been truly differentially expressed in patients with RCC relative to healthy controls, but were missed following an overnight incubation that resulted in severe artifactually induced downregulation of a subset of transcripts in both populations of blood samples. It is clear that immediate processing is the optimal approach, but for large multicenter trials this remains a practical impossibility since many of the sites lack the necessary staff or equipment to execute these purifications. In this instance, then, overnight shipping of collected blood samples and processing by a central laboratory is the method of choice, since this method treats all samples similarly. Nonetheless, our laboratory and others are constantly evaluating alternative approaches to blood collection, stabilization, and preparation for the purposes of minimizing artifactual *ex vivo* changes in gene expression profiles determined from *in vivo* samples. The issues and caveats associated with blood profiling are presented in this book in greater detail in Chapter 2.

In addition to the *ex vivo* effects of storage prior to processing, depending on the disease state of the individual in question, activated neutrophils and other polymorphonuclear leukocytes can differentially migrate through density gradients designed to enrich for mononuclear cells, further confounding analyses of PBMC transcriptional profiles.[41] While these alterations may be very relevant to the overall disease process rather than a simple technical artifact (neutrophil activation may be correlated with tumor aggressiveness, etc.), these variabilities in cellular composition nonetheless present difficulties associated with analysis and interpretation of transcriptional profiling data in PBMCs.

Our laboratory now routinely employs ANCOVA approaches to account for variations in cell populations, and this approach greatly assists in delineation of transcriptional differences in PBMCs that appear related to differences in cell populations vs. differences that are independent of cell populations and therefore likely represent bona fide alterations in transcriptional regulation. Nonetheless, newer technologies like PaxGene (Qiagen, Valencia, CA) and other whole-blood stabilization technologies afford an opportunity to minimize variation due to sample handling and processing. While they are associated with their own challenges (for instance, PaxGene-purified RNA contains excess hemoglobin RNA that can dramatically reduce the sensitivity of RNA profiles measured on oligonucleotide arrays), these advances in technology are continuing to evolve and striving to provide a more reliable alternative to enable surrogate tissue profiling with greater reproducibility. For example, globin reduction protocols have been developed, and other non-RNAse-based methods for globin reduction are currently under development, for the removal of globin from whole-blood samples prior to expression profiling. Further studies are required to determine whether profiling of surrogate tissues such as PBMCs will provide transcriptional markers with meaningful applicability in solid tumor diseases, but given the abundant evidence for the interaction between the immune system and tumor cells trying to evade it, the possibility remains an attractive hypothesis for exploration.

4.8 SUMMARY

In summary, initial encouraging data have provided support for the incorporation of transcriptional profiling of surrogate tissues in clinical studies. The promise afforded by genome-wide transcriptional profiling technology is great and will be poised for realistic evaluation in upcoming years as pharmaceutical companies continue to employ these strategies in decision making during the drug development process. It will be important in the future to determine (1) the disease-specificity of transcriptional patterns in PBMCs in patients with solid tumors; (2) the overlap in common between transcriptional responses in PBMCs to the presence of solid tumors; and (3) most importantly, whether transcriptional profiles in peripheral blood may serve as indicators of tumor aggressiveness or responsiveness to therapy. More research is required to determine whether surrogate tissues will prove capable of providing prognostic or theranostic information in given disease(s) in the field of oncology.

ACKNOWLEDGMENTS

The author thanks all of the patients who have contributed samples to foster research in the field of clinical pharmacogenomics.

REFERENCES

1. Perou, C.M., Jeffrey, S.S., van de Rijn, M., Rees, C.A., Eisen, M.B., Ross, D.T., Pergamenschikov, A., Williams, C.F., Zhu, S.X., Lee, J.C.F., Slano, D., Brown, P.O., and Bottstein, D. Distinctive gene expression patterns in human mammary epithelial cells and breast cancers. *Proc. Natl. Acad. Sci. U.S.A.*, 96, 9212–9217, 1999.
2. Sgroi, D.C., Teng, S., Robinson, G., LeVangie, R., Hudson, J.R., Jr., and Elkahloun, A.G. *In vivo* gene expression profile analysis of human breast cancer progression. *Cancer Res.*, 59, 5656–5661, 1999.
3. Bertucci, F., Houlgatte, R., Benziane, A., Granjeaud, S., Adelaide, J., Tagett, R., Loriod, B., Jacquemier, J., Viens, P., Jordan, B., Birnbaum, D., and Nguyen, C. Gene expression profiling of primary breast carcinomas using arrays of candidate genes. *Hum. Mol. Genet.,* 9, 2981–2991, 2000.
4. Perou, C.M., Sorlie, T., Eisen, M.B., van de Rijn, M., Jeffrey, S.S., Rees, C.A., Pollack, J.R., Ross, D.T., Johnsen, H., Akslen, L.A., Fluge, O., Pergamenschikov, A., Williams, C., Zhu, S.X., Lonning, P.E., Borresen-Dale, A.L., Brown, P.O., and Botstein, D. Molecular portraits of human breast tumours. *Nature,* 406, 747–752, 2000.
5. Sorlie, T., Perou, C.M., Tibshirani, R., Aas, T., Geisler, S., Johnsen, H., Hastie, T., Eisen, M.B., van de Rijn, M., Jeffrey, S.S., Thorsen, T., Quist, H., Matese, J.C., Brown, P.O., Botstein, D., Eystein Lonning, P., and Borresen-Dale, A.L. Gene expression patterns of breast carcinomas distinguish tumor subclasses with clinical implications. *Proc. Natl. Acad. Sci. U.S.A..,* 98, 10869–10874, 2001.
6. West, M., Blanchette, C., Dressman, H., Huang, E., Ishida, S., Spang, R., Zuzan, H., Olson, J.A., Jr., Marks, J.R., and Nevins, J.R. Predicting the clinical status of human breast cancer by using gene expression profiles. *Proc. Natl. Acad. Sci. U.S.A.*, 98, 11462–11467, 2001.
7. van de Vijver, M.J., He, Y.D., van't Veer, L.J., Dai, H., Hart, A.A., Voskuil, D.W., Schreiber, G.J., Peterse, J.L., Roberts, C., Marton, M.J., Parrish, M., Atsma, D., Witteveen, A., Glas, A., Delahaye, L., van der Velde, T., Bartelink, H., Rodenhuis, S., Rutgers, E.T., Friend, S.H., and Bernards, R. A gene-expression signature as a predictor of survival in breast cancer. *N. Engl. J. Med.*, 347, 1999–2009, 2002.
8. van't Veer, L.J., Dai, H., van de Vijver, M.J., He, Y.D., Hart, A.A., Mao, M., Petersen, H.L., van der Kooy, K., Marton, M.J., Witteveen, A.T., Schreiber, G.J., Kerkhoven, R.M., Roberts, C., Linsley, P.S., Bernards, R., and Friend S.H. Gene expression profiling predicts clinical outcome of breast cancer. *Nature*, 415, 530–536, 2002.
9. Chang, J.C., Wooten, E.C., Tsimelzon, A., Hilsenbeck, S.G., Gutierrez, M.C., Elledge, R., Mohsin, S., Osborne, C.K., Chamness, G.C., Allred, D.C., and O'Connell, P. Gene expression profiling for the prediction of therapeutic response to docetaxel in patients with breast cancer. *Lancet*, 362, 362–369, 2003.
10. Ayers, M., Symmans, W.F., Stec, J., Damokosh, A.I., Clark, E., Hess, K., Lecocke, M., Metivier, J., Booser, D., Ibrahim, N., Valero, V., Royce, M., Arun, B., Whitman, G., Ross, J., Sneige, N., Hortobagyi, G.N., and Pusztai, L. Gene expression profiles predict complete pathologic response to neoadjuvant paclitaxel and fluorouracil, doxorubicin, and cyclophosphamide chemotherapy in breast cancer. *J. Clin. Oncol.*, 22, 2284–2293, 2004.
11. Rockett, J.C., Burczynski, M.E., Fornace, A.J., Jr., Hermann, P.C., Krawetz, S.A., and Dix, D.J. Surrogate tissue analysis: monitoring toxicant exposure and health status of inaccessible tissues through the analysis of accessible tissues and cells. *Tox. Appl. Pharmacol.*, 194, 189–199, 2004.

12. Park, J.W., Kerbel, R.S., Kelloff, G.J., Barrett, J.C., Chabner, B.A., Parkinson, D.R., Peck, J., Ruddon, R.W., Sigman, C.C., and Slamon, D.J. Rationale for biomarkers and surrogate endpoints in mechanism-driven oncology drug development. *Clin. Cancer Res.,* 10, 3885–3896, 2004.
13. Yarden, Y. and Sliwkowski, M.X. Untangling the ErbB signalling network. *Nat. Rev. Mol. Cell Biol.*, 2, 127–137, 2001.
14. Baselga, J. and Hammond, L.A. HER-targeted tyrosine-kinase inhibitors. *Oncology,* 63(Suppl. 1), 6–16, 2002.
15. Slamon, D.J., Clark, G.M., Wong, S.G., Levin, W.J., Ullrich, A., and McGuire WL. Human breast cancer: correlation of relapse and survival with amplification of the HER-2/neu oncogene. *Science*, 235, 177–182, 1987.
16. Ravdin, P.M. and Chamness, G.C. The c-erbB-2 proto-oncogene as a prognostic and predictive marker in breast cancer: a paradigm for the development of other macromolecular markers — a review. *Gene,* 159, 19–27, 1995.
17. Revillion, F., Bonneterre, J., and Peyrat, J.P. ERBB2 oncogene in human breast cancer and its clinical significance. *Eur. J. Cancer*, 34, 791–808, 1998.
18. Harries, M. and Smith, I. The development and clinical use of trastuzumab (Herceptin). *Endocr. Relat. Cancer*, 9, 75–85, 2002.
19. Lugo, T.G., Pendergast, A.M., Muller, A.J., and Witte, O.N. Tyrosine kinase activity and transformation potency of bcr-abl oncogene products. *Science*, 247, 1079–1082, 1990.
20. Druker, B.J. Imatinib alone and in combination for chronic myeloid leukemia. *Semin. Hematol.*, 40, 50–58, 2003.
21. Deininger, M.W., O'Brien, S.G., Ford, J.M., and Druker, B.J. Practical management of patients with chronic myeloid leukemia receiving imatinib. *J. Clin. Oncol.*, 21, 1637–1647, 2003.
22. Simon, R. Diagnostic and prognostic prediction using gene expression profiles in high dimensional microarray data. *Br. J. Cancer*, 9, 1599–1604, 2003.
23. Simon, R., Radmacher, M.D., and Dobbin, K. Pitfalls in the use of DNA microarray data for diagnostic and prognostic classification. *J. Natl. Cancer Inst.,* 95, 14–18, 2003.
24. Halapi, E. Oligoclonal T cells in human cancer. *Med. Oncol*., 15, 203–211, 1998.
25. Zusman, I., Gurevich, P., Gurevich, E., and Ben-Hur, H. The immune system, apoptosis and apoptosis-related proteins in human ovarian tumors. *Int. J. Oncol.*, 18, 965–972, 2001.
26. Lin, E.Y. and Pollard, J.W. Role of infiltrated leucocytes in tumour growth and spread. *Br. J. Cancer.*, 90, 2053–2058, 2004.
27. Landis, S.H., Murray, T., Bolden, S., and Wingo, P.A. Cancer statistics, 1999. *CA Cancer J. Clin.*, 49, 8–31, 1999.
28. Moch, H.., Gasser, T., Amin, M.B., Torhorst. J., Sauter. G., and Mihatsch, M.J. Prognostic utility of the recently recommended histologic classification and revised TNM staging system of renal cell carcinoma: a Swiss experience with 588 tumors. *Cancer,* 89, 604–614, 2000.
29. Minasian, L.M., Motzer, R.J., Gluck, L., Mazumdar, M., Vlamis, V., and Krown, S.E. Interferon alfa-2a in advanced renal cell carcinoma: treatment results and survival in 159 patients with long-term follow-up. *J. Clin. Oncol.*, 11, 1368–1375, 1993.
30. Slonim, D.K. Transcriptional profiling in cancer: the path to clinical pharmacogenomics. *Pharmacogenomics*, 2, 123–136, 2001.

31. Young, A.N., Amin, M.B., Moreno, C.S., Lim, S.D., Cohen, C., Petros, J.A., Marshjall, F.F. and Neish, A.S. Expression profiling of renal epithelial neoplasms: a method for tumor classification and discovery of diagnostic molecular markers. *Am. J. Pathol.*, 158, 1639–1651, 2001.
32. Twine, N.C, Stover, J.A., Marshall, B., Dukart, G., Hidalgo, M., Stadler, W., Logan, T., Dutcher, J., Hudes, G., Dorner, A.J., Slonim, D.K., Trepicchio, W.L., and Burczynski, M.E. Disease-associated expression profiles in peripheral blood mononuclear cells from patients with advanced renal cell carcinoma. *Cancer Res.,* 63, 6069–6075, 2003.
33. Xu, T., Shu, C.T., Purdom, E., Dang, D., Ilsley, D., Guo, Y., Weber, J., Holmes, S.P., and Lee, P.P. Microarray analysis reveals differences in gene expression of circulating CD8+ T cells in melanoma patients and healthy donors. *Cancer Res.*, 64, 3661–3667, 2004.
34. Golub, T.R., Slonim, D.K., Tamayo, P., Huard, C., Gaasenbeek, M., Mesirov, J.P., Coller, H., Loh, M.L., Downing, J.R., Caligiuri, M.A., Bloomfield, C.D., and Lander, E.S. Molecular classification of cancer: class discovery and class prediction by gene expression monitoring. *Science*, 286, 531–537, 1999.
35. Slonim, D.K., Tamayo, P., Mesirov, J.P., Golub, T.R., and Lander, E.S. Class prediction and discovery using gene expression data. *Proc. Fourth Annual Conference on Computational Molecular Biology*, 263–272, 2000.
36. Burczynski, M.E., Twine, N.C., Dukart, G., Marshall, B., Hidalgo, M., Stadler, W.M., Logan, T., Dutcher, J., Hudes, G., Trepicchio, W.L., Strahs, A., Immermann, F., Slonim, D.K., and Dorner, A.J. Transcriptional profiles in peripheral blood mononuclear cells prognostic of clinical outcomes in patients with advanced renal cell carcinoma. *Clin. Cancer Res.*, 11:1181–1189, 2005.
37. Panelli, M.C., Wang, E., Phan, G., Puhlmann, M., Miller, L., Ohnmacht, G.A., Klein, H.G., and Marincola, F.M. Gene-expression profiling of the response of peripheral blood mononuclear cells and melanoma metastases to systemic IL-2 administration. *Genome Biol.,* 3, research0035, 2002.
38. DePrimo, S.E., Wong, L.M., Khatry, D.B., Nicholas, S.L., Manning, W.C., Smolich, B.D., O'Farrell, A.M., and Cherrington, J.M. Expression profiling of blood samples from an SU5416 Phase III metastatic colorectal cancer clinical trial: a novel strategy for biomarker identification. *BMC Cancer*, 3, 3–14, 2003.
39. Debey, S., Schoenbeck, U., Hellmich, M., Gathof, B.S., Pillai, R., Zander, T., and Schultze, J.L. Comparison of different isolation techniques prior gene expression profiling of blood derived cells: impact on physiological responses, on overall expression and the role of different cell types. *Pharmacogenomics J.*, 4, 193–207, 2004.
40. Baechler, E.C., Batliwalla, F.M., Kaypis, G., Gaffney, P.M., Moser, K., Ortmann, W.A., Espe, K.J., Balasubramanian, S., Hughes, K.M., Chan, J.P., Begovich, A., Chang, S.Y., Gregersen, P.K., and Behrens, T.W. Expression levels for many genes in human peripheral blood cells are highly sensitive to *ex vivo* incubation. *Genes Immun.*, 5, 347–353, 2004.
41. Schmiealau, J. and Finn, O.J. Activated granulocytes and granulocyte-derived hydrogen peroxide are the underlying mechanism of suppression of T-cell function in advanced cancer patients. *Cancer Res.*, 61, 4756–4760, 2002.

Chapter 5

Blood-Derived Transcriptomic Profiles as a Means to Monitor Levels of Toxicant Exposure and the Effects of Toxicants on Inaccessible Target Tissues

John C. Rockett

CONTENTS

5.1 INTRODUCTION

The postgenomic era has seen the emergence of new molecular biological techniques and the development of new disciplines as these techniques have been integrated into more traditional fields of study. One such discipline is toxicogenomics, which uses contemporary genomic and proteomic techniques to elucidate mechanisms of toxicant action. One of the primary tenets of toxicogenomics is that the effects of toxicants on cellular functions are mediated by gene expression changes, or at least cause gene changes to occur as secondary effects. In most cases these gene changes occur prior to clinical manifestation of toxicity, which affords a possible window of opportunity for preclinical diagnosis of toxic end points that may arise as a result of the exposure. Such a diagnosis could employ, among other things, the use of gene expression profiling (GEP), either on a global or restricted scale. GEP offers the potential to classify toxicant exposures (Burczynski et al., 2000; Bartosiewicz et al., 2001; Thomas et al., 2001; Hamadeh et al., 2002a, 2002b) and predict the clinical outcome of such exposures (Waring et al., 2001a; Hamadeh et al., 2002c; Kier et al., 2004), as well as informing on the mechanism of action (Waring et al., 2001b).

Where humans are concerned, the use of GEP to determine toxicant exposures or predict possible outcomes is largely limited to the use of accessible biospecimens. Although there are a number of such biospecimens available from humans (see Chapter 1), blood is currently the most practical choice. Its benefits are several:

1. It is available from almost all people and is taken routinely for monitoring or diagnostic purposes.
2. It is a source of live, nucleated cells — primarily leukocytes — which can provide the high-quality RNA necessary for GEP.
3. Just a few hundred microliters of blood can provide a sufficient quantity of DNA or RNA on which to conduct DNA adduct or GEP analysis.

With this in mind, many investigators are turning to blood as a surrogate tissue for monitoring exposure and effect in both the whole body and specific target tissues. Indeed, the astute reader will already have recognized that most of the work described in this book has been conducted on blood. The use of blood as a surrogate in toxicity studies can be broadly categorized into two areas: (1) as a means to measure whole-body toxicant exposure levels; and (2) to evaluate molecular events that are occurring in specific inaccessible target tissues following a toxicant exposure.

5.2 BLOOD GENE EXPRESSION AS A BIOMARKER OF WHOLE-BODY TOXICANT EXPOSURE

Current methods for measuring toxicant exposure levels require foreknowledge of the chemical (or metabolite thereof) to which a person has been exposed. This approach cannot therefore be used in cases where general monitoring would be advantageous, as, for example, in the case of agricultural workers, who may be at elevated risk of developing occupationally related diseases because of seasonal or chronic exposure to pesticides or other toxic chemicals used in their workplace. In most cases it is impractical to monitor these workers routinely for the multiple exposures and potential health effects they face. This means that disease development as a result of such exposures, particularly those affecting internal organs or those which occur over a period of months or even years, are likely to go unnoticed until manifestation of clinical symptoms. At this point, medical intervention becomes remedial or palliative rather than preventative. Since in any given agricultural area there may be a number of potentially toxic compounds and mixtures in use, some of which are persistent and bioaccumulative, it is a difficult proposition to monitor for body burden of these various possible exposures using current methods. Such methods usually measure metabolite levels of a single compound in blood or urine. An emerging alternative for monitoring toxicant exposure might be to use blood gene expression profiles to search for a specific gene expression "fingerprint," which is indicative of exposure to a specific chemical or chemicals or chemical class or classes, and may even be predictive of toxicity-associated disease development.

There have already been multiple *in vivo* rodent studies, referred to earlier, demonstrating how gene expression profiles can be used to classify toxicant exposures in specific tissues such as liver. However, few studies have been conducted to determine whether accessible biospecimens such as peripheral blood lymphocytes (PBLs) can also be used in this way. Of those that have, the two most commonly reported models have been exposure of cell lines and *ex vivo* PBLs to ionizing radiation (IR). For example, Amundson et al. (2000) found that the induction of DDB2, CDKN1A, and XPC in human *ex vivo* PBLs showed a linear dose–response relationship between 0.2 and 2 Gy of IR at 24 and 48 h after exposure, but with less linearity at earlier or later times. Although the magnitude of mRNA induction generally decreased over time, the expression of many of these genes was still significantly elevated up to 72 h after irradiation. Gene expression changes also occur at low doses of IR (0.002 to 0.05 Gy) — Amundson et al. (2003) demonstrated a linear induction for multiple stress genes in the human p53-wt myeloblastic leukemia (ML)-1 line. Clustering of these data indicated two distinct groups of responder genes: one group was induced in a dose rate-dependent fashion (e.g., GADD45, CDKN1A), and the other in a dose rate-independent fashion (e.g., MDM-2). Genes belonging to the former group may prove particularly useful as a measure of human radiation exposure. To test this, a study was conducted on samples from patients undergoing total body irradiation for allogenic or autologous hematopoietic stem cell transplantation (Amundson et al., 2004); it was found that stress-gene induction in the *in vivo* samples generally agreed with those obtained from the *ex vivo* experiments.

Other studies support the idea that gene expression changes in PBLs can be used as a biomarker for IR exposure. Blakely et al. (2002) exposed human PBLs to 25 to 100 cGy of x-ray radiation, and using Northern blot analysis found a dose-dependent elevation in the Haras gene expression levels 17 h after exposure. Kang et al. (2003) used cDNA microarrays to identify highly expressed genes in *ex vivo* human PBLs following exposure to IR. At 12 h after irradiation they found a linear dose–response relationship between 0.5 and 4 Gy for the expression of TRAIL receptor 2, FHL2, and cyclin G; however, there was less linearity at later times. Together, these findings indicate that the dose, dose rate, and elapsed time since ionizing radiation exposure result in variations in the response of stress genes, and suggest that gene expression signatures may be informative markers of radiation exposure.

Use of PBL-derived gene expression profiles as a means to measure exposure levels might also apply to chemical toxicants. As early as 1991, Cosma et al. found metallothionein (MT) was induced in rat PBLs following intraperitoneal cadmium exposure. Ganguly et al. (1996) subsequently found that levels of MT mRNA in the PBLs of workers occupationally exposed to high levels of cadmium were more than twice that of unexposed individuals. Similarly, Lu et al. (2001) reported that MT mRNA levels were elevated in the PBLs of humans living in a cadmium-contaminated area. Together, these studies indicate the potential value of PBL MT mRNA expression as a biomarker of cadmium exposure.

Other chemical toxicant exposures that can be detected via gene expression changes in the blood have also been reported. For example, Ember et al. (1998) found elevation in p53 and N-ras mRNA levels in the PBLs of individuals occupationally exposed to ethylene oxide compared to controls. Lang et al. (1998) exposed a human cell line and primary cultures of cells to a range of 2,3,7,8-tetrachlorodibenzo-*p*-dioxin (TCDD) doses, and found that CYP1A1 mRNA levels were dose-dependently increased in bronchoepithelial cells and PBLs following exposure.

It is clear that the expression of some or all of the genes mentioned above may be similarly changed by other toxicant exposures. Nevertheless, each single gene whose expression is a biomarker of exposure to a certain toxicant is still useful in that it could be included in a diagnostic "identification panel" of genes, the complement of which would be unique for each type of toxicant. For example, diagnosis of exposure to IR might be determined by analyzing expression of genes in an "IR-exposure panel," which might include Ddb2, Cdkn1a, Xpc, Gadd-45, Mdm-2 (Amundson et al., 2000, 2003), Haras (Blakely et al., 2002), TRAIL receptor 2, Fhl2, and cyclin G (Kang et al., 2003).

Clearly, there are several issues to resolve if the use of GEP as a diagnostic tool for toxicant exposure is going to be used. These include characterizing for each potential biomarker the effects of dose level, time since exposure, and the effects of simultaneous exposure to other chemicals/toxicants, as well as biological variables such as genetics, age, diet, and health.

Thus, although application of this approach in routine clinical practice is not yet realized, the use of PBL gene expression profiling to determine levels of toxicant exposure is a very real possibility. The main advantage of this approach would be that foreknowledge of possible type of exposure would not be required. Theoretically, a clinical worker would simply take a small amount of whole blood from a subject, determine the gene expression profile of the sample, and from that data diagnose the nature and (possibly) level of any toxicant exposure that might have taken place. It is quite plausible that small sets of just a few tens or hundreds of biomarker genes will provide all the necessary information to distinguish among numerous toxicants. However, the ability to conduct such a sophisticated diagnosis will not be possible without the development of validated biomarkers of exposure and, more importantly, extensive databases that can house data against which to compare patient samples.

5.3 BLOOD AS A SURROGATE TISSUE FOR MONITORING GENE EXPRESSION CHANGES IN AN INACCESSIBLE TARGET TISSUE

The use of GEP to monitor for toxicant exposure or subclinical disease development in inaccessible human tissues is a difficult prospect, since direct biopsy of such tissues is not feasible unless strong medical reason (usually provoked by clinical symptoms) dictates otherwise. A less invasive method must therefore be developed if diagnostic procedures or monitoring programs are to be developed based on toxicogenomic analysis. One possible solution is to use surrogate tissues. A number of companies and institutions are pursuing the idea that gene expression changes in accessible (surrogate) tissues might reflect those in inaccessible (target) tissues, thus offering a convenient biomonitoring method to provide insight into the effects of environmental toxicants on target tissues.

5.3.1 The Evolution of Blood-Based Surrogate Tissue Analysis

The use of blood as a surrogate tissue is not a new concept. Indeed, there have been a number of published studies, which, although not necessarily expounding on the wider utility of surrogate tissue analysis (STA), nevertheless provided initial proof of concept and helped to shape current thinking. For example, Nesnow et al. (1993) showed that DNA adduct formation, a potential method of measuring exposure to environmental genotoxicants, exhibited a similar pattern in rat lung, liver, and PBLs following exposure to polycyclic hydrocarbons, and that this was detectable at least 56 days after treatment. These findings strongly suggested that PBLs might offer a convenient method of assessing exposure to, and effect of, genotoxic agents on internal, inaccessible tissues and organs.

More recently, PBLs have been analyzed as possible surrogates in studies involving gene expression analysis. Hukkannen et al. (1997) used RT-PCR to compare levels of mRNA expression of a number of xenobiotic-metabolizing cytochrome P450s (CYPs) in lung and PBLs. A similar study, comparing CYP expression in liver and PBLs was later published by Finnstrom et al. (2001). Disappointingly, both studies concluded that differences in the CYP gene expression patterns between PBLs and target tissues were too great to use PBL gene expression as a surrogate for examining gene expression in these particular target tissues, at least for the specific CYPs tested. At about the same time, researchers at the university of Pecs (Hungary) published the results of two studies in which rats were exposed to the carcinogenic chemicals 1-nitropyrene (Ember et al., 2000), and 7,12-dimethylbenz(*a*)anthracene (Gyongyi et al., 2001). At 24 or 48 h after exposure the animals were necropsied and RNA extracted from PBLs and certain internal target tissues (lung, liver, lymph nodes, kidneys, spleen). Dot blots were then used to detect expression of two oncogenes (c-myc, H-ras) and a suppressor gene (p53). Results for both chemicals suggested that expression of H-ras and p53 in PBLs correlated with that in several of the target tissues. The authors thus concluded: (1) the expression of H-ras and p53 in PBLs might be potential early biomarkers of exposure to the tested chemicals; and (2) that PBLs may be effective surrogates for certain internal target tissues.

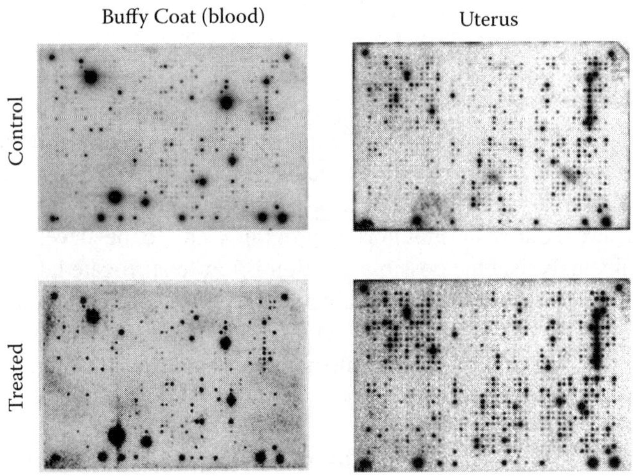

Figure 5.1 Representative images of microarrays following hybridization of RNA from the buffy coat (blood) and uterus of control and estradiol-treated adult female rats. RNA from buffy coat and uterine samples of control or estradiol-treated ovariectomized rats was used to produce ^{32}P-labeled cDNA probes. Probes were hybridized overnight to Clontech Atlas Rat 1.2 membrane Arrays, washed, and the hybridization image captured using a phosphorimaging screen (see Rockett et al., 2002, for further details).

Despite the mixed findings and the small number of genes analyzed, these initial studies were nevertheless pioneering in that they helped highlight the idea of using gene expression profiling of blood/PBLs in STA. Others have since advanced the concept, and elsewhere in this book the reader can see how gene expression profiling of PBLs using powerful microarray technology has been used to identify the occurrence of non-neoplastic disease (Chapter 3) and to provide diagnostic and prognostic information for oncology patients (Chapter 4). In this chapter we discuss further how blood might be used as a surrogate tissue in toxicology studies.

5.3.2 Use of DNA Arrays to Monitor Gene Expression in Rat Blood and Uterus Following 17-β-Estradiol Exposure — Biomonitoring Environmental Effects Using Surrogate Tissues

In an *in vivo* study designed to investigate the potential utility of STA in identifying perturbations in the endocrine system, researchers at the U.S. Environmental Agency compared gene expression changes in PBLs and uteri of adult rats to identify genes whose expression was altered in both tissues following estradiol treatment (Rockett et al., 2002). Ovariectomized rats were treated with either 17-β-estradiol or vehicle control for 3 days. PBL and uterine RNAs from these animals were then hybridized to Clontech rat toxicology 1.2 arrays (Figure 5.1) containing 1185 genes. In all, 193 of these genes were detectable in both leukocytes and uterus, 18 of which were significantly altered in both tissues (Table 5.1). The changes in eight of these

Table 5.1 Genes That Are Significantly Changed in Both the PBL (Buffy Coat) Fraction and Uterus 3 Days after Estradiol Treatment of Ovariectomized Rats

GenBank Gene Name	Direction in Blood/Uterus	Blood Proportion Change (Treated/Control)	Uterus Proportion Change (Treated/Control)
Jun-D	+/+	2.03	2.60
Neuropilin	+/+	2.45	2.41
NGF-inducible anti-proliferative secreted protein	+/+	2.63	2.57
Phospholipase A2, cytosolic	+/+	1.76	1.47
Synaptotagmin XI	+/+	2.51	1.75
Thymidine kinase, cytosolic	+/+	3.58	2.23
Dipeptidase	–/–	0.34	0.46
Tissue inhibitor of metalloproteinase 2	–/–	0.63	0.62
Insulin-like growth factor binding protein 1	–/–	0.21	0.50
Adenine nuleotide translocator, mitochondrial	–/+	0.62	1.41
Beta-arrestin 2	–/+	0.64	1.29
Beta-actin, cytoplasmic	–/+	0.69	1.54
GTP-binding protein (G-alpha-i2)	–/+	0.74	1.25
H(+)-Transporting ATPase	–/+	0.60	1.46
Macrophage migration inhibitory factor	–/+	0.51	1.88
Microglobulin	–/+	0.58	1.73
5-Hydroxytryptamine (serotonin) receptor 5B	+/–	1.82	0.67
Sky	+/–	2.36	0.40

Note: For the top 8 genes the treatment effect (i.e., direction and degree of change) is similar in both tissues. For the lower 10 genes there is some difference between the treatment effect for the two tissues, i.e., direction of gene change is opposite or degree of gene change is greater than twofold.

Source: Adapted from Rockett et al. (2002).

genes appeared to be treatment specific, rather than tissue specific (i.e., the genes demonstrated a similar degree and direction of expression change in both tissues). This group of genes appears to offer the best opportunity for identifying shared mechanistic changes in the target and surrogate tissues. Changes in the other 10 genes appeared tissue specific, rather than treatment specific. This means that either there was greater than a twofold difference between the tissues in the degree of change, or the change was in the opposite direction. These genes may be less useful for identifying shared mechanisms between target and surrogate tissues, but if the changes are consistent over dose and time, they may be useful in fingerprinting types of exposure.

Although the number of coregulated genes discovered here (18) might appear rather limited, it should be remembered that the arrays used in this pilot study contained only 1185 genes. Given that the latest estimates are that the mouse and human genome contain in excess of 35,000 genes, and the reasonable assumption that the rat will have a similar number, it is possible that there may be in excess of 500 genes in total that are coregulated in this PBL-uterus estrogen exposure model. Furthermore, gene expression differences were assessed at only one time point. Since genes expression is a dynamic process, it is probable that additional genes will also change at different times during or following an exposure. Thus, there is likely to be an abundant pool of target genes expressed in both tissues from which to derive candidates for biomonitoring exposure and/or effect.

This proof-of-concept study thus provided initial supportive evidence for toxicogenomics-based STA of toxicant exposure/effect. More specifically, it demonstrated that PBLs might be appropriate surrogates for observing gene expression changes in the uterus following changes in steroid hormone levels induced by age, disease, or toxicant exposure.

5.4 CHALLENGES TO THE USE OF BLOOD AS A SURROGATE TISSUE

Like all new methods and approaches, there are likely to be a number of challenges to overcome before it can be determined where and when STA is both applicable and appropriate. Some challenges that have been identified so far in relation to gene expression profiling in blood include inter-individual variation in gene expression and technologically induced variation in gene expression

5.4.1 Inter-Individual Variation in Gene Expression

Expression of any given gene can be changed by multiple environmental or genetic factors associated with the regulation and function of that gene, or the genetic network of which it forms part. When one considers the 35,000+ genes that are expressed in the human body, it is not difficult to see how there can be a large range in both the specific genes expressed in each tissue and the degree of that expression, even in normal, resting, healthy individuals. Thus, before it can be determined whether certain gene expression profiles are indicative of some toxicant exposure or effect, the range of genes expressed in the blood of normal, healthy individuals must first be characterized. Studies have already been published that show that multiple biological variables can affect whole-blood gene expression, including age, gender, and circadian rhythm (e.g., Whitney et al., 2003). Much of this biological variation is due to the relative proportions of nucleated cells in the blood. The white blood cells (leukocytes), which are the largest RNA-containing fraction of blood cells, include multiple cell types such as lymphocytes, neutrophils, monocytes, eosinophils, and granulocytes. The proportion of each type varies between individual. Radich et al. (2004) reported that whole-blood gene expression tends to remain constant within an individual from month to month. However, it is known that relative levels of each leukocyte subpopulation can change depending on such factors as physical condition, disease status, diet, etc., which in turn affects the relative abundance of different mRNAs in the whole-blood RNA population.

5.4.2 Technologically Induced Variation in Gene Expression

Layered on top of biological sources of differences in gene expression are differences that can be introduced by collection, transport, storage, processing, and hybridizing the samples. To analyze blood, one of the first challenges is to develop appropriate methods for collection, storage, and transportation of tissues at and between sites of collection and analysis. "Appropriate" means that:

1. Sufficient specimen must be collected to enable extraction of reasonable amounts (e.g., > 500 ng) of good-quality total RNA.
2. The collection, transportation, and storage procedures must inhibit RNA degradation.
3. The population of RNAs (the "transcriptome") in each specimen must not change between obtaining the specimen from the patient at the field site and extraction of RNA from the specimen in the laboratory.

Methods have been developed to overcome these challenges and are discussed in detail in Chapter 2.

5.5 SUMMARY

Gene expression profiling has the potential to revolutionize clinical monitoring of toxicant exposures by providing information that can be used to identify mechanisms of action and discern different types of exposure, even when the nature of an exposure is unknown. Furthermore, unlike other methods used for measuring levels of toxicant exposure, it has been shown that gene expression profiles can be used to predict future adverse outcomes that arise as a result of that exposure.

Unfortunately, many targets of toxicant action are internal tissues or organs and are therefore inaccessible in terms of obtaining biopsy samples for gene expression profiling. Consequently, there is extensive interest in using accessible tissues that can act as surrogates for these inaccessible tissues. In this context, an ideal surrogate tissue is obtainable from all individuals, yields good-quality RNA in sufficient quantities to conduct GEP, and has a rich complement of gene expression networks that reflect, at least in part, those that exist in inaccessible tissues following toxicant exposure. Peripheral blood has proved to be the most popular surrogate tissue for use in most STA studies. Although to date there have been relatively few toxicology studies conducted using STA, those that have suggest that in some cases gene expression profiles in peripheral blood can be used either as a surrogate for measuring whole-body toxicant exposure or as a surrogate to monitor molecular changes in inaccessible target tissues.

Nevertheless, there is still much work to be done before blood can be widely used as a surrogate in toxicological studies or diagnoses. This includes (1) resolving technical issues related to collection and processing of blood and target tissue(s), and the generation and analysis of gene expression data; (2) fully documenting what constitutes normal gene expression in both animal models and humans; (3) identifying gene expression changes that constitute biomarkers of exposure and effect; and (4) developing and upkeeping databases to house these data. The last two issues are most problematic, and how they are addressed will probably determine whether the use of STA in toxicology will become a widespread phenomenon or relegated to a minor role.

REFERENCES

Amundson, S.A., Do, K.T., Shahab, S., Bittner, M., Meltzer, P., Trent, J., and Fornace, A.J., Jr. (2000). Identification of potential mRNA biomarkers in peripheral blood lymphocytes for human exposure to ionizing radiation. *Radiat. Res.*, 154, 342–346.

Amundson, S.A., Lee, A., Koch-Paiz, C.A., Bittner, M.L., Meltzer, P., Trent, J.M., and Fornace, A.J., Jr. (2003). Differential responses of stress genes to low dose-rate γ irradiation. *Mol. Cancer Res.*, 1, 445–452.

Amundson, S.A., Grace, M.B., McLeland, C.B., Epperly, M.W., Yeager, A., Zhan, Q., Greenberger, J.S., and Fornace, A.J., Jr. (2004). Human *in vivo* radiation-induced biomarkers: gene expression changes in radiotherapy patients. *Cancer Res.*, 64(18), 6368–6371.

Bartosiewicz, M., Penn, S., and Buckpitt, A. (2001). Applications of gene arrays in environmental toxicology: fingerprints of gene regulation associated with cadmium chloride, benzo(*a*)pyrene, and trichloroethylene. *Environ. Health Perspect.*, 109, 71–74.

Blakely, W.F., Miller, A.C., Luo, L., Lukas, J., Hornby, Z.D., Hamel, C.J., Nelson, J.T., Escalada, N.E., and Prasanna, P.G. (2002). Nucleic acid molecular biomarkers for diagnostic biodosimetry applications: use of the fluorogenic 5′-nuclease polymerase chain reaction assay. *Mil. Med.*, 167(2 Suppl.), 16–19.

Burczynski, M.E., McMillian, M., Ciervo, J., Li, L., Parker, J.B., Dunn, R.T., II, Hicken, S., Farr, S., and Johnson, M.D. (2000). Toxicogenomics-based discrimination of toxic mechanism in HepG2 human hepatoma cells. *Toxicol. Sci.*, 58(2), 399–415.

Chipping Forecast II, 2002. *Nat Gen.* Supplement, Vol. 32, December.

Cosma, G.N., Currie, D., Squibb, K.S., Snyder, C.A., and Garte, S.J. (1991). Detection of cadmium exposure in rats by induction of lymphocyte metallothionein gene expression. *J. Toxicol. Environ. Health*, 34(1), 39–49.

Ember, I., Kiss, I., Gombkoto, G., Muller, E., and Szeremi, M. (1998). Oncogene and suppressor gene expression as a biomarker for ethylene oxide exposure. *Cancer Detect. Prev.*, 22(3), 241–245.

Ember, I., Kiss, I., Gyongyi, Z., and Varga, C.S. (2000). Comparison of early onco/suppressor gene expressions in peripheral leukocytes and potential target organs of rats exposed to the carcinogen 1-nitropyrene. *Eur. J. Cancer Prev.*, 9, 439–442.

Finnstrom, N., Thorn, M., Loof, L., and Rane, A. (2001). Independent patterns of cytochrome P450 gene expression in liver and blood in patients with suspected liver disease. *Eur. J. Clin. Pharmacol.*, 57, 403–409.

Ganguly, S., Taioli, E., Baranski, B., Cohen, B., Toniolo, P., and Garte, S.J. (1996). Human metallothionein gene expression determined by quantitative reverse transcription-polymerase chain reaction as a biomarker of cadmium exposure. *Cancer Epidemiol. Biomarkers Prev.*, 5, 297-301.

Gyongyi, Z., Ember, I., Kiss, I., and Varga, C. (2001). Changes in expression of onco- and suppressor genes in peripheral leukocytes — as potential biomarkers of chemical carcinogenesis. *Anticancer Res.*, 21(5), 3377–3380.

Hamadeh, H.K., Bushel, P.R., Jayadev, S., DiSorbo, O., Bennett, L., Li, L., Tennant, R., Stoll, R., Barrett, J.C., Paules, R.S., Blanchard, K., and Afshari, C.A. (2002a). Prediction of compound signature using high density gene expression profiling. *Toxicol. Sci.*, 67(2), 232–240.

Hamadeh, H.K., Bushel, P.R., Jayadev, S., Martin, K., DiSorbo, O., Sieber, S., Bennett, L., Tennant, R., Stoll, R., Barrett, J.C., Blanchard, K., Paules, R.S., and Afshari, C.A. (2002b). Gene expression analysis reveals chemical-specific profiles. *Toxicol. Sci.*, 67(2), 219–231.

Hamadeh, H.K., Knight, B.L., Haugen, A.C., Sieber, S., Amin, R.P., Bushel, P.R., Stoll, R., Blanchard, K., Jayadev, S., Tennant, R.W., Cunningham, M.L., Afshari, C.A., and Paules RS. (2002c). Methapyrilene toxicity: anchorage of pathologic observations to gene expression alterations. *Toxicol. Pathol.*, 30(4), 470–482.

Hukkanen, J., Hakkola, J., Anttila, .S, Piipari, R., Karjalainen, A., Pelkonen, O., and Raunio, H. (1997). Detection of mRNA encoding xenobiotic-metabolizing cytochrome P450s in human bronchoalveolar macrophages and peripheral blood lymphocytes. *Mol. Carcinog.*, 20(2), 224–230.

Kang, C.M., Park, K.P., Song, J.E., Jeoung, D.I., Cho, C.K., Kim, T.H., Bae, S., Lee, S.J., and Lee, Y.S. (2003). Possible biomarkers for ionizing radiation exposure in human peripheral blood lymphocytes. *Radiat. Res.*, 159(3), 312–319.

Kier, L.D., Neft, R., Tang, L., Suizu, R., Cook, T., Onsurez, K., Tiegler, K., Sakai, Y., Ortiz, M., Nolan, T., Sankar, U., and Li, A.P. (2004). Applications of microarrays with toxicologically relevant genes (tox genes) for the evaluation of chemical toxicants in Sprague Dawley rats *in vivo* and human hepatocytes *in vitro. Mutat. Res.*, 549(1–2), 101–113.

Lang, D.S., Becker, S., Devlin, R.B., and Koren, H.S. (1998). Cell-specific differences in the susceptibility of potential cellular targets of human origin derived from blood and lung following treatment with 2,3,7,8-tetrachlorodibenzo-*p*-dioxin (TCDD). *Cell Biol. Toxicol.*, 14(1), 23–38.

Lu, J., Jin, T., Nordberg, G., and Nordberg, M. (2001). Metallothionein gene expression in peripheral lymphocytes from cadmium-exposed workers. *Cell Stress Chaperones*, 6(2), 97–104.

Nesnow, S., Ross, J., Nelson, G., Holden, K., Erexson, G., Kligerman, A., and Gupta, R.C. (1993). Quantitative and temporal relationships between DNA adduct formation in target and surrogate tissues: implications for biomonitoring. *Environ. Health Perspect.,* 101(Suppl. 3), 37–42.

Radich, J.P., Mao, M., Stepaniants, S., Biery, M., Castle, J., Ward, T., Schimmack, G., Kobayashi, S., Carleton, M., Lampe, J., and Linsley, P.S. (2004). Individual-specific variation of gene expression in peripheral blood leukocytes. *Genomics*, 83(6), 980–988.

Rockett, J.C., Kavlock, R.J., Lambright, C.R., Parks, L.G., Schmid, J.E., Wilson, V., Wood, C., and Dix, D.J. (2002). DNA arrays to monitor gene expression in rat blood and uterus following 17-beta-estradiol exposure: biomonitoring environmental effects using surrogate tissues. *Toxicol. Sci.*, 69(1), 49–59.

Rockett, J.C., Burczynski, M.E., Fornace, A.J., Herrman, P.C., Krawetz, S.A., and Dix, D.J. (2004). Surrogate tissue analysis: monitoring toxicant exposure and health status of inaccessible tissues through the analysis of accessible tissues and cells. *Toxicol. Appl. Pharmacol.*, 194(2), 189–199.

Thomas, R.S., Rank, D.R., Penn, S.G., Zastrow, G.M., Hayes, K.R., Pande, K., Glover, E., Silander, T., Craven, M.W., Reddy, J.K., Jovanovich, S.B., and Bradfield, C.A. (2001). Identification of toxicologically predictive gene sets using cDNA microarrays. *Mol. Pharmacol.*, 60(6), 1189–1194.

Waring, J.F., Jolly, R.A., Ciurlionis, R., Lum, P.Y., Praestgaard, J.T., Morfit, D.C., Buratto, B., Roberts, C., Schadt, E., and Ulrich, R.G. (2001a). Clustering of hepatotoxins based on mechanism of toxicity using gene expression profiles. *Toxicol. Appl. Pharmacol.*, 175(1), 28–42.

Waring, J.F., Ciurlionis, R., Jolly, R.A., Heindel, M., and Ulrich, R.G. (2001b). Microarray analysis of hepatotoxins *in vitro* reveals a correlation between gene expression profiles and mechanisms of toxicity. *Toxicol. Lett.*, 120(1–3), 359–368.

Whitney, A.R., Diehn, M., Popper, S.J., Alizadeh, A.A., Boldrick, J.C., Relman, D.A., and Brown, P.O. (2003). Individuality and variation in gene expression patterns in human blood. *Proc. Natl. Acad. Sci. U.S.A.*, 100(4), 1896–1901.

CHAPTER 6

Spermatozoal RNAs as Surrogate Markers of Paternal Exposure

G. Charles Ostermeier and Stephen A. Krawetz

CONTENTS

6.1 INTRODUCTION

Monitoring the toxicological load applied to inaccessible tissues is overtly challenging. To overcome this challenge, it has been proposed that surrogate tissue analysis (STA) be employed to monitor the health and condition of the inaccessible "target" tissue using markers that are indicative of insult. For example, it has been observed that following exposure to polycyclic hydrocarbons, similar groups of DNA adducts appear in rat peripheral blood leukocytes (PBLs), lungs and liver.[1] These modifications are maintained for up to 56 days after exposure. This clearly shows that PBLs can be used as surrogates for assessing the influence of genotoxic agents on remote internal organs or tissues. STAs have also been applied to reproductive tissues. For example, the levels of alpha-4 and beta-3 integrins in PBLs have been used to monitor and correctly predict the receptivity of the uterine endometrium to embryonic implantation[2] (see Chapter 8). In this example, the benefits of using PBLs as a surrogate for endometrial biopsies are evident, as drawing blood is relatively trouble-free, reducing the opportunity for intrauterine infection with little trauma. Although some of the fractions isolated from peripheral blood, e.g., leukocytes, have received the most attention as possible surrogates for tissue analysis, several other candidates including hair and sperm are well suited to this task.

The toxicology of the male reproductive system has received increased interest in recent years. In part, this has been fueled by the growing controversy of falling spermatozoal counts and rising reproductive disorders in the human population.[3–6] The pathogenesis of reduced male fecundity can often be traced to aberrant spermatogenesis caused by an impediment of male germ plasm differentiation. This situation can be compounded by pituitary disorders, testicular cancer, germ cell aplasia, and varicocele. Testicular biopsies are generally employed to directly assess the impact of these factors on the quality of spermatogenesis. The anxiety and pain associated with this procedure necessitate the use of a local anesthetic[7] and the procedure typically yields only a small portion of testicular parenchyma. This procedure is subject to several limitations that can result in poor-quality data, and produce local hematomas that can lead to a further decrease in fertility.[7,8] To overcome these hindrances, many have turned to using spermatozoa as surrogates for spermatogenic evaluation. With this approach the anxiety associated with the surgical necessity of testicular biopsy is removed, a broader sample representing the functional status of both testes is obtained, and postsampling injury is negated. However, the optimal method for utilizing spermatozoa to investigate spermatogenic function and environmental insult still remains to be established.

Semen analysis is the most common means to address spermatogenic competence of the male gonad.[9,10] This typically encompasses a morphological screen that includes determining the percentage of spermatozoa that are viable and motile, as well as acrosome status. However, there is increasing evidence that these subjective measurements are relatively poor indicators of testicular function as they rely on physiological and morphological criteria.[11–14] The need for objective assessment, like that provided by genetic profiling,[15–17] is thus evident.

6.2 RNA IN SPERM: INITIAL OBSERVATIONS

Despite the general acceptance that ejaculate spermatozoa are transcriptionally inert,[18] it is well documented that these cells contain a complex yet specific population of RNAs.[16–17,19–26] Initially, spermatozoal mRNAs were identified within the condensed chromatin of the fern *Scolopendrium.*[25] RNAs within mature rat spermatozoa were subsequently visualized using an RNase colloidal gold assay.[24] The first specific spermatozoal RNAs to be identified were the rat U1 and U2 snRNAs,[20] while the first specific mRNA identified was the murine proto-oncogene c-myc.[21] Reverse transcription-polymerase chain reaction amplification (RT-PCR) was later used to detect leukocyte antigen class I-G and -B mRNAs in human sperm.[19] Differential display clearly illustrated the complexity of the human spermatozoal

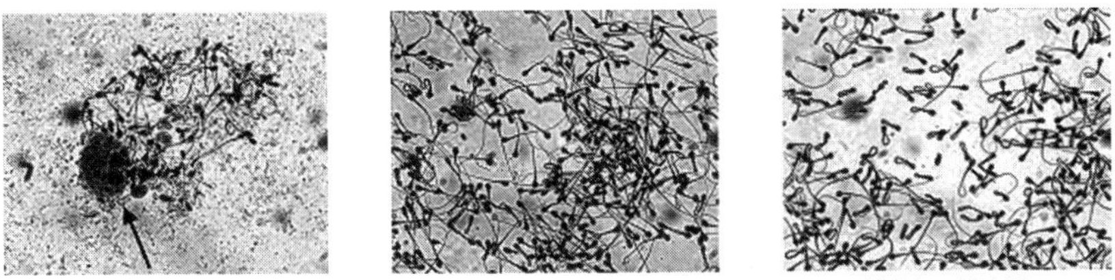

Figure 6.1 Photomicrographs of sperm during the various stages of purification. Spermatozoa from the ejaculate (panel 1), were prepared using sequential centrifugations through 40:80 discontinuous Percoll gradients. The pellet from the first centrifugation (panel 2) was processed through the second centrifugation then treated with Triton-X 100. The aliquots were stained with H&E for evaluation. As shown (panel 3), essentially pure spermatozoa were obtained without centrifugation throught Percoll gradients after treatment with Triton-X 100. (Reprinted from *The Lancet*, vol. 360, Ostermeier et al., Spermatozoal RNA profiles of normal fertile men, p. 774, 2002, with permission from Elsevier.)

RNA population.[23] These data were independently corroborated by *in situ* hybridization, which showed the presence of β-*actin* mRNA as well as the mRNAs corresponding to all three members of the coordinately haploid expressed *PRM1*→*PRM2*→*TNP2* domain in spermatozoa in both mice and humans.[26,27] Further support for a specific population of spermatozoal RNAs came when a testis cDNA library was probed with total spermatozoal RNA.[22] When randomly selected cDNA clones were sequenced, it was firmly established that spermatozoa contain a wealth of both known and unknown protein-encoding and noncoding RNAs.

As shown in Figure 6.1, stringent precautions must be taken to ensure spermatozoal purity prior to RNA extraction. This includes the use of discontinuous 40:80 Percoll gradients, followed by somatic cell lysis and washing in mild detergent solutions. These washing steps have proved so effective that Percoll gradient centrifugation is no longer required (right panel). Interestingly, when RNAs are isolated from humans and then compared by electrophoretic analysis, a broad distribution of various sized RNAs is revealed as shown in Figure 6.2. Unlike the RNA isolated from somatic cells or testis, there is a virtual absence of spermatozoal rRNAs. As shown by PCR these preparations are essentially free of contaminating DNA. Repeated isolations from many different individuals have established that the yield of RNA per sperm cell varies and can average as little as 10–80 fg per cell. This is similar to the total amount of actin mRNA within an ovulated mouse egg.[28] If the average size of each RNA is estimated as 1500 nucleotides, then on average, each sperm would contain 100,000 RNA molecules or approximately 10 copies of each of the different RNAs per cell.

Collectively, these studies have shown that a specific suite of gene transcripts accumulates during spermatogenesis. A host of these transcripts is maintained as the round spermatid differentiates into the mature spermatozoa. From plant to humans, this suite of transcripts is then carried by the spermatozoa through completion of their journey for delivery upon fertilization.

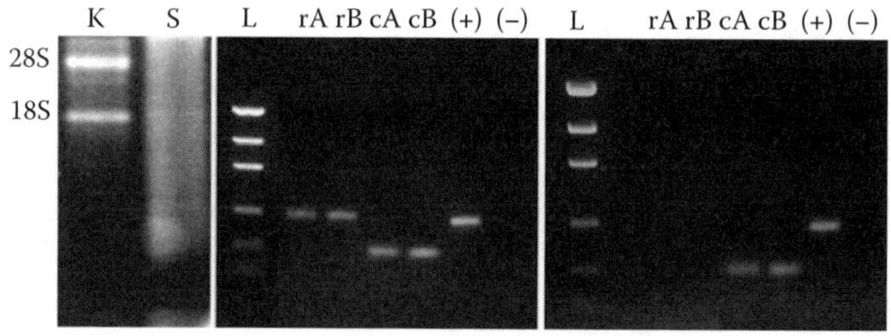

Figure 6.2 Distribution of RNAs in human sperm and quality control PCR analysis. (Left panel) Total RNA from human kidney (K) and ejaculated spermatozoa (S). The spermatozoal RNAs exhibit a wide distribution of sizes and lack 28S and 18S ribosomal bands. Subsequent to RNA isolation, samples were treated with RNase-free DNase I and subjected to quality control using sets of intron spanning PCR primers. (Middle panel) Quality control results after a single DNase treatment; (Right panel) The results following two DNase treatments. Note that genomic contamination is completely removed by a second DNase treatment. L = 100-bp ladder; rA, rB = PCR amplified human Protamine 2 from isolated sperm RNA samples; cA, cB = RT-PCR products of human Protamine 2 from cDNAs of same samples; (+) = human genomic DNA, positive control; (–) negative control, no template. (Reprinted from *The Lancet*, vol. 360, Ostermeier et al., Spermatozoal RNA profiles of normal fertile men, p. 774, 2002, with permission from Elsevier.)

6.3 TRANSCRIPT SURVEY TECHNIQUES

Early attempts to build gene expression databases relied on Northern blotting, RNase protection, RT-PCR, *in situ* hybridization, and nuclease protection strategies. For the most part, these methodologies are limited by low throughput and their labor-intensive nature. More recent strategies have included the use of subtractive hybridization and differential display. However, these protocols are labor intensive and cost prohibitive. They also suffer from varied false positive rates and a general lack of sensitivity, rendering it difficult to identify a subset of genes and gene products involved in processes key to normal or aberrant development. To date, these technical factors have precluded a genome-wide scan to elucidate the sperm transcriptome. However, techniques that afford genome-wide scans, such as Serial Analysis of Gene Expression (SAGE), permit one to identify genomic heterogeneity that underlies the developmental pathways specific to individual cells, tissues, and organ systems.[29] While in some cases SAGE may provide a sensitive means of detecting RNA species, the sequences defined by SAGE can be unknown "snapshots" of a complete sequence. Even though this is a powerful technology, this sequencing-intensive process is comparatively slow and relatively expensive, and thus has not become widely employed.

The potential of microarrays to address key issues of development and differentiation was immediately realized.[30–32] In a single experiment, this technology permits the simultaneous determination of the expression of thousands of genes. This enables the construction of detailed expression and genetic profiles.[33–35] Recent microarray-based ovarian and breast cancer studies have demonstrated both the potential diagnostic and prognostic value of this method.[36,37] It is also clear that this technology

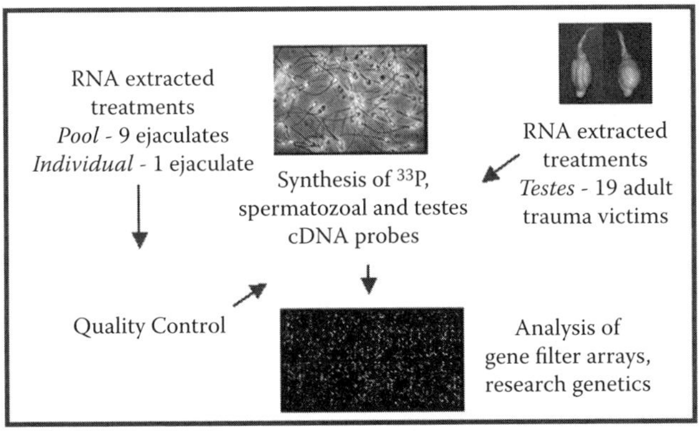

Figure 6.3 Experimental design. Human ejaculates were obtained from 10 healthy volunteers of proven fertility. Nine of the samples were pooled then purified using two sequential centrifugations through 40:80 discontinuous Percoll gradients. The purity and integrity of both preparations of spermatozoal RNAs were electrophoretically assessed and verified by RT-PCR using intron spanning protamine-2 primers. RNA from pooled histologically normal human testes was purchased from Clontech Laboratories, Inc. (Palo Alto, CA, USA). Probes were prepared from the testes and spermatozoal RNAs by reverse-transcription from the total, poly(A+) sperm RNAs and total testes RNA, then individually hybridized to the arrays.

is well suited to the toxicological arena[15] and a new field of toxicogenomics has been created.[38] One of the early applications of this technology was the classification of toxicants based on the responsive profile of the transcriptome.[39] This has since been reviewed.[40]

6.4 DEFINING THE NORMAL FERTILE MALE

It was initially suggested that the mRNAs observed in mature ejaculate spermatozoa were remnants of untranslated spermatogenic stores, and that they would provide a historic record or fingerprint of spermatogenesis.[41] Indeed, spermatozoal RNAs are concordant with those found in testes as determined by microarray analysis.[16,42,43] As summarized in Figure 6.3, initial characterization utilized a pool of testes cDNAs from 19 trauma victims, while the ejaculate spermatozoal samples were analyzed as a pool of 9 individuals. The spermatozoal pool was prepared by two density gradient centrifugations followed by poly(A+) RNA isolation. In addition, total spermatozoal RNA was isolated by simply lysing the somatic cells then washing the sperm with a series of mild detergents. As summarized in Table 6.1, the corresponding cDNA probes simultaneously interrogated 27,016 unique expressed sequence tags (ESTs). The spermatozoal sequences identified were a discrete subset of those identified within the testes. This showed that a specific population of RNAs within mature ejaculate human spermatozoa echoed spermatogenic gene expression. Without constructing or sequencing a spermatozoal cDNA library the genetic fingerprint of those transcripts present in spermatozoa was

Table 6.1 Testis and Spermatozoal RNAs Overlap

Probe	ESTs Interrogated	ESTs Identified	ESTs Shared with Testes Probe
Testes	27,016	7,157	—
Pooled ejaculate	27,016	3,281	3,281
Individual ejaculate	27,016	2,784	2,784

Note: Testes = pool of 19 trauma victims. Pooled ejaculate = RNA isolated from the ejaculates of 9 men. Individual = RNA isolated from a single ejaculate. ESTs = expressed sequence tags.

defined. This strategy has also been employed to characterize the distribution of transcripts in human testes and is likely to be applicable to any previously uncharacterized population of cells.[37]

Interestingly, within the pooled ejaculate probe all but four of the ESTs from the individual probe were identified. This suggested that among normal fertile men minimal spermatozoal RNA variation exists. To assess spermatozoal transcript variation among men, spermatozoal RNAs from three different individuals were isolated then compared using the Clontech Atlas Human Toxicology 1.2 Array. As shown in Figure 6.4, this comparison established that transcript variation among normal fertile men is minimal and that a core set of invariant fertile transcripts could be identified.

The size of the cohort that is necessary to saturate the identification of spermatozoal RNAs can be estimated based on our genome containing at least 30,000 unique genes. A subset of ~10,000 to 15,000 genes is likely expressed in each cell type.[48] We expect that all transcripts present in the spermatozoa will be derived from those expressed during spermatogenesis, since spermatozoa are transcriptionally inert.[18] A total of 7157 unique testes transcripts were identified from a pool of 19 different individuals. This corresponds to approximately one half of the total number of transcripts expected per cell if the testis were to contain 15,000 different transcripts. Interestingly, approximately one half of the 7157 testes transcripts identified all of the 3281 sperm transcripts. Accordingly, as a lower limit, spermatozoan transcripts may only represent one half of the transcripts present in testes. Thus, the number of transcripts in sperm should be in the range of 5000 to 7500. If this estimate is correct, then at least 48%, or possibly even 75%, of the total number of different transcripts present in sperm has been identified. This may already provide sufficient discriminatory power to describe the normal fertile male given that breast cancer prognosis can be reliably based on as few as 70 specific ESTs that were derived from an initial genome-wide survey of 25,000 cDNAs.[37]

6.5 DATA MINING SPERM mRNAS

Characterizing expression fingerprints like those of the normal fertile male can be undertaken using ontological classification.[43] This affords the global organization of data into several biological groups and has been applied to address the nature of

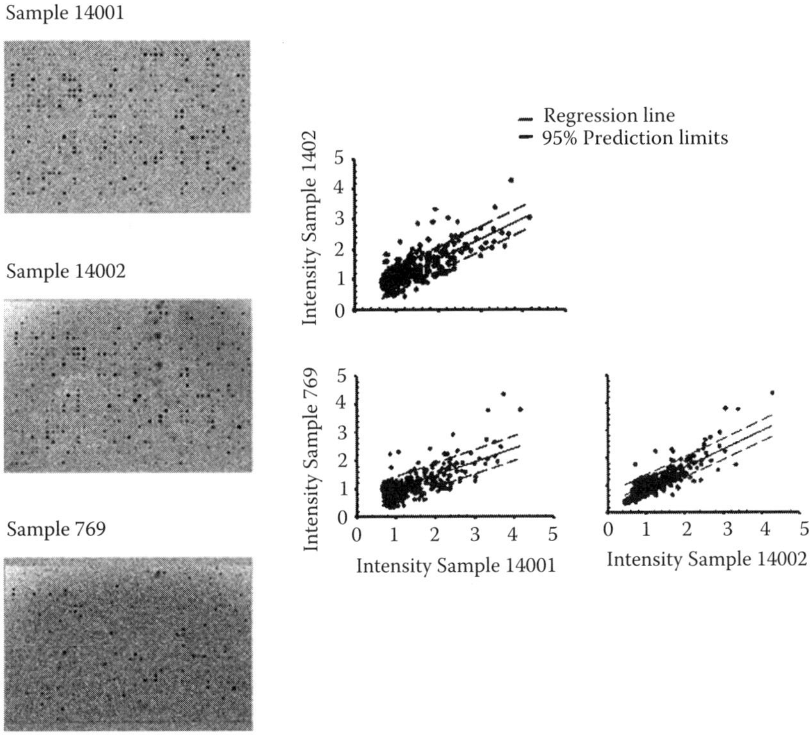

Figure 6.4 Spermatozoal transcript variation among men. Spermatozoal RNAs were isolated from the ejaculates of three different men and array specific cDNA labeled probes constructed. Each probe was hybridized to a Clontech Atlas Human Toxicology 1.2 Array. To assess variation, the standardized intensities from each of the 1176 unique ESTs were compared among the men. The linear regression equation is shown as a solid line, while the dotted lines delineate 99% prediction limits. The data show minimal variation, as greater than 99% of the data are positioned along the major diagonal in each plot.

the spermatozoa's complement of mRNAs. As shown in Figure 6.5, the largest groups of spermatozoal RNAs of known function participate in signal transduction, oncogenesis, and cell proliferation. As expected, the majority of this collection corresponds to nuclear proteins and plasma membrane proteins. They are similarly distributed in testes, pooled sperm, or sperm from a single individual. This attests to the robustness of this data and lack of variation among normal fertile men. Interestingly, upon biological classification a series of transcripts that are key to various stages of fertilization and early embryo development is highlighted. As shown in Figure 6.6 this included FOXG1B and WNT5A. They are present in testis and sperm, and ontologically classified as members of the embryogenesis and morphogenesis pathways. This has led to the introduction of the concept that spermatozoal RNAs may be part of a suite of transcripts that are functionally required in the early fertilized egg. For example, FOXG1B is a member of the family of fork head domain transcription factors restricted to the fetal brain and adult testis.[45] Similarly, WNT5

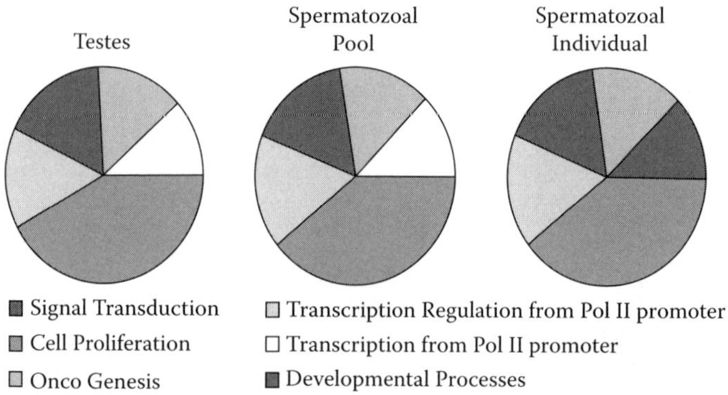

Figure 6.5 Spermatozoal RNA ontogeny. The biological processes of the proteins that represent each expressed sequence tag identified by the testes, pooled- and individual-ejaculate spermatozoal cDNA was data-mined using Onto-Express.[64] The biological process delineates the biological "objective" to which the protein contributes. The five categories having the largest representation by each of the probes are reported. The different sections of the pie chart represent the proportion of proteins identified to have the biological process that is indicated by shading in the legend.

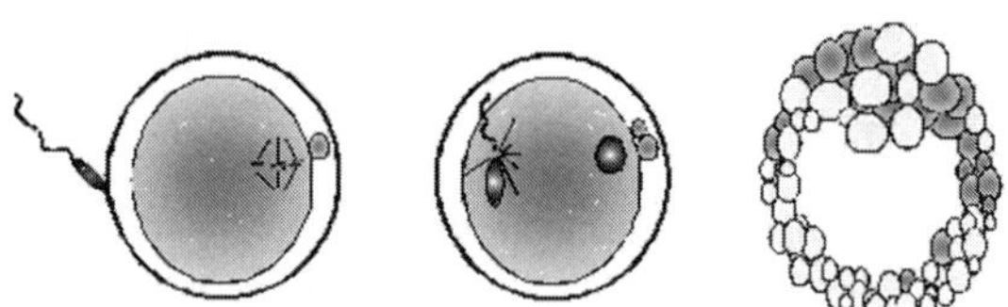

Fertilization	Stress Response	Embryogenesis Morphogenesis
Clusterin	HSF2	MID1
Calmegin	HSPA1L	NLVCF
AKAP4	DNAJB1	CYR61
Oscillin	HSBP1	EYA3
PRM2	DUSP5	FOXG1B
		WNT5A
		WHSC1
		SOX13

Figure 6.6 Paternally derived transcripts implicated in early development. A set of expressed genes common to testis and spermatozoa was obtained by microarray analysis. The concordant genes were grouped into functional ontological categories using Onto-Express[64] and compared with SAGE databases of oocyte-expressed genes.[65,66] This revealed a set of candidate spermatozoa-specific transcripts implicated in fertilization, stress response, and zygotic and embryonic development.

appears to be restricted to fetal heart and lung and to adult testis and germ-line tumors. Homologues of the WNT5A family of proto-oncogenic signaling molecules participate in embryological and morphogenetic patterning associated with cellular differentiation.[46] Recently, these sperm RNAs have been shown to be delivered to the egg upon fertilization.[47] They also include a group of micro-RNAs.[48]

6.6 SPERM AS A SURROGATE TISSUE

The body of work discussed above clearly demonstrates that a specific and relatively large population of mRNAs exists within mature ejaculated human spermatozoa and that they reflect the gene expression of spermatogenesis. These findings have clear implications for the use of spermatozoa as surrogates for spermatogenic tissue. Since it is believed that gene transcription does not occur in ejaculated spermatozoa, inter- as well as intra-individual differences in spermatozoal transcriptomes must have originated during spermatogenesis. If an individual is infertile or subfertile, the implication is that genetic aberrations or some toxicant adversely altered the genetic program of spermatogenesis. Such an action would negatively affect the differentiative processes within spermatogenesis, yielding spermatozoa that are unable to fertilize oocytes and/or orchestrate embryonic development. Accordingly, since specific patterns of gene expression can be associated with exposure to definitive classes of toxicants[39] or diseases,[49] RNA profiles obtained from ejaculate spermatozoa should be well suited to identifying toxicological exposures.

6.7 APPLICATION

Mature spermatozoa provide a key repository of genetic information that can be used to determine paternal exposure to environmental factors. This is evidenced by the observed sensitivity of the male gamete to environmental exposures of a chemical, thermal, or biological nature. Spermatozoa ultimately determine the paternal genetic load our children will bear. In effect, spermatozoa are a useful model for understanding the importance of the genetic complement passed from parent to offspring as part of the environment in which the molecular genetic processes are carried out.

It is widely held that there has been a decline in human male fertility within the past few decades.[50,51] The direct cause or causes of this reduction remain controversial, although, in part, it may reflect a trend toward decreased family size in the Western world. However, concurrent with the growing decrease in male infertility, there has been a corresponding increase in the incidence of testicular cancer and cryptorchidism.[52] Several theories have been put forward to help explain the causes of increased male infertility, including increased environmental and systemic exposure to pesticides, herbicides, estrogenic compounds, heavy metals, and reactive oxygen species.[53–55]

Establishing the use of the spermatozoal RNA genetic fingerprint as a molecular biomarker for exposure should prove quite valuable in risk assessment, forming public policy, and predicting individual health outcomes. In this respect, the identity of environmental toxins suspected of playing a role in decreasing male fertility and evaluating the reproductive toxicity of newly discovered environmentally significant compounds could be addressed. For the first time, this could provide the means to intercede before environmental/toxicological repercussions in new generations become apparent.

Toward this goal, microarrays offer the powerful ability of multiple analyses and simultaneous evaluations with objective markers and statistical correlations. Their ability to identify co-regulated genes, genes whose products are interconnected in specific biologically relevant mechanistic pathways, and genes that play key roles in specific diseases and genetic disorders has been used to provide insight into both normal and diseased states.[12] This is well exemplified in the heterozygous CREM⁻ male. This individual presents as subfertile and can be classified as oligozoospermic. Based on current mouse CREM⁻ microarray expression data (http://www.dkfz-heidelberg.de/tbi/crem/affydiff.html), this phenotype is characterized by the greater than fivefold upregulation of 16 genes, including laminin, beta 3, C-Ros proto-oncogene, spermidine/spermine *N*1-acetyltransferase, smooth muscle calponin gene, and acidic epididymal glycoprotein, and greater than fivefold downregulation of 119 genes including STAT4, RAR-related orphan receptor alpha, outer dense fiber of sperm tails 1, inositol polyphosphate-1-phosphatase, and fibrous sheath component 1. The up- and downregulation of each member of the affected pathway demarcates the molecular lesion. As expected, at the point of the lesion, messages before the affected member of the pathway were upregulated and those after the affected member of the pathway were downregulated. The results of this simple profiling study yield potential management strategies targeted to the various affected pathway members.

As articulated in the "Recommendations for the Future" outlined in the published article "Exposure to Hazardous Substances and Male Reproductive Health: A Research Framework,"[56] the development of standardized biomarkers of paternal environmental exposure for clinical application is critical. This challenge can be addressed using microarray analyses of paternally derived sperm RNAs as biomarkers of environmental exposure. A demonstration project has been initiated to address an area of acute concern to those individuals who consume sport-caught fish.[57,58] They account for up to 90% of the individuals with a tenfold increase (~20 ppb) in their load of organochlorinated compounds.[59] The burden it places presents as reduced sperm motility,[60] decreased *in vitro* fertilization rates,[61] decreased sperm counts,[62] and atrophy of the germinal epithelium.[63] These symptoms reflect the significant reproductive risk to the conception of a healthy child.

Paternal toxicological screening could provide the means to intercede before repercussions originating from paternal exposures become apparent in the next generation. Because spermatogenesis is a process of continuous self-renewal, a transcriptome-based assay system such as a microarray provides the means to monitor and diagnose exposures, as well as provide a history of previous exposure. For example, collection of samples at regular intervals for 60 to 80 days (time to complete one round of spermatogenesis) and comparison of their RNA profiles to a "normal fingerprint" could be used to establish the type and severity of an exposure and subsequent detoxification. The value in our ability to identify, screen, and intercede is not limited to current environmental exposure. The significance and need to develop this capability are now reinforced by the threat of an unwarranted biological and chemical terrorist attack on the mass population. With this new diagnostic capacity we could ensure the fitness of the paternal contribution to our next generation.

ACKNOWLEDGMENTS

The authors thank David J. Dix from the Reproductive Toxicology Division (MD-72) of the National Health and Environmental Effects Research Laboratory at the U.S. Environmental Protection Agency for his critical review of this manuscript. Support of this research program from the Michigan Economic Development Corporation and the Michigan Technology Tri-Corridor is gratefully acknowledged.

REFERENCES

1. Nesnow, S., et al., Quantitative and temporal relationships between DNA adduct formation in target and surrogate tissues: implications for biomonitoring. *Environ. Health Perspect.,* 101(Suppl. 3), 37–42, 1993.
2. Reddy, V.R., Gupta, S.M., and Meherji, P.K., Expression of integrin receptors on peripheral lymphocytes: correlation with endometrial receptivity. *Am. J. Reprod. Immunol.,* 46(3), 188–195, 2001.
3. Adamopoulos, D.A., et al., Seminal volume and total sperm number trends in men attending subfertility clinics in the greater Athens area during the period 1977–1993. *Hum. Reprod.*, 11(9), 1936–1941, 1996.
4. Fisch, H., et al., Semen analyses in 1,283 men from the United States over a 25-year period: no decline in quality. *Fertil. Steril.*, 65(5), 1009–1014, 1996.
5. Irvine, S., et al., Evidence of deteriorating semen quality in the United Kingdom: birth cohort study in 577 men in Scotland over 11 years. *Br. Med. J.*, 312(7029), 467–471, 1996.
6. Benshushan, A., et al., Is there really a decrease in sperm parameters among healthy young men? A survey of sperm donations during 15 years. *J. Assist. Reprod. Genet.*, 14(6), 347–353, 1997.
7. Goldstein, M., Surgical management of male infertility and other scrotal disorders, in *Campbell's Urology,* 8th ed., P. Walsh, et al., Eds. New York: W.B. Saunders, 2002.
8. Moldenhauer, J.S., et al., Diagnosing male factor infertility using microarrays. *J. Androl.*, 24(6), 783–789, 2003.
9. Andrade-Rocha, F.T., Semen analysis in laboratory practice: an overview of routine tests. *J. Clin. Lab. Anal.*, 17(6), 247–258, 2003.
10. Brugh, V.M., III and Lipshultz, L.I., Male factor infertility: evaluation and management. *Med. Clin. North Am.*, 88(2), 367–385, 2004.
11. Ostermeier, G.C., et al., Relationship of bull fertility to sperm nuclear shape. *J. Androl.*, 22(4), 595–603, 2001.
12. Menkveld, R., et al., Semen parameters, including WHO and strict criteria morphology, in a fertile and subfertile population: an effort towards standardization of *in-vivo* thresholds. *Hum. Reprod.*, 16(6), 1165–1171, 2001.
13. Chia, S.E., Tay, S.K., and Lim, S.T., What constitutes a normal seminal analysis? Semen parameters of 243 fertile men. *Hum. Reprod.*, 13(12), 3394–3398, 1998.
14. Duty, S.M., et al., Phthalate exposure and human semen parameters. *Epidemiology*, 14(3), 269–277, 2003.
15. Rockett, J.C., et al., Surrogate tissue analysis: monitoring toxicant exposure and health status of inaccessible tissues through the analysis of accessible tissues and cells. *Toxicol. Appl. Pharmacol.*, 194(2), 189–199, 2004.

16. Ostermeier, G.C., et al., Spermatozoal RNA profiles of normal fertile men. *Lancet*, 360(9335), 772–777, 2002.
17. Wang, H., et al., A spermatogenesis-related gene expression profile in human spermatozoa and its potential clinical applications. *J. Mol. Med*, 82, 317–324, 2004.
18. Kierszenbaum, A. and Tres, L.L., Structural and transcriptional features of the mouse spermatid genome. *J. Cell Biol.*, 65, 258–270, 1975.
19. Chiang, M.H., et al., Detection of human leukocyte antigen class I messenger ribonucleic acid transcripts in human spermatozoa via reverse transcription-polymerase chain reaction. *Fertil. Steril.*, 61(2), 276–280, 1994.
20. Concha, II, V et al., U1 and U2 snRNA are localized in the sperm nucleus. *Exp. Cell. Res.*, 204(2), 378–381, 1993.
21. Kumar, G., Patel, D., and Naz, R.K., c-MYC mRNA is present in human sperm cells. *Cell Mol. Biol. Res.*, 39(2), 111–117, 1993.
22. Miller, D., et al., A complex population of RNAs exists in human ejaculate spermatozoa: implications for understanding molecular aspects of spermiogenesis. *Gene*, 237(2), 385–392, 1999.
23. Miller, D., et al., Differential RNA fingerprinting as a tool in the analysis of spermatozoal gene expression. *Hum. Reprod.*, 9(5), 864–869, 1994.
24. Pessot, C.A., et al., Presence of RNA in the sperm nucleus. *Biochem. Biophys. Res. Commun.*, 158(1), 272–278, 1989.
25. Rejon, E., et al., RNA in the nucleus of a motile plant spermatozoid: characterization by enzyme-gold cytochemistry and *in situ* hybridization. *Mol. Reprod. Dev.*, 1(1), 49–56, 1988.
26. Wykes, S.M., Visscher, D.W., and Krawetz, S.A., Haploid transcripts persist in mature human spermatozoa. *Mol. Hum. Reprod.*, 3(1), 15–19, 1997.
27. Wykes, S.M., Miller, D., and Krawetz, S.A., Mammalian spermatozoal mRNAs: tools for the functional analysis of male gametes. *J. Submicrosc. Cytol. Pathol.*, 32(1), 77–81, 2000.
28. Giebelhaus, D.H., Weitlauf, J.J., and Schultz, G.A., Actin mRNA content in normal and delayed implanting mouse embryos. *Dev. Biol.*, 107, 407–413, 1985.
29. Velculescu, V.E., et al., Serial analysis of gene expression. *Science,* 270, 484–487, 1995.
30. Ramsay, G., DNA chips: state-of-the-art. *Nat. Biotechnol.,* 16, 40–44, 1998.
31. Behr, M.A., Wilson, M.A., Gill, W.P., Salamon, H., Schoolnik, G.K., Rane, S., and Small, P.M., Comparative genomics of BCG vaccines by whole-genome DNA microarray. *Science,* 284, 1520–1523, 1999.
32. Iyer, V.R., Eisen, M.B., Ross, D.T., Schuler, G., Moore, T., Lee, J.C.F., Trent, J.M., Staudt, L.M., Hudson, J., Jr., Boguski, M.S., Lashkari, D., Shalon, D., Botstein, D., and Brown, P.O., The transcriptional program in the response of human fibroblasts to serum. *Science,* 283, 83–87, 1999.
33. Perou, C.M., Sorlie, T., Eisen, M.B., van de Rijn, M., Jeffrey, S.S., Rees, C.A., Pollack, J.R., Ross, D.T., Johnsen, H., Aksien, L.A., Fluge, O., Pergamenschikov, A., Williams, C., Zhu, S.X., Lonning, P.E., Borresen-Dale, A.L., Brown, P.O., and Botstein, D., Molecular portraits of human breast tumours. *Nature,* 406, 747–752, 2000.
34. Clark, E.A., Golub, T.R., Lander, E.S., and Hynes, R.O., Genomic analysis of metastasis reveals an essential role for RhoC. *Nat. Biotechnol.*, 406, 532–535, 2000.

35. Bittner, M., Meltzer, P., Chen, Y., Jiang, Y., Seftor, E., Hendrix, M., Radmacher, M., Simon, R., Yakhini, Z., Ben-Dor, A., Sampas, N., Dougherty, E., Wang, E., Marincola, F., Gooden, C., Lueders, J., Glatfelter, A., Pollock, P., Carpten, J., Gillanders, E., Leja, D., Dietrich, K., Beaudry, C., Berens, M., Alberts, D., Sondak, V., Hayward, N., and Trent, J., Molecular classification of cutaneous malignant melanoma by gene expression profiling. *Nat. Biotechnol.*, 406, 536–540, 2000.
36. Welcsh, P.L., Lee, M.K., Gonzalez-Hernandez, R.M., Black, D.J., Mahadevappa, M., Swisher, E.M., and Warrington, J.A.M.-C., K., BRCA1 transcriptionally regulates genes involved in breast tumorigenesis. *Proc. Natl. Acad. Sci. U.S.A.*, 99, 7560–7565, 2002.
37. van't Veer, L.J., Dai, H., van de Vijver, M.J., He, Y.D., Hart, A.A., Mao, M., Peterse, H.L., van der Kooy, K., Marton, M.J., Witteveen, A.T., Schreiber, G.J., Kerkhoven, R.M., Roberts, C., Linsley, P.S., Bernards, R., and Friend, S.H., Gene expression profiling predicts clinical outcome of breast cancer. *Nature,* 415, 530–536, 2002.
38. Nuwaysir, E.F., Bittner, M., Trent, J., Barrett, J.C., and Afshari, C.A., Microarrays and toxicology: the advent of toxicogenomics. *Mol. Carcinog.,* 243, 153–159, 1999.
39. Thomas, R.S., Rank, D.R., Penn, S.G., Zastrow, G.M., Hayes, K.R., Pande, K., Glover, E., Silander, T., Craven, M.W., Reddy, J.K., Jovanovich, S.B., and Bradfield, C.A., Identification of toxicologically predictive gene sets using cDNA microarrays. *Mol. Pharmacol.,* 60, 1189–1194, 2001.
40. Lash, L.H., Hines, R.N., Gonzalez, F.J., Zacharewski, T.R., and Rothstein, M.A., Genetics and susceptibility to toxic chemicals: do you (or should you) know your genetic profile? *J. Pharmacol. Exp. Ther.,* **305**, 403–409, 2003.
41. Kramer, J.A. and Krawetz, S.A.,RNA in spermatozoa: implications for the alternative haploid genome. *Mol. Hum. Reprod.*, 3(6), 473–478, 1997.
42. Ostermeier, G.C., Dix, D.J., and Krawetz, S.A., A bioinformatic strategy to rapidly characterize cDNA libraries. *Bioinformatics,* 18(7), 949–52, 2002.
43. Khatri, P., et al., Profiling gene expression using onto-express. *Genomics,* 79(2), 266–270, 2002.
44. Martins, R.P., Leach, R.E., and Krawetz, S.A., Whole body gene expression by data mining. *Genomics*, 72, 34–42, 2001.
45. Granadino, B., Arlas-de-la-Fuente, C., Perez-Sanchez, C., Parraga, M., Lopez-Fernandez, L.A., del Mazo, J., and Rey-Campos, J., Fhx (Foxj2) expression is activated during spermatogenesis and very early in embryonic development. *Mech. Dev.*, 97, 157–160, 2000.
46. Yamaguchi, T.P., Bradley, A., McMahon, A.P., and Jones, S., A Wnt5a pathway underlies outgrowth of multiple structures in the vertebrate embryo. *Development*, 126, 1211–1123, 1999.
47. Ostermeier, G.C., Miller, D., Huntriss, J.D., Diamond, M.P., and Krawetz, S.A., Reproductive biology: Delivering spermatozoan RNA to the oocyte. *Nature,* 429, 154, 2004.
48. Ostermeier, G.C., Goodrich, R.J., Moldenhauer, J.S., Diamond, M.P., and Krawetz, S.A., A suite of novel human spermatozoal RNA's. *J. Androl.*, 26, 70–74, 2005.
49. Copland, J.A., et al., The use of DNA microarrays to assess clinical samples: the transition from bedside to bench to bedside. *Recent Prog. Horm. Res.*, 58, 25–53, 2003.
50. Carlsen, E., Giwercman, M., Keiding, N., and Skakkebaek, N.E., Evidence for decreasing quality of semen during past 50 years. *Br. Med. J.*, 305, 609–613, 1992.
51. Parazzini, F., Bortolotti, A., and Colli, E., Declining sperm count and fertility in males: an epidemiological controversy. *Arch. Androlol.*, 41, 27–30, 1998.

52. Skakkebaek, N.E., Rajpert De Meyts, E., Jorgensen, N., Carlsen, E., Petersen, P.M., Giwercman, A., Andersen, A.G., Jensen, T.K., Andersson, A.M., and Muller, J., Germ cell cancer and disorders of spermatogenesis: an environmental connection? *Apmis*, 106, 3–11, 1998.
53. Telisman, S., Cvitkovic, P., Jurasovic, J., Pizent, A., Gavella, M., and Rocic, B., Semen quality and reproductive endocrine function in relation to biomarkers of lead, cadmium, zinc, and copper in men. *Environ. Health Perspect.*, 108, 45–53, 2000.
54. Lindbohm, M.L., Effects of occupational solvent exposure on fertility. *Scand. J. Work Environ. Health*, 25, 44–46, 1999.
55. Aitken, R.J. and Clarkson, J.S., Cellular basis of defective sperm function and its association with the genesis of reactive oxygen species by human spermatozoa. *J. Reprod. Fertil.*, 81, 459–469, 1987.
56. Moline, J.M., Golden, A.L., Bar-Chama, N., Rauch, E.M.E., Chapin, R.E., Perreault, S.D., Schrader, S.M., Suk, W.A., and Landrigan, P.J., Exposure to hazardous substances and male reproductive health: a research framework. *Environ. Health Perspect.*, 108, 803–813, 2000.
57. Kimbrough, R., Polychlorinated biphenyls (PCBs) and human health: an update. *Crit. Rev. Toxicol.*, 25, 133–163, 1995.
58. Kearney, J.P., Cole, D.C., Ferron, L.A., and Weber, J.P., Blood PCB, p,p1-DDE, and mirex levels in Great Lakes fish and waterfowl consumers in two Ontario communities. *Environ. Res.*, 80, S138–S149, 1999.
59. Falk, C., Hanrahan, L., Anderson, H.A., et al., Body burden levels of dioxin, furans and PCBs among frequent consumers of Great Lakes sport fish. The Great Lakes Consortium. *Environ. Res.*, 80, S19–S25, 1999.
60. Bush, B., Bennett, A.H., and Snow, J.T., Polychlorobiphenyl congeners, p,p1-DDE, and sperm function in humans. *Arch. Environ. Contam. Toxicol.,* 15, 333–341, 1986.
61. Tielemans, E., van Kooij, R., te Velde, E.R., Burdorf, A., and Heederik, D., Pesticide exposure and decreased fertilisation rates *in vitro. Lancet,* 354, 484–485, 1999.
62. Goldsmith, J.R., Potashnik, G., and Israeli, R., Reproductive outcomes in families of DBCP-exposed men. *Arch. Environ. Health*, 39, 85–89, 1984.
63. Potashnik, G. and Yanai-Inbar, I., Dibromochloropropane (DBCP): an 8-year reevaluation of testicular function and reproductive performance. *Fertil. Steril.,* 47, 317–323, 1987.
64. Draghici, S., Khatri, P., Martins, R.P., Ostermeier, G.P., and Krawtz, S.A., Global functional profiling of gene expression. *Genomics*, 81, 98–104, 2003.
65. Stanton, J.L., Bascard, M., Fisher, L., Quinn, M., Macgregor, A., and Green, D.P.L., Gene expression profiling of human GV oocytes: an analysis of a profile obtained by serial analysis of gene expression (SAGE). *J. Reprod. Immunol*, 53, 193–201, 2001.
66. Stanton, J.L., Macgregor, A.B., and Green, D.P.L., Using expressed sequence tag databases to identify ovarian genes of interest. *Mol. Cell. Endocrinol.*, 191, 11, 2002.

Section III

Proteomic Approaches

CHAPTER 7

Proteomic Analysis of Surrogate Tissues: Mass Spectrometry-Based Profiling of the Circulatory Proteome for Cancer Detection and Stratification

Emanuel F. Petricoin III, Katherine R. Calvo, Julia Wulfkuhle, and Lance A. Liotta

CONTENTS

7.1 CLINICAL CANCER BIOMARKERS: IS THE PIPELINE DRIED UP?

Despite the urgent need for biomarkers that can improve cancer clinical outcome through early detection, risk stratification, and therapy optimization, relatively few new cancer biomarkers have been advanced to routine clinical use.[1] The poor yield of clinically useful biomarkers is not for lack of trying by thousands of scientists worldwide. Gene and protein array data have revealed that each malignancy may have a different molecular portrait.[2–5] Unfortunately, discovery of cancer-specific markers has proved much harder than was initially anticipated. The three major impediments are (1) molecular heterogeneity between histologically identical appearing tumors; (2) prevalence of noncancer diseases that reduce biomarker specificity for cancer; and (3) low biomarker concentrations (especially in early stage disease), which reduce sensitivity.

In addition to these impediments, two other significant roadblocks to clinical biomarker development are the complexity of the clinical studies needed to uncover them and the number of subjects required for such studies to achieve adequate statistical power. A cancer biomarker validation trial must be designed specifically to address sensitivity and specificity delimited to the *intended use*. Examples of intended use categories are as follows: (1) early detection in the general population, (2) high-risk screening, (3) secondary screening in combination with other modalities, and (4) recurrence monitoring. Depending on the intended use, the clinical trial design can be vastly different. In fact, while general population screening for early-stage detection (indication 1 above) would no doubt have significant public health impact, one could argue that population screening should be the last "clinical intended use" that is explored when identifying and characterizing a candidate cancer biomarker. One reason for this is demonstrated in the example of ovarian cancer screening. In the general population, it is estimated that 1 in 2500 women will develop ovarian cancer at any point in time.[6] Consequently, if 2500 women are screened using a candidate ovarian cancer biomarker that theoretically achieves 99% sensitivity and 99% specificity, approximately 50 women will be misdiagnosed (false negative or false positive) for every one cancer that is detected correctly. Even if the candidate biomarker is 100% sensitive and 100% specific, if one of the statistical requirements of a screening trial is to identify 200 cancer events, then the number of patients needed to successfully conduct the trial will be approximately 200 times 2500, equaling 500,000 women! These 500,000 women would likely have to be monitored over many years to obtain the necessary prospective data. Thus, the total number of individual blood specimens collected under such a trial would number in the millions, with each requiring appropriate collection, processing, and storage, in addition to detailed phenotypic evaluation that significantly influences the trial economics.

In contrast to population screening, pursuing the indication of risk stratification may provide a new opportunity for biomarker evaluation and potentially a faster route to clinical use. While diagnostic imaging technologies are rapidly advancing due to computer sophistication, these methods at present have poor specificity and remain too expensive at this time to be used for general population-based screening. The combination of a validated blood-based biomarker in conjunction with an appropriately specific diagnostic imaging modality could save unnecessary surgical procedures while retaining the ability to detect early-stage disease. The current chapter describes (1) the rationale for mass spectrometry (MS)-based profiling of surrogate tissues for the identification of cancer biomarkers, (2) approaches that can be applied in the profiling of surrogate tissues for the identification of novel cancer biomarkers, (3) issues in the application of surface-enhanced laser desorption/ionization time-of-flight (SELDI-TOF)-based methods for these purposes, and (4) the effect of carrier protein sequestration on potential low-molecular-weight biomarkers that can be exploited to enhance the sensitivity and specificity of assays designed to detect these surrogate-tissue-based biomarkers in humans.

7.1.1 Abandoning Old Assumptions about Cancer Biomarker Biology

To improve the sensitivity and specificity of cancer biomarkers, the assumption that a single cancer biomarker exists for one or multiple types of cancer may have to be abandoned. Cancer cells are genetically deranged normal cells, not exogenous infectious agents. Consequently, biomarkers associated with cancer can logically be expected to be quantitatively, but not qualitatively, different from normal cellular molecules. Indeed, all clinically employed circulating cancer biomarkers to date appear to be present in both malignant and nonmalignant conditions.[1] In an attempt to improve specificity and sensitivity, investigators have been evaluating the use of multiple panels of (1) identified markers,[7] or (2) panels of unidentified and uncharacterized molecules. The rationale for this approach is based on a new appreciation of the tumor microenvironment. Tumor cells are involved in complex and poorly understood interactions with surrounding organ parenchyma, local stroma, vasculature, and immune cell populations. This biochemical cross talk is hypothesized to generate a cascade of specific and sensitive biomarkers produced directly from the tumor cell population, indirectly from the interacting nontumor cells or extracellular molecules, or to be a specific product of the microecology. The most specific cancer biomarkers may turn out to be chemically modified molecules derived from this last category. Molecules that normally play a nonmalignant role in physiology may be cleaved, phosphorylated, glycosylated, or otherwise altered in a manner that provides an ongoing and specific biomarker record of the pathophysiology of the tumor–host microenvironment.[8,9] Larger proteins that may be unable to cross the endothelial vascular wall due to their size, and are thus excluded from the circulatory proteome, may in fact be represented in the blood by smaller isoforms. Thus, fragments in the circulatory proteome may represent almost every tissue protein and provide a fountainhead of new diagnostic information.

7.2 A RICH POTENTIAL SOURCE OF CANDIDATE BIOMARKERS IN THE LOW-MOLECULAR-WEIGHT REALM

Investigators have recently evaluated MS as a means to detect modified proteins derived from the tumor–host microenvironment, even though the identity of such molecules was not known ahead of time.[10–22] Past attempts to discover new biomarkers, using hypothesis-generating discovery approaches such as two-dimensional gel electrophoresis, would have ignored a biomarker archive that comprised many of these small modified and clipped molecules. This deficiency arises from the fact that gel-based separation methods have poor resolution in the low-molecular-weight (LMW) range of the proteome and often undersample small components due to loss during fixation and analysis. MS, on the other hand, has its best sensitivity and resolution within the LMW range of the proteome. MS also has a unique advantage as a discovery tool because it does not require prior knowledge of protein characteristics or the development of specific capture agents (i.e., antibody) to separate and profile fractions of the proteome of interest. Initially, our laboratory used MS to determine whether tissue cell lysates obtained with laser capture microdissection contained a proteomic portrait that could discriminate different tumor types, differences between primary and metastatic disease, or early-stage premalignant lesions.[23] The results of these and other later studies[24] indicated that there were significant differences in the proteomic fingerprints of the tumor cells themselves, especially in the LMW range (mass-to-charge ratio below 40,000). In fact, MS-based tissue profiling or imaging MS may be able to rapidly identify tissue-borne proteomic fingerprints that can both be prognostic for outcome[25] and used to evaluate response to molecular therapeutics.[26] Based on the compendium of this information and the demonstration of the existence of this uncharted LMW information archive, the obvious next question was: How much of this new information can be captured from a blood sample?

Based on these previous findings, ourselves and others set out to test the hypothesis that the LMW range of the circulatory proteome contained previously unknown diagnostic information. Work by the groups of Goodacre[27] and Lay[28] indicated that it was possible to combine mass spectral data with pattern recognition methodology to identify fingerprints that could discriminate bacterial species without prior knowledge of the molecules themselves. This type of approach provided a facile means to query mass spectral information, without even knowing if diagnostic information existed in this low-molecular-mass region. Using a variety of different pattern recognition methods, high-throughput MS, and a variety of disease states, we and others have generated data that appear to indicate that discriminatory information can be found within the study sets employed.[10–22] The results of these many independent studies appear to support the initial hypothesis that there does indeed exist a rich source of previously unknown biomarker information in the circulatory proteome.

7.2.1 Prospecting Approaches

The new LMW information archive residing in the circulatory proteome is being actively mined by investigators using two separate avenues of translational investigation. The two approaches are complementary rather than exclusive. One avenue of investigation uses the fingerprints, or patterns, of mass spectral information as the diagnostic itself, even without knowledge of the identity or sequence of the molecules contributing to the mass spectra. Investigators using this approach are examining collections of mass spectral amplitude values and then evaluating whether the combined relative intensities of the m/z values can be used to classify disease states accurately. The other avenue is to sequence the proteins comprising this new set of candidate biomarkers directly. Once each candidate biomarker is identified, the next objective is to develop capture reagents (e.g., antibodies) that can be used to measure multiplexed panels of analytes consisting of subsets of the candidate biomarkers. Both of these avenues have significant advantages, disadvantages, and roadblocks ahead. It is not readily apparent at this time which approach will ultimately have the earliest impact at the bedside. Our stated opinion is that both avenues should be explored concomitantly since any method that could achieve patient benefit warrants rigorous and serious investigation.[29,30]

What are the impediments facing investigators choosing between these approaches? With regard to an approach based on patterns of unidentified molecules, the major challenge is one of reproducibility across platforms, time, and laboratories. Since MS platforms are in a constant state of technologic evolution, with new improved systems coming online every year, no common platform or standard operating procedure has yet been adopted by the scientific community. Lack of agreement on the utilization and type of reference standards further complicates this issue. Additionally, since the molecules that underpin the pattern are not known, a further impediment is the difficulty of assuring that experimental bias is not a contributor to the discrimination. Experimental bias can occur due to differences in how the cases and control specimens are collected and processed, or from the procedure and process of generating the mass spectra itself. In addition to the problems of experimental bias, investigators must recognize the further challenges presented by high-dimensional data analysis. Rigorous validation based on blinded study sets are absolutely required to guard against overfitting. Nevertheless, MS profiling remains an attractive and very rapid analytical approach, well suited for commercialization. This discovery-based approach does not require the lengthy development and validation of antibody reagents and immunoassay-based systems. The difficulty of constructing and validating calibrators and controls suitable for CLIA/ASCP/CAP certification, not to mention formal validation and licensure, is a very formidable task.

In contrast to direct MS profiling of blood or tissue, sequencing and characterization of the underlying constituents is a very laborious process. In fact, the cycle time for protein sequencing, characterization, antibody (or analyte-specific ligand) development, validation in clinical research study sets, and immunoassay development is the biggest impediment for the direct characterization approaches. The obvious advantage of this path is that once characterized, reproducibility of measurements of the analytes using well-tested and validated immunoassay platforms is not an issue. Additionally, once the molecules are identified, bias and overfitting can be assessed directly.

7.3 POINTS TO CONSIDER FOR MASS SPECTROMETRY-BASED PROFILING STUDIES

While investigators have used a variety of different bioinformatic algorithms for pattern discovery, the most common analytical MS profiling platform today comprises a Protein Chip Biomarker System-II (PBS-II, a low-resolution TOF mass spectrometer). Herein samples are ionized by surface enhanced laser desorption/ionization (SELDI), a protein chip array-based chromatographic retention technology that allows for direct MS analysis of analytes retained on the array (Figure 7.1). Only a subset of the proteins in the serum bind to the chromatographic surface of the chip, and the unbound proteins are washed away. The bait region containing individual captured serum protein samples is overlaid with a coating of an organic

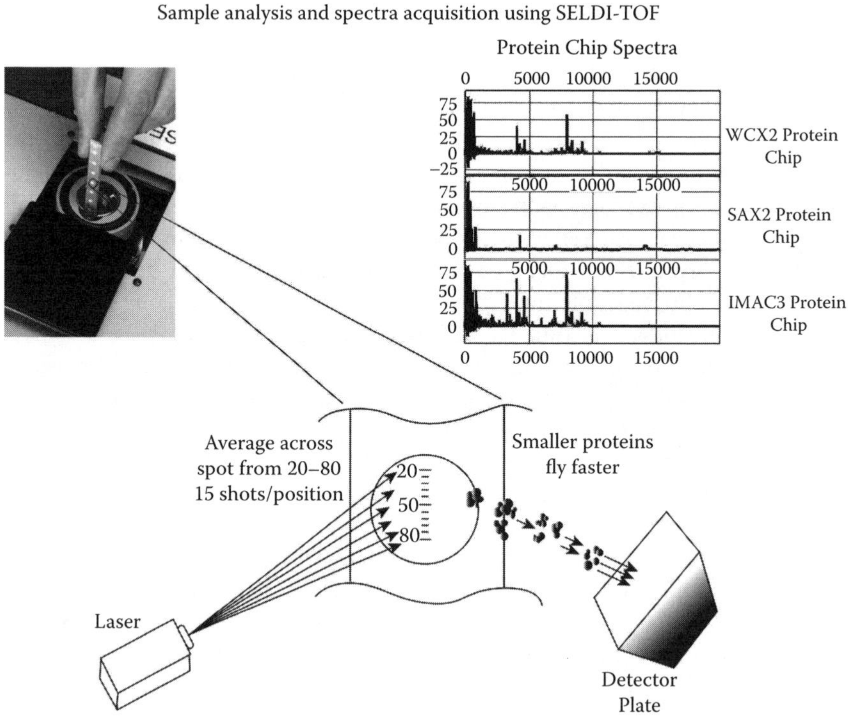

Figure 7.1 MS as a diagnostic tool. Surface-enhanced laser desorption/ionization TOF (SELDI-TOF) MS is one type of proteomic analytical tool and is a class of MS instrument useful in high-throughput proteomic fingerprinting of serum. Depending on the surface chemistry used (WCX2 = weak cation exchange surface; SAX2 = strong anion exchange surface; IMAC3 = immobilized metal affinity surface), a subset of the proteins in the sample bind to the surface of the chip with unbound proteins washed off after incubation. The bound proteins are treated with a MALDI matrix, washed, and dried. The chip, containing multiple patient samples, is inserted into high (ABI Qstar) or low (Ciphergen PBS) resolution mass spectrometers and analyzed by laser desorption/ioization. The TOF of the ion prior to detection by an electrode is a measure of the mass to charge (*m*/*z*) value of the ion, with most ions being singly charged. The mass spectra can then be analyzed by various pattern recognition software to discover potential diagnostic differences based not on one molecule, but on a pattern of multiple decreases and increases in ion amplitudes.

acid matrix (e.g., α-cyano-5-hydroxycinnamic acid), which crystallizes, and then the entire chip is inserted into a vacuum chamber and a laser beam is fired at each spot. The organic acid matrix serves as an energy transfer medium for protein ionization whereby the kinetic energy from the laser causes protein desorption/ionization.[31] The mass-to-charge value of each ion is estimated from the time it takes for the launched ion to reach the electrode; small ions travel faster. Consequently, the spectrum provides a TOF signature of ions ordered by size. Recently, this concept has been extended to a high-resolution MS employing a hybrid quadrupole TOF MS (QSTAR pulsar *i*, Applied Biosystems, Inc., Framingham, MA) fitted with a ProteinChip array interface (Ciphergen Biosystems, Inc., Fremont, CA). As a point or

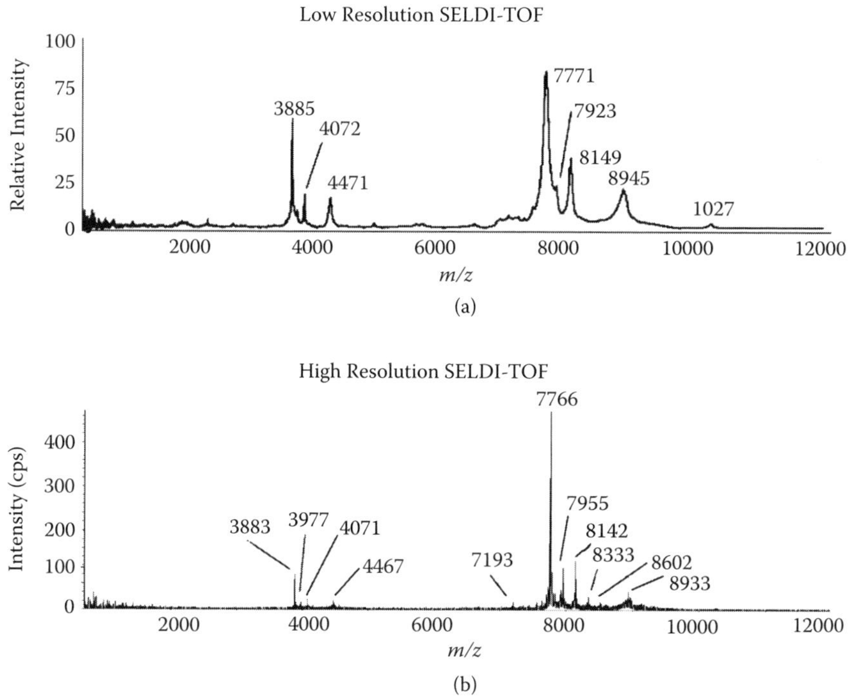

Figure 7.2 Comparison between low resolution and high resolution SELDI-TOF mass spectra. Spectra from the same weak cation exchange chip (queried at the same spot on the same chip) were generated on either a PBS IIc (low-resolution instrument Ciphergen Biosystems, Inc.) (Panel A) or on a QSTAR Pulsar *i* high-resolution instrument (Applied Biosystems, Inc.) (Panel B).

analytical comparison, the Qq-TOF MS (routine resolution ~ 8000) can completely resolve species differing in *m/z* of only 0.375 (e.g., at *m/z* 3000) whereas complete resolution of species with the Ciphergen PBS-II TOF MS (routine resolution ~ 150) is only possible for species that differ by *m/z* of 20 (Figure 7.2).[32] Moreover, the spectral resolution of the lower-resolution instrumentation may not be able to separate specific ions that are close in mass/charge and which can coalesce multiple specific discrete ions into a single peak. Of course, whether or not any low-resolution or high-resolution mass spectrometer is ever used as a routine clinical diagnostic platform remains to be seen, as the field is just in its infancy. Clinical utility is not just predicated based on clinical performance. The entire process will need to be evaluated for each step of the process: sample handling, archiving, database and processing standard operating procedures, sample application and robotic handling procedures, MS cGMP and ISO9001 performance, protein chip and/or MALDI plate performance, software validation, and database management.

In a clinical setting where a fingerprint-based test comprising unknown molecules could be eventually employed as a diagnostic, it will be important to determine overall spectral quality and develop spectral release specifications such that variances introduced into the process can be evaluated and monitored. Day-to-day, lot-to-lot, and machine-to-machine variances brought in from sample handling/storage and shipping conditions will need to be evaluated and understood as well as the mass spectrometer itself. An important component of the analysis will be assessment of in-process controls and calibrators, The use of a reference standard sample, such as that which can be obtained from NIST (SRM–1951A), can be employed and randomly applied to one spot on each protein array as a quality control for overall process integrity, sample preparation, and mass spectrometer function. Additionally, for spectral quality control, quality assurance, and spectral release specification, all spectra should be subjected to a suite of statistical measures such as total ion current (total record count), average/mean and standard deviation of amplitude, chi-square and *t*-test analysis of each ion or bin, and quartile plotting measures. Process measures can then be checked by analyzing the statistical plots of the serum reference standard, and spectra that fail statistical checks for homogeneity are eliminated from in-depth modeling and analysis. This type of upfront analysis is critical so that it is possible to compare the total analytical variance obtained from a constant reference sample with the variance of the clinical sample populations. The total variance of the reference sample should be no less than that for the clinical specimens. Reproducibility of the MS profiling type approaches above is currently being evaluated.

A typical low-resolution SELDI-TOF proteomic profile will have up to 15,500 data points that comprise the recordings of data between 500 and 20,000 m/z, with a high-resolution mass spectrometer generating upwards of 1,000,000 data points. A multitude of downstream pattern recognition systems exist, and all may show very good reliability at detecting and discovering sets of classifying ion features. To reduce complexity of high-resolution data, one simple approach is to bin the data from the spectra, for example, by using a simple ppm binning equation that gradually increases as a function of the resolution capacity of the machine. If one uses a 400-ppm binning function, one can reduce the number of data points from 350,000 to a little over 7000 points per sample.[17] The binning function should be based on the estimate of what the mass drift of the MS machine routinely obtains by external and internal calibration results. The data are then normalized (necessary since MS is inherently nonquantitative) and then randomly separated into equal groups for training and testing. Data normalization is an important element of pattern recognition so as to ensure a commonality in the spectra itself and assess potential for bias (e.g., introduced by protein chip quality, mass spectrometer instrumentation and operator variance, sample collection, sample handling and storage) and which can effect overall spectral performance and introduce potential nondisease-related artifact into the spectra. It is likely that different data normalization procedures will generate different ions selected, especially in a clustering algorithm where multiple ion features are used as the pattern. Since MS is not inherently quantitative, scalar intensity changes may be apparent, yet the overall pattern may not changed. Normalization can be achieved by a variety of means, such as dividing the spectra by the total ion current, amplitude value sums, or average.

7.4 BIOMARKER AMPLIFICATION VIA CARRIER PROTEIN SEQUESTRATION: UNDERPINNINGS OF THE MASS SPECTRAL INFORMATION

Based on the need for identification, our own laboratory research efforts have centered on the identification and sequencing of the underlying discriminatory information that exists in the mass range profiled by direct MS profiling work. Some have argued that only high-abundance molecules are represented in the mass spectral read-out, and that, as such, these molecules can be only nonspecific epiphenomena.[33] Relevant to this concern, we are beginning to understand some of the mechanisms by which low-abundance biomarkers can be amplified biologically to detectable concentrations. As we sought to understand the source and identity of the molecules, we realized, and experimentally demonstrated, that a vast majority of the LMW biomarkers under study were actually complexed with high-abundance circulating carrier proteins.[8,30,34,35] Accumulation of LMW biomarkers in association with circulating carrier proteins greatly amplifies the total serum/plasma concentration of the measurable biomarker. This enrichment of detectable LMW components is clearly demonstrated (Figure 7.3), where the MS profile of neat serum is compared to that of LMW components associated with albumin. This enrichment is due to the biomarker elimination half-life taking on the half-life of the carrier protein.[30] Such association drives equilibrium toward the plasma compartment, even if the association constant is low. This is because the carrier protein exists in concentrations many orders of magnitude greater than the biomarker. In fact, noncovalent association with albumin has been shown to extend the half-life of short-lived proteins introduced into the circulation.[36,37]

These findings now shift the focus of biomarker analysis from a serum-wide analysis to the just the carrier protein and its biomarker content. The discriminatory molecules are likely to be metabolic products, enzymatically generated fragments, and modified protein fragments. In fact, the most important biomarkers may be normal host proteins that are aberrantly clipped, modified, or reduced in abundance. Until now, conventional protocols for biomarker discovery discard the abundant "contaminating" high-molecular-mass proteins to focus on the low mass range. Unfortunately, this procedure may remove most of the important diagnostic biomarkers — the carrier protein-bound LMW molecules.[8] We can now develop new tools, created at the intersection of proteomics and nanotechnology, whereby nanoharvesting agents can be instilled into the circulation (e.g., derivatized gold particles) or into the blood collection device to act as "molecular mops" that soak up and amplify biomarkers via accumulation (Figure 7.4). These nanoparticles, with their bound diagnostic cargo, can be directly analyzed by MS and the LMW and enriched biomarker signatures revealed. Coupling this method with ultrahigh-resolution MS (e.g., Fourier transform ion cyclotron resonance mass spectrometry[38,39]) will allow for rapid protein identification and diagnostic analysis at the same time with the same machine.

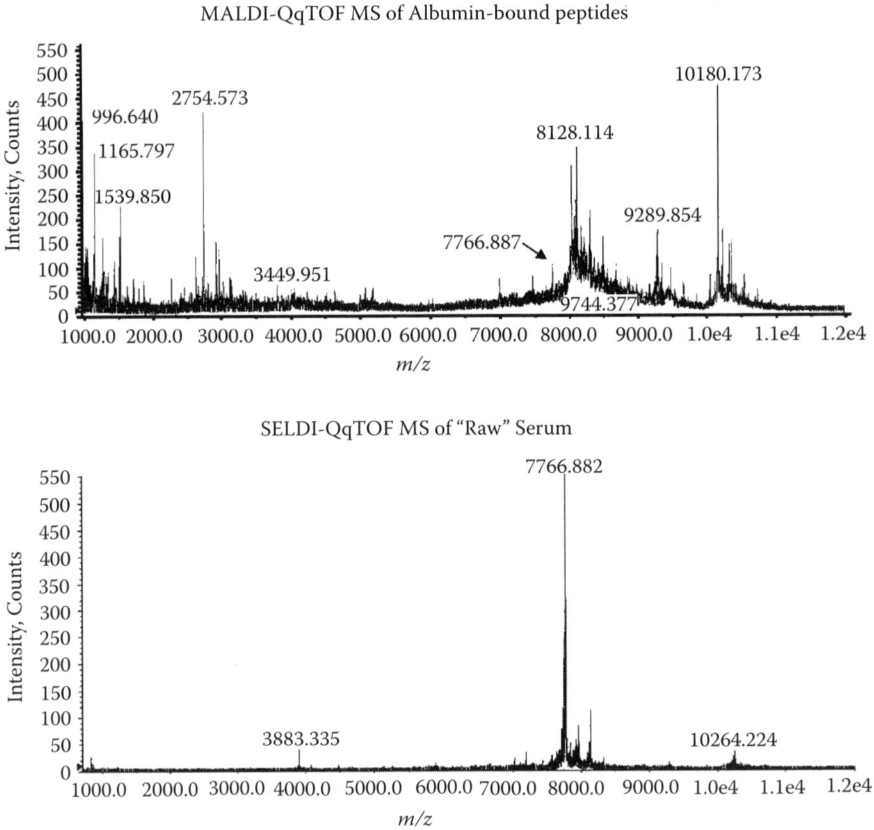

Figure 7.3 Enrichment of LMW (1,000 to 12,000 *m/z*) proteomic information via carrier protein binding and amplification. Mass spectral comparison of the same serum where the input was either the albumin-bound fraction (Upper panel) or the native total sample (Lower panel). The spectra are scaled equally so that a direct comparison can be made. MALDI-QqTOF = matrix-assisted laser desorption ionization hybrid quadrupole time-of-flight; SELDI-TOF = surface-enhanced laser desorption ionization time-of-flight; MS = mass spectrometry.

7.5 CONCLUDING REMARKS AND A VIEW TO THE FUTURE

Recognition that cancer is a product of the proteomic tissue microenvironment has important clinical implications from both an early detection and therapeutic targeting point of view. The tissue microenvironment can spawn entirely new biomarker cascades of LMW information that are amplified by subtle changes at the earliest times of tumor growth and invasion. The exchange of information in the communication linkages at the invasion interface can give rise to changes that are reflected in specific alterations of the proteome of the circulation. The LMW component of the circulatory proteome offers an exciting, untapped, and unexplored source of potentially useful diagnostic information. The two major paths to utilizing

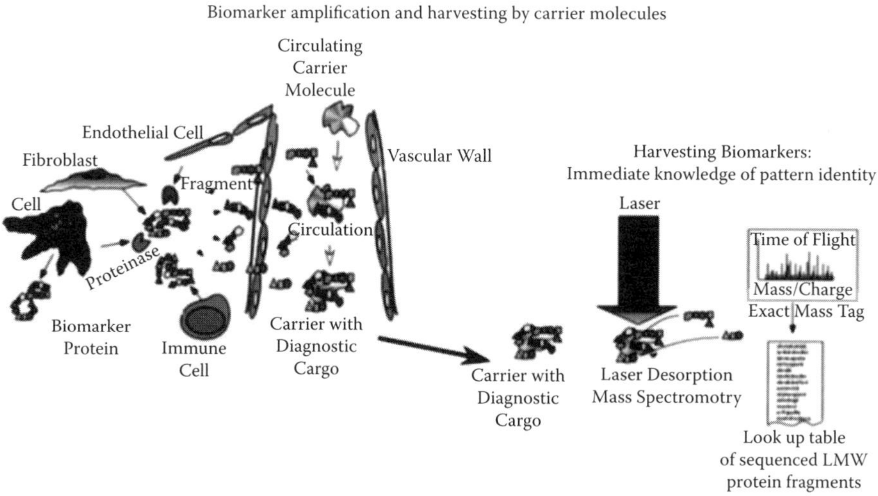

Figure 7.4 Biomarker amplification and harvesting by carrier molecules. LMW peptide fragments, produced at the interface of the diseased cell and the tissue microenvironment, permeate through the endothelial cell wall barrier and trickle into the circulation. Here, these fragments are immediately bound with circulating high-abundance carrier proteins such as albumin and protected from rapid kidney clearance. The sequestration of the low-abundance biomarkers by the carrier protein pool over time results in the net effect of an enrichment and amplification of the biomarker fragments. In the future, harvesting nanoparticles, engineered with high affinity for binding, can be instilled into the collected body fluids or perhaps even injected directly into the circulation. These nanoparticles and their bound biomarkers can then be collected, filtered over engineered nanofilters, and directly queried by high-resolution MS. A lookup table, where the exact identities of each of the peaks will be compared against the accurate mass tag of each of the peaks within the spectra (e.g., through the use of FTICR-type systems) will soon enable the simultaneous identification of each entity within the pattern as well as the discovery of the diagnostic pattern itself.

this archive — patterns of unknown and unidentified MS-generated ions, or a multiplex immunoassay of known and sequenced molecules — are being explored concomitantly at this time. These two paths will intersect at some point in the near future as we understand the identities of each molecule displayed by MS. It will be critical to the field that demonstration of reproducibility across time and between laboratories using a pattern-based approach where the underlying identities of the molecules are unknown, is achievable. Unfortunately, some publications[40] have based conclusions about lack of reproducibility based on publicly posted mass spectral data sets (http://home.ccr.cancer.gov/ncifdaproteomics/) where each data set was derived from a purposefully altered experimental condition as methods were being optimized and sources of variability identified and measured. These conclusions provide an inaccurate picture of the state of the science, since it was expected that each posted data set would be different and that the data should not be used as a reproducibility study.[30] Importantly, however, it appears that MS profiling-based approaches are beginning to demonstrate inter- and intralaboratory reproducibility when robustness is an actual stated goal of the study.[41]

The ability of carrier protein sequestration to provide a rich and untapped source of biomarker information may soon populate the biomarker pipeline with interesting candidate molecules available in surrogate tissues such as the circulatory proteome. Upon rigorous and extensive qualification and scientific validation using large clinical study sets, multiplexed measurements of some of theses candidate molecules may eventually reach clinical utility. This same information archive, in a complex fashion, appears to underpin serum mass spectral profiles. Thus, a list of sequence-identified proteins or peptides that reside in the mass range encompassed by a mass spectral profile becomes a facile conduit between profile-based approaches and multiplexed immunoassay based systems. Such a marriage of publicly available identities to MS patterns, we believe, should expedite translation of this knowledge to the bedside, independent of the analytical method employed.

ACKNOWLEDGMENTS

The views expressed here are expressed solely by the authors and should not be construed as representative of those of the Department of Health and Human Services and the U.S. Food and Drug Administration. Moreover, aspects of the topics discussed have been filed as U.S. Government-owned patent applications. Drs. Petricoin and Liotta are co-inventors on these applications and may receive royalties provided under U.S. law.

REFERENCES

1. Anderson, N.L. and Anderson, N.G. The human plasma proteome: history, character, and diagnostic prospects. *Mol. Cell Proteomics* 1(11):845–867, 2002.
2. Wright, G., Tan, B., Rosenwald, A., Hurt, E.H., Wiestner, A., and Staudt, L.M. A gene expression-based method to diagnose clinically distinct subgroups of diffuse large B cell lymphoma. *Proc. Natl. Acad. Sci. U.S.A.* 100(17), 9991–9996, 2003. Epub 2003 Aug 04.
3. Ramaswamy, S., Tamayo, P., Rifkin, R., Mukherjee, S., Yeang, C.H., Angelo, M., et al. Multiclass cancer diagnosis using tumor gene expression signatures. *Proc. Natl. Acad. Sci. U.S.A.* 98(26), 15149–15154, 2001. Epub 2001 Dec 11.
4. Wulfkuhle, J.D., Aquino, J.A., Calvert, V.S., Fishman, D.A., Coukos, G., Liotta, L.A., and Petricoin, E.F., III. Signal pathway profiling of ovarian cancer from human tissue specimens using reverse-phase protein microarrays. *Proteomics* 3(11), 2085–2090, 2003.
5. Petricoin, E.F., Bichsel, V.E., Calvert, V.S., Espina, V., Winters, M., Young, L., et al. Mapping molecular networks using proteomics: a vision for patient-tailored combination therapy. *J. Clin. Oncol.*, 23(15), 3614–3621, 2005.
6. Jacobs, I.J. and Menon, U. Progress and challenges in screening for early detection of ovarian cancer. *Mol. Cell Proteomics* 3(4), 355–366, 2004. Epub 2004 Feb 05.

7. Skates, S.J., Horick, N., Yu, Y., Xu, F.J., Berchuck, A., Havrilesky, L.J., et al. Preoperative sensitivity and specificity for early-stage ovarian cancer when combining cancer antigen CA-125II, CA 15-3, CA 72-4, and macrophage colony-stimulating factor using mixtures of multivariate normal distributions. *J. Clin. Oncol.* 22(20), 4059–4066, 2004. Epub 2004 Sep 20.
8. Liotta, L.A., Ferrari, M., and Petricoin, E. Clinical proteomics: written in blood. *Nature* 425(6961), 905, 2003.
9. Liotta, L.A. and Kohn, E.C. The microenvironment of the tumour-host interface. *Nature* 411(6835), 375–179, 2001.
10. Petricoin, E.F., III, Ornstein, D.K., Paweletz, C.P., Ardekani, A., Hackett, P.S., Hitt, B.A., et al. Serum proteomic patterns for detection of prostate cancer. *J. Natl. Cancer Inst.* 94(20), 1576–1578, 2002.
11. Petricoin, E.F., Ardekani, A.M., Hitt, B.A., Levine, P.J., Fusaro, V.A., Steinberg, S.M., et al. Use of proteomic patterns in serum to identify ovarian cancer. *Lancet* 359(9306), 572–577, 2002.
12. Ornstein, D.K., Rayford, W., Fusaro, V.A., Conrads, T.P., Ross, S.J., Hitt, B.A., et al. Serum proteomic profiling can discriminate prostate cancer from benign prostates in men with total prostate specific antigen levels between 2.5 and 15.0 ng/ml. *J. Urol.* 172(4 Pt. 1), 1302–1305, 2004.
13. Wadsworth, J.T., Somers, K.D., Cazares, L.H., Malik, G., Adam, B.L., Stack, B.C., Jr., et al. Serum protein profiles to identify head and neck cancer. *Clin. Cancer Res.* 10(5), 1625–1632, 2004.
14. Vlahou, A., Laronga, C., Wilson, L., Gregory, B., Fournier, K., McGaughey, D., et al. A novel approach toward development of a rapid blood test for breast cancer. *Clin. Breast Cancer* 4(3), 203–209, 2003.
15. Villanueva, J., Philip, J., Entenberg, D., Chaparro, C.A., Tanwar, M.K., Holland, E.C., and Tempst, P. Serum peptide profiling by magnetic particle-assisted, automated sample processing and MALDI-TOF mass spectrometry. *Anal. Chem.* 76(6), 1560–1570, 2004.
16. Hingorani, S.R., Petricoin, E.F., Maitra, A., Rajapakse, V., King, C., Jacobetz, M.A., et al. Preinvasive and invasive ductal pancreatic cancer and its early detection in the mouse. *Cancer Cell* 4(6), 437–450, 2003.
17. Conrads, T.P., Fusaro, V.A., Ross, S., Johann, D., Rajapakse, V., Hitt, B.A., et al. High-resolution serum proteomic features for ovarian cancer detection. *Endocr. Relat. Cancer* 11(2), 163–178, 2004.
18. Malyarenko, D.I., Cooke, W.E., Adam, B.L., Malik, G., Chen, H., Tracy, E.R., et al. Enhancement of sensitivity and resolution of surface-enhanced laser desorption/ionization time-of-flight mass spectrometric records for serum peptides using time-series analysis techniques. *Clin. Chem.,* 51(1), 65–74, 2004.
19. Grizzle, W.E., Semmes, O.J., Basler, J., Izbicka, E., Feng, Z., Kagan, J., et al. The early detection research network surface-enhanced laser desorption and ionization prostate cancer detection study: a study in biomarker validation in genitourinary oncology. *Urol. Oncol.* 22(4), 337–343, 2004.
20. Koopmann, J., Zhang, Z., White, N., Rosenzweig, J., Fedarko, N., Jagannath, S., et al. Serum diagnosis of pancreatic adenocarcinoma using surface-enhanced laser desorption and ionization mass spectrometry. *Clin. Cancer Res.* 10(3), 860–868, 2004.
21. Zhang, Z., Bast, R.C., Jr., Yu, Y., Li, J., Sokoll, L.J., Rai, A.J., et al. Three biomarkers identified from serum proteomic analysis for the detection of early stage ovarian cancer. *Cancer Res.* 64(16), 5882–5890, 2004.

22. Li, J., Zhang, Z., Rosenzweig, J., Wang, Y.Y., and Chan, D.W. Proteomics and bioinformatics approaches for identification of serum biomarkers to detect breast cancer. *Clin. Chem.* 48(8), 1296–1304, 2002.
23. Paweletz, C.P., Gillespie, J.W., Ornstein, D.K., Simone, N.L., Brown, M.R., Cole, K.A., et al. Biomarker profiling of stages of cancer progression directly from human tissue using a protein biochip. *Drug Dev. Res.* 49, 34–42, 2000.
24. Stoeckli, M., Chaurand, P., Hallahan, D.E., and Caprioli, R.M Imaging mass spectrometry: a new technology for the analysis of protein expression in mammalian tissues. *Nat. Med.* 7(4), 493–496, 2001.
25. Yanagisawa, K., Shyr, Y., Xu, B.J., Massion, P.P., Larsen, P.H., White, B.C., Roberts, J.R., Edgerton, M., Gonzalez, A., Nadaf, S., Moore, J.H., Caprioli, R.M., and Carbone, D.P. Proteomic patterns of tumour subsets in non-small-cell lung cancer. *Lancet* 362(9382), 433–439, 2003.
26. Reyzer, M.L., Caldwell, R.L., Dugger, T.C., Forbes, J.T., Ritter, C.A., Guix, M., Arteaga, C.L., and Caprioli, R.M. Early changes in protein expression detected by mass spectrometry predict tumor response to molecular therapeutics. *Cancer Res.* 64(24), 9093–9100, 2004.
27. Goodacre, R., Neal, M.J., Kell, D.B., Greenham, L.W., Noble, W.C., and Harvey, R.G. Rapid identification using pyrolysis mass spectrometry and artificial neural networks of *Propionibacterium acnes* isolated from dogs. *J. Appl. Bacteriol.* 76(2), 124–134, 1994.
28. Holland, R.D., Wilkes, J.G., Rafii, F., Sutherland, J.B., Persons, C.C., Voorhees, K.J., and Lay, J.O., Jr. Rapid identification of intact whole bacteria based on spectral patterns using matrix-assisted laser desorption/ionization with time-of-flight mass spectrometry. *Rapid Commun. Mass Spectrom.* 10(10), 1227–1232, 1996.
29. Petricoin, E.F., Fishman, D.A., Conrads, T.P., Veenstra, T.D., and Liotta, L.A. Lessons from Kitty Hawk: from feasibility to routine clinical use for the field of proteomic pattern diagnostics. *Proteomics* 4(8), 2357–2360, 2004.
30. Liotta, L.A., Lowenthal, M., Conrads, T.P., Veenstra, T.D., Fishman, D.A., and Petricoin, E.F., III. Importance of communication between producers and consumers of publicly available experimental data. *JNCI*, 97(4), 1–8, 2005.
31. Hillenkamp, F. and Karas, M. Mass spectrometry of peptides and proteins by matrix-assisted ultraviolet laser desorption/ionization. *Methods Enzymol.* 193, 280–295, 1990.
32. Conrads, T.P., Zhou, M., Petricoin, E.F., III, Liotta, L., and Veenstra, T.D. Cancer diagnosis using proteomic patterns. *Expert Rev. Mol. Diagn.* 3(4):411–420, 2003.
33. Diamandis, E.P. Analysis of serum proteomic patterns for early cancer diagnosis: drawing attention to potential problems. *J. Natl. Cancer. Inst.* 96(5), 353–356, 2004.
34. Mehta, A.I., Ross, S., Lowenthal, M.S., Fusaro, V., Fishman, D.A., Petricoin, E.F., III, and Liotta, L.A. Biomarker amplification by serum carrier protein binding. *Dis. Markers* 19(1), 1–10, 2003–2004.
35. Zhou, M., Lucas, D.A., Chan, K., Issaq, H.J., Petricoin, E.A., III, Liotta, L.A., Veenstra, T.D., and Conrads, T. P^{2004} investigation into the human serum interactome. *Electrophoresis* 25(9), 1289–1298, 2004.
36. Dennis, M.S., Zhang, M., Meng, Y.G., Kadkhodayan, M., Kirchhofer, D., Combs, D., et al. Albumin binding as a general strategy for improving the pharmacokinetics of proteins. *J. Biol. Chem.* 277, 35035–35043, 2002.

37. Yeh, P., Landais, D., Lemaitre, M., Maury, I., Crenne, J.Y., Becquart, J., et al. Design of yeast-secreted albumin derivatives for human therapy: biological and antiviral properties of a serum albumin-CD4 genetic conjugate. *Proc. Natl. Acad. Sci. U.S.A.* 89, 1904–1908, 1992.
38. Shen, Y., Tolic, N., Masselon, C., Pasa-Tolic, L., Camp, D.G., II, Hixson, K.K., Zhao, R., Anderson, G.A., and Smith, R.D. Ultrasensitive proteomics using high-efficiency on-line micro-SPE-nanoLC-nanoESI MS and MS/MS. *Anal. Chem.* 76(1), 144–154, 2004.
39. Shen, Y., Tolic, N., Zhao, R., Pasa-Tolic, L., Li, L., Berger, S.J., Harkewicz, R., Anderson, G.A., Belov, M.E., and Smith, R.D. High-throughput proteomics using high-efficiency multiple-capillary liquid chromatography with on-line high-performance ESI FTICR mass spectrometry. *Anal. Chem.* 73, 3011–3021, 2001.
40. Baggerly, K.A., Morris, J.S., and Coombes, K.R. Reproducibility of SELDI-TOF protein patterns in serum: comparing datasets from different experiments. *Bioinformatics* 20(5), 777–785, 2004.
41. Semmes, O.J., Feng, Z., Adam, B.L., Banez, L.L., Bigbee, W.L., Campos, D., Cazares, L.H., Chan, D.W., Grizzle, W.E., Izbicka, E., Kagan, J., Malik, G., McLerran, D., Moul, J.W., Partin, A., Prasanna, P., Rosenzweig, J., Sokoll, L.J., Srivastava, S., Srivastava, S., Thompson, I., Welsh, M.J., White, N., Winget, M., Yasui, Y., Zhang, Z., and Zhu, L. Evaluation of serum protein profiling by surface-enhanced laser desorption/ionization time-of-flight mass spectrometry for the detection of prostate cancer: I. Assessment of platform reproducibility. *Clin. Chem.* 51(1), 102–112, 2005.

CHAPTER 8

Lymphocyte Integrins: Potential Surrogate Biomarkers for Evaluation of Endometrial Receptivity

K.V.R. Reddy, S.M. Gupta, and P.K. Meherji

CONTENTS

8.1 INTRODUCTION

Human endometrium undergoes a remarkable series of developmental changes during the menstrual cycle in preparation for embryonic implantation.[1] During a limited period of time called the "window of implantation," the uterus is ready to accept the implanting embryos; before and after this time it may be either indifferent or hostile to the embryo.[2,3] It is during this critical period that a proper dialogue can be established between an intrusive blastocyst and a receptive endometrium. If for any reason this dialogue is not established or is perturbed, the embryo is aborted. The identity of the molecular repertoire that makes the endometrium receptive to implantation and/or leads to menstruation is now being revealed and broadly includes cytokines, adhesion molecules, and matrix metalloproteases.[4]

The endometrium is composed of the uterine epithelium and stroma. The stroma contains many cellular elements such as fibroblasts, vascular components, endothelial cells, smooth muscle cells that coat the vessels of the endometrium, and a dynamic array of immune cells.[5] These include polymorphonuclear leukocytes, natural killer cells (NK cells), and large granular lymphocytes and macrophages.[6] The role of immune cells in the endometrium has been the subject of considerable interest in recent years. Normal eutopic endometrium contains numerous leukocytes in both stromal and intraepithelial locations.[7] The number of leukocytes in the endometrial stroma increases from the proliferative phase to the late secretory phase where 20 to 25% of the stromal cells are leukocytes.[7,8] This large increase in the endometrial leukocyte population is due to an increase in the phenotypically unusual population of $CD56^+CD16^-$ endometrial granulated lymphocytes.[7] During the luteal phase, there is a dense mucosal infiltration of these $CD56^+$ and NK cells. NK cells comprise the major leukocyte population at implantation sites, accounting for 70% of the total number of cells during the first trimester. Once pregnancy progresses into the second trimester, the number of these cells greatly decreases.

It is now apparent that immunologic implantation failure is more than likely mediated through the activation of NK cells, which along with macrophages and T cells produce a variety of TH-1 cytokines — tumor necrosis factor alpha (TNF-α), interferon-γ, and interleukins IL-1 and IL-2 — and TH-2 cytokines — IL-3, IL-4, IL-6, IL-7, IL-8, IL-11, and IL-12. An orderly, controlled release of TH-1 cytokines, occurring in association with an appropriate production of TH-2 cytokines, is vital to proper implantation, decidualization, and placentation. This TH-1/TH-2 homeostasis creates an environment fostering implantation and optimal intrauterine development.[8] In contrast, excessive release of TH-1 cytokines, particularly TNF-α and interferon-γ, is cytotoxic to the trophoblast and endometrial glandular cells causing unregulated apoptosis and subsequent failed implantation.[9]

8.2 ENDOMETRIAL BIOMARKERS IN IMPLANTATION

Psychoyos and Nikas[10] demonstrated the presence of specialized surface protrusions on the uterine luminal epithelium called pinopodes, appearing between days 19 and 21 of the normal menstrual cycle, i.e., during the period of maximal endometrial receptivity. Further, Lessey et al.[11] reported stage-dependent changes in pinopode formation during normal and stimulated menstrual cycles.

It is well documented that many of the physiological events that are crucial to successful implantation are driven by cyclic changes in the ovarian steroid hormonal milieu and that both morphological and functional maturation of the endometrium are mediated by these hormones.[12] In addition, endometrial receptors for estrogen and progesterone are essential for the establishment of receptive phases of implantation and for the expression of some of the endometrial biomarkers. Both receptors show maximal expression in the glandular epithelium and stroma during late proliferative and early secretory phase. After day 19, there is an abrupt disappearance of these receptors from the glands due to the effect of progesterone, although they do persist in the stroma.[13,14] Besides morphological and physiological markers, several biomarkers have been identified in the human endometrium that seem to participate in tissue remodeling and the implantation process.

8.2.1 Leukemia Inhibitory Factor

Leukemia inhibitory factor (LIF) is an important biomarker that has been heavily implicated in the implantation process. Embryos from transgenic mice with no LIF expression are unable to implant but show normal development *in vitro*.[15] In humans, LIF has been found in the endometrium at the time of implantation with maximal expression between days 19 and 25 of an ideal cycle. These findings indicate that LIF may be an important regulator of human embryonic implantation, modulating the differentiation of trophoblast.[11]

8.2.2 Interleukin-1 Receptor Type I

IL-1 receptor type I (IL-1RI) is expressed in the epithelial and stromal cells during the entire menstrual cycle with maximum levels during early and late luteal phases.[16] The binding of IL-1 to maternal IL-1R is a necessary step in implantation. Abundant expression of this receptor through the luminal epithelium is required for adequate embryo attachment.[17]

8.2.3 Mucin-1

Mucin-1 (MUC-1), a highly glycosylated cell surface and secretory mucus of endometrial epithelium, which has been described as an inhibitor of blastocyst attachment, is specifically expressed in the uterine epithelium of rodents, rabbits, pigs, baboons, and humans. MUC-1 inhibits the initial phases of implantation by steric hindrance or promotes cell–cell interaction.[18]

8.2.4 Mouse Ascites Golgi

Mouse ascites Golgi (MAG) is normally expressed in the glandular Golgi on day 5. Its secretion begins on day 16, appears on the apical surface of the human luminal epithelium on day 17, and lasts until day 19. Abnormalities in expression of MAG are known to be associated with unexplained infertility.[19]

8.2.5 Adhesion Molecules (Integrins)

The most important biomarkers are adhesion molecules of the integrin family. The complex structure of the endometrium requires an array of distinct molecules that contribute to cell distribution, adhesion, trafficking, and signaling with matrix proteins of the endometrial meshwork.[20] Over the past decade, insights into the mechanisms underlying cell migration, adherence to extracellular matrix and cell-to-cell attachment have been greatly expanded with the discovery of cell surface molecules known as *integrins*, which represent a possible key for linking these ostensibly unrelated phenomena.[21]

8.3 INTEGRINS AND ENDOMETRIAL FUNCTION

Integrins are a large family of cation-dependent heterodimeric (α-chain and β-chain), transmembrane glycoprotein receptors that consist of different α and common β subunits. They interact with a variety of ligands including extracellular matrix (ECM) glycoproteins and several other cell surface molecules.[22,23] Many of the integrin receptors recognize the tripeptide, arginine-glycine-aspartic acid (RGD) sequence, which commonly appears in ECM components. To date, 18 α and 8 β subunits have been identified and these subunits form 24 known heterodimers (Figure 8.1). These molecules appear attractive markers to study implantation defects since they are continuously expressed from the gamete right through to birth. Their roles in adhesion, migration, invasion, and a multitude of intracellular effects on organization of the cytoskeleton as

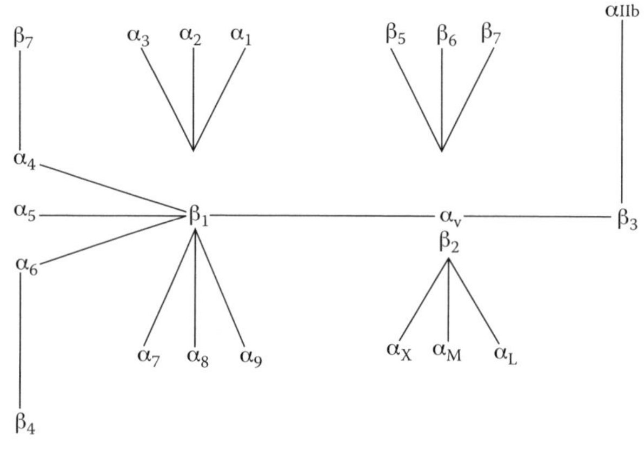

Figure 8.1 Integrin subunit association.

Table 8.1 Distribution Pattern of Various Integrin Subunits in Normal Endometrium during the Menstrual Cycle

	Integrin Subunits									
Phase of the Cycle	α1	α2	α3	α4	α5	α6	αv	β1	β4	β3
Proliferative Phase (day 8/9)										
Epithelial cells	–	+	+	–	–	+	–	+	+	–
Stromal cells	–	+	+	–	–	+	–	+	+	–
Ovulatory Phase (day 14/15)										
Epithelial cells	+	+	+	+	–	+	–	+	+	–
Stromal cells	+	+	+	+	–	+	–	+	+	–
Mid-luteal Phase (day 19/20)										
Epithelial cells	+	+	+	++	–	++	+++	+	+	+++
Stromal cells	+	+	+	++	–	++	+++	+	+	+++
Menstrual Phase (day 26/27)										
Epithelial cells	–	–	+	+	–	+	–	+	+	–
Stromal cells	–	–	+	–	–	+	–	+	+	–

well as their ability to respond to intracellular and extracellular signals make integrins attractive potential participants in the complex events of fertilization, implantation, decidualization, and placentation. It has been suggested that integrins also play a crucial role in the reproductive process.[20,24]

Integrins are regulated spatially and temporally within the uterus throughout the reproductive cycle and early pregnancy.[25–28] A defined temporal and spatial expression of specific integrins, which include α1β1, αvβ3, α4β1, α6β4, is considered to play a key role in endometrial receptivity for embryo implantation.[29,30] To date, 14 integrin subunits have been identified in the endometrium of human,[31] 10 subunits in the baboon,[28] and 7 in the pig.[18] In humans, integrin subunits α2, α3, α6, α9, αv, β1, β3, β4, β5, and β6 have been identified on the luminal epithelium of the uterus. With the exception of β5 and β6, all of the integrins listed are also expressed in the glandular epithelium[25,32] (Table 8.1). Thus, the possible known heterodimers available at the uterine luminal surface include α1β1, α2β1, α3β1, α4β1, α6β1, α9β1, αvβ1, αvβ3, and αvβ5. The α4β1 integrin is absent from the proliferative endometrium[25,31] and appears on glandular epithelial cells just after ovulation (day 14) and disappears on day 24 of the cycle when the period of maximum uterine receptivity ends. The α1β1 molecule is expressed by the epithelial cells between days 14 and 28 of the cycle. The αvβ1 integrin is expressed by glandular epithelial cells after cycle day 19/20, when the window of implantation opens. During this period, αvβ3 appears for the first time on uterine luminal epithelial cells. This period coincides with the onset of maximum uterine receptivity.[31] The loss of α4β1 integrin heralds the closure of this window.[20] It seems likely that the co-expression of all four of these integrins during this period is crucial for embryo–maternal recognition and successful implantation of the blastocyst.[1,32] After recognition and penetration of the uterine epithelium, the blastocyst invades the uterine stroma. During this time, at least four newly expressed integrins (α1, α2, α6, α7) appear on the blastocyst surface. These integrin subunits are required for blastocyst–stromal interactions.

8.4 EMBRYONIC INTEGRINS AND IMPLANTATION

Implantation of the human blastocyst requires proper development of the uterine endometrium to a state of receptivity synchronized with the developmental stage of the embryo.[17] Cell–cell interactions during pre-implantation development are likely to require the expression of a variety of cell adhesion molecules on the embryo surface.[33] Some of these molecules are developmentally regulated, either appearing or becoming redistributed at specific stages of development. Integrin $\alpha 1$ and $\beta 1$ and CD44 molecules are located on the surface of the human oocyte.[34] It has been shown that integrin $\alpha 6\beta 1$ on the egg surface facilitates fertilization by interacting with fertilin on the spermatozoa.[35,36] The activation of $\alpha 6\beta 1$ on the oocyte may lead to intracellular signals that could aid in the development of embryo. Once the embryo has attached to the uterine epithelium, it invades the uterine stroma. The interaction between embryo and uterine epithelium is similar to the leukocyte endothelial interactions and metastatic processes where integrins play a major role in the adhesion process.[11,37]

8.5 INTEGRINS AND REPRODUCTIVE DYSFUNCTION

Studies have indicated that adhesion molecules play a critical role in the migration of PBLs and their recruitment into the endometrium.[38] Through their binding to specific receptors, cytokines may activate molecular changes in the expression pattern of adhesion and anti-adhesion molecules (MUC-1) that are essential for the adhesion of endometrial epithelial cells (EECs). During the adhesion phases, direct contact occurs between the lateral borders of EECs and the trophectoderm.[39] Finally, during invasion, the embryonic trophoblast penetrates the basal membrane and invades the uterine stroma. This involves different trophoblast lineages and several endometrial cell types, such as stromal cells, endothelial cells, and resident immune cell types.

Aberrant $\alpha v\beta 3$ integrin expression has been associated with endometriosis and may identify some women with decreased cycle fecundity due to defects in uterine receptivity.[40,41] Integrins and E-cadherins are involved in the shedding of endometrial tissue during menstruation and in the attachment of endometrial tissue fragments to the peritoneum.[30,40] Retrograde menstruation is considered an important factor in the development of endometriosis. Meyer et al.[42] demonstrated that inflammatory hydrosalpinges adversely affect endometrial receptivity. The expression of $\alpha v\beta 3$ integrin was less in the mid-luteal-phase endometrium of women with hydrosalpinges[42] and of recurrent spontaneous abortors.[43] In patients with polycystic ovarian syndrome (PCOS), the expression of $\alpha v\beta 3$ was either delayed or absent in the endometrium.[44] Lack or poor expression of some of the integrins in endometrial cells may lead to failure of embryo–endometrial interaction and implantation.[20] This disruption of integrin expression may be associated with certain types of infertility in women including endometriosis, anovulation, luteal phase dysfunction, and unexplained infertility.[29]

8.6 INTEGRINS AND INFERTILITY

Infertility affects approximately 2 to 4 million couples annually and, despite a widening in the number of infertility diagnosis, many of the molecular defects associated with female infertility are still not understood. Recurrent spontaneous abortion and failure of implantation due to defects in uterine receptivity may contribute to 20% of these cases.[43] One of the methods used earlier to monitor the defects in implantation is based on dating of the endometrial biopsy. The classic work of Noyes et al.[45] on endometrial morphology that has served clinicians so well for 50 years is losing its power in the evaluation of human endometrium in normal and abnormal circumstances, due to conflicting views on the timing and interpretation of endometrial biopsies.[46]

Endometrial biopsies allow morphological and ultrastructural assessment, but these procedures are invasive and disrupt the synchrony of the peri-implantation program at critical times. Moreover, the procedure is traumatic to the patients and can also lead to intrauterine infections. This hampers the assessment of defects in endometrial function. Therefore, a simple and less invasive procedure may be of immense help in the evaluation of luteal adequacy in infertile couples.

Recent studies indicated that lack or delayed expression of $\alpha4\beta1$ and $\alpha v\beta3$ integrins on endometrial cells of infertile women may provide a new diagnostic modality for the evaluation of endometrial defects.[20] These integrins appear to be promising, but the method is based on the evaluation of endometrial biopsy. Hence, there is an ongoing search to find an alternative diagnostic tool that does not require endometrial biopsies. An accurate marker for uterine receptivity during blastocyst implantation, along with a better definition of the mechanisms regulating this event, is urgently needed.[47]

With the advent of mAbs that react with specific types of cells, it has been determined that the majority of endometrial stromal cells are of lymphoid origin[48] and that the tissue is specialized in the process of T-lymphocyte selection, maturation, and expansion that occur in the context of its microenvironment.[49] Immunostaining for leukocyte common antigen (LCA) has demonstrated that PBLs are the major cell population (55 to 60%) in the human endometrium,[38] followed by macrophages. The absolute number of these cells is reported to vary throughout the menstrual cycle and according to the stage of pregnancy.[38] A positive correlation has been shown between abnormal function of leukocyte subsets such as cytotoxic T lymphocytes (CTLs, $CD8^+$) and natural killer (NK, $CD56^+$) cells, and implantation failure in infertile women.[6] It has been demonstrated that intravenous administration of thymocytes, especially $CD4^+$ lymphocytes derived from immature nonpregnant female mice, significantly promoted embryo implantation in recipient mice on pseudopregnancy day 2.[50] Further, intra-endometrial injection of splenocytes prepared from mice in the early stages of pregnancy enhanced embryo implantation,[51] suggesting that immune cells possess information about the presence of embryo and facilitate embryo–endometrial interaction.[52]

8.7 ROLE OF PERIPHERAL BLOOD LYMPHOCYTES IN ENDOMETRIAL FUNCTION

PBLs are resident in the spleen, liver, and uterus. Some of the signaling pathways transduced by integrins facilitate the recruitment of PBLs to the site of implantation.[43] It is now believed that endometrial lymphocytes and PBLs play a vital role in implantation and maintenance of pregnancy.[32] If the functional activities of these cells are suboptimal, the establishment of early pregnancy may be impaired. Consequently, utero-placental function may be associated not only with miscarriage but also with later pregnancy complications including pre-eclampsia and intrauterine growth retardation.

A significant correlation has been shown between PBL dysfunction and subclinical embryo loss in recurrent aborters.[53] Morphometric analysis of endometrium from nonpregnant women who have experienced several recurrent spontaneous abortions shows evidence of defective maturation of glandular function.[54] Recurrent spontaneous abortion (RSA) is one of the most severe complications of pregnancy, and about 15% pregnancies end spontaneously within the first trimester. Immunological disturbance due to lymphocyte dysfunction accounts for almost 50% of abortions. Studies have shown that these women have a high concentration of CTLs and NKa cells, and raised levels of TH-1 cytokines (embryotoxic factor) in their decidual supernatants, as well as in the peripheral blood. In addition, it has been reported that the endometrial cells express lymphocyte function antigen-3 (LFA-3) on their surface, implicating physiological interaction between endometrial cells and T lymphocytes through the menstrual cycle into pregnancy.[55] The immunoendocrine function of these cells is reported to vary among the fertile and infertile women.[56,57]

8.8 CORRELATION BETWEEN ENDOMETRIAL CELL AND PERIPHERAL LYMPHOCYTE INTEGRINS

Expression of integrins on endometrial cells has been compared with that on PBLs with an aim of identifying a marker for assessing implantation failure in infertile women. We have shown that changes in expression of $\alpha4\beta1$ and $\alpha v\beta3$ integrins on PBLs mirror the changes in endometrial cells during implantation and that this may be helpful in assessing endometrial functional defects.[32] Immunocytochemical and immunofluorescence studies revealed that the expression of both $\alpha4\beta1$ and $\alpha v\beta3$ integrins was significantly decreased in infertile women compared with those who were fertile (Figure 8.2 through Figure 8.5). Considering the relationship of PBLs with endometrial cells, we believe that failure of the immune system to support pregnancy through production of various integrins might be responsible for the demise of embryo. PBLs were once considered as mere target cells for various hormones, and not involved in cell–cell signal transduction mechanisms. However, the evidence now indicates that lymphocytes are endocrine glands with autocrine and paracrine functions.[58] They synthesize and secrete various hormones, specifically prolactin (PRL), follicle stimulating hormone (FSH), and luteinizing hormone (LH), and are involved in cell–cell interactions.[56] The identification

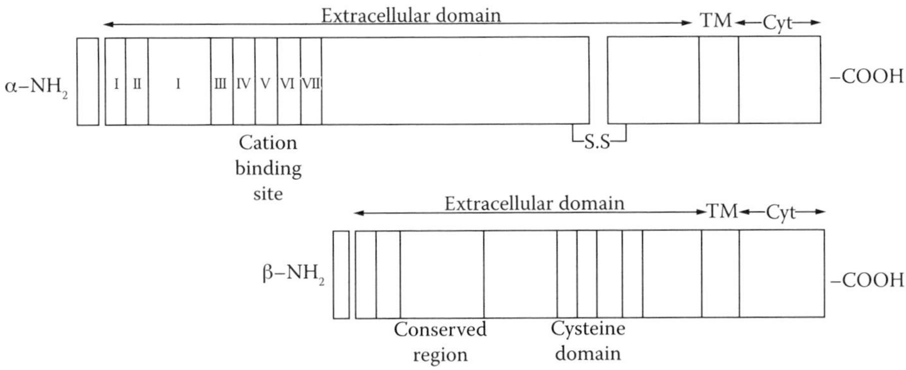

Figure 8.2 Structure of integrin and subunits. $-NH_2$ = aminoterminal; Cyt = cytoplasmic domain; Tm = transmembrane domain; i = insert-domain; S–S = disulfide linkage; –COOH = carboxy terminal.

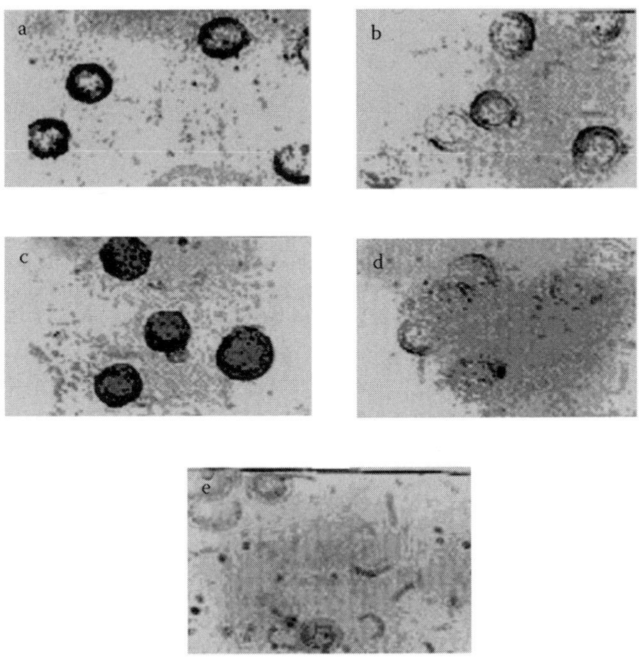

Figure 8.3 Immunocytochemical localization of integrins on PBL of fertile (a) and infertile (unexplained infertile) women (b) during midluteal phase (day 19/20). The expression of $\alpha 4\beta 1$ (a, b) and $\alpha v\beta 3$ (c, d) was significantly low in infertile cases. In negative control (e) (with out primary antibody) the antibody did not react with the cells.

of α4, α6, and 3 integrin (ITG) receptors on PBL supports these studies. It is therefore possible that poor expression of integrins on endometrial cells and PBLs in the majority of infertile patients observed in our study may be attributed to the downregulation of signal transduction across the T-cell membrane, leading to dysfunction of PBLs as suggested by earlier workers.[56] If these observations withstand

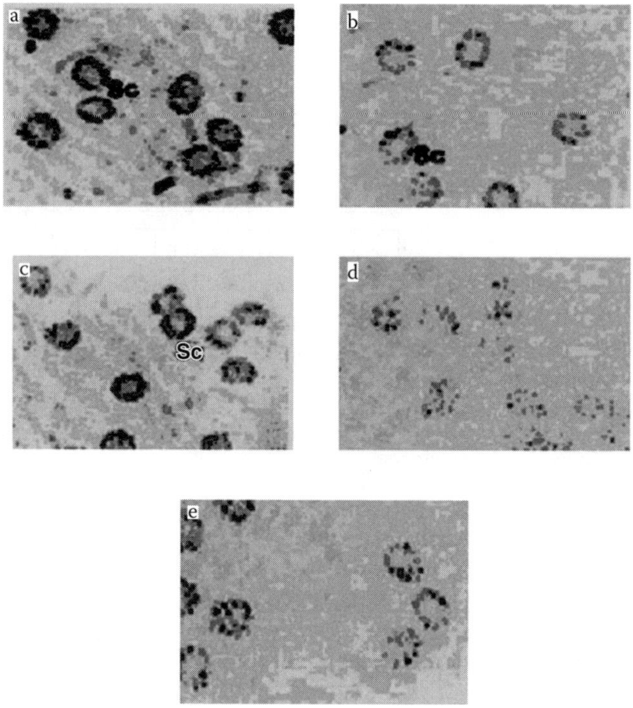

Figure 8.4 Immunocytochemical localization of integrins on endometrial stromal cells (sc) of fertile (a) and women with unexplained infertility (b) during midluteal phase (day 19/20). The expression of $\alpha4\beta1$ (a, b) and $\alpha v\beta3$ (c, d) was significantly low in infertile cases. In negative control (e) (with out primary antibody) the antibody did not react with the cells.

further scientific scrutiny, one of the causes of infertility could be identified and further investigated to enable medical intervention.

In general, our studies on the expression of integrins on PBLs have demonstrated that dynamic alterations in lymphocyte integrin expression accompany the endometrial integrin expression that characterizes the endometrial cycle and that lymphocytes appear to be an alternative diagnostic tool to assess the uterine function.

8.9 SUMMARY AND CONCLUSIONS

The multidimensional nature of integrins, as well as the redundancy in integrin expression on reproductive cell types and their ability to bind to more than one integrin, is likely to make them important participants in the process of reproduction. The embryo and endometrium use the language of integrins for the early stages of communication. As mediators of attachment and signal transduction, their expression offers clues to understanding the regulation of uterine receptivity. Aberrant expression of integrins by the endometrium is associated with an adverse effect on blastocyst implantation, and this could be one explanation for impaired

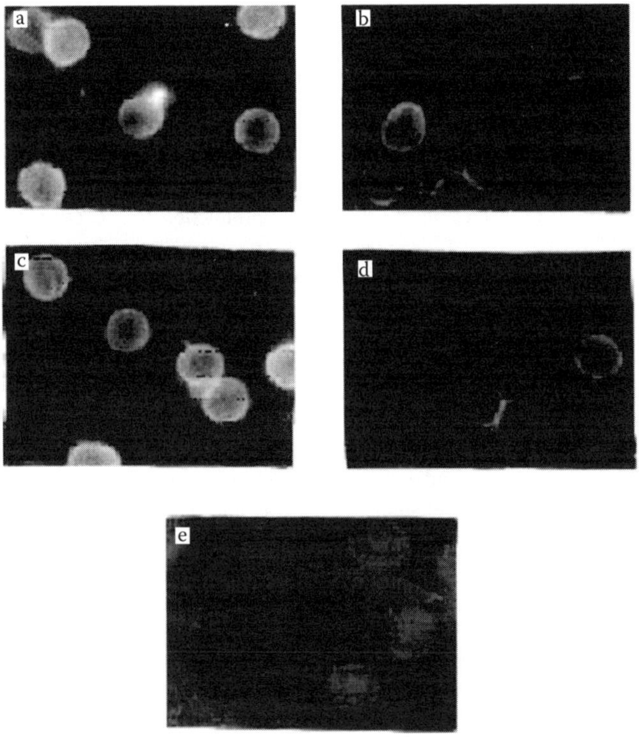

Figure 8.5 **(Color figure follows p. 138.)** Immunofluorescence localization of integrins on PBL of fertile (a) and women with unexplained infertility (b) during mid-luteal phase. The expression of α4β1 (a, b) and αvβ3 (c, d) was significantly low in infertile cases (b, d). In negative controls (e) (without primary antibody), the antibody did not react with the cells.

fertility. Collection of endometrial biopsies is an invasive procedure, and due to ethical restrictions, obtaining frequent biopsies from a woman during the different phases of menstrual cycle is not permissible, thus hampering the diagnosis of infertility. Lymphocytes may be used as an alternative source material to endometrial stromal cells to assess the defects in endometrial function, since the expression of integrins on lymphocytes correlates well with the expression on endometrial cells. Moreover, frequent sampling of blood is advantageous over repeated endometrial biopsies, as the former approach is easier, nontraumatic, and avoids intrauterine infections.

ACKNOWLEDGMENTS

The authors are grateful to Dr. Chander P. Puri, Director, for his consistent encouragement throughout the study. This work was supported by Indian Council of Medical Research (Reference No. NIRRH/MS/ 30 /2004).

REFERENCES

1. Tabibzadeh, S. Pattern of expression of integrin molecules in human endometrium throughout the menstrual cycle. *Hum. Reprod.* 7, 876, 1992.
2. Psychoyos, A. The implantation window. Basic and clinical aspects. In *Prospectives on Assisted Reproduction*, Mori, T., Aono, T., Tominaga, T. and Hiroi, M., Eds. Ares-Serono Symposia, Rome, 1995, 57.
3. Klentzeris, L.D. Adhesion molecules in reproduction. *Br. J. Obst. Gynaecol.* 104, 401, 1997.
4. Tabibzadeh, S. VLA-1 associated with endometrial biopsies. *Fertil. Steril.* 54, 624, 1990.
5. Bowen, J.A. and Hunt, J.S. The role of integrins in reproduction. *Proc. Soc. Exp. Biol. Med.* 223, 331, 2000.
6. Klentzeris, L.D. et al. Lymphoid tissue in the endometrium of women with unexplained infertility: morphometric and immunohistochemical aspects. *Hum. Reprod.* 9, 646, 1994.
7. Bulmer, J.N. et al. Granulated lymphocytes in human endometrium: histochemical and immunohistochemical studies. *Hum. Reprod.* 6, 791, 1991.
8. Jones, R.K., Bulmer, J.N., and Searle, R.F. Immunohistochemical characterization of stromal leucocytes in ovarian endometriosis: comparison of eutopic and ectopic endometrium with normal endometrium. *Fertil. Steril.* 66, 81, 1996.
9. Pace, D., Longfellow, M., and Bulmer, J.N. Characterization of intraepithelial lymphocytes in human endometrium. *J. Reprod. Fertil.* 91, 165, 1991.
10. Psychoyos, A. and Nikas, G. Uterine pinopodes as markers of uterine receptivity. *Asst. Reprod. Rev.* 4, 26, 1994.
11. Lessey, B.A. et al. Integrin adhesion molecules in the human endometrium. *J. Clin. Invest.* 90, 180, 1992.
12. Somkuti, S.G. et al. Epidermal growth factor and sex steroids dynamically regulate a marker of endometrial receptivity in Ishikawa cells. *J. Clin. Endocrinol. Metab.* 82, 2192, 1997.
13. Garcia, E. et al. Use of immunocytochemistry of progesterone and estrogen receptors for endometrial dating. *J. Clin. Endocrinol. Metab.* 67, 80, 1988.
14. Edger, E.D.J. and Ar, D.H. Estrogen and human implantation. *Hum. Reprod.* 10, 223, 1995.
15. Stewart, C.L. et al. Blastocyst implantation depends on maternal expression of leukemia inhibitory factor. *Nature* 359, 70, 1992.
16. Chegini, N. and Williams, R.S. Cytokines and growth factors network in human endometrium from menstruation to embryo implantation. In *Cytokines and Reproduction*. Hill, J., Ed., John Wiley & Sons, New York, 1998.
17. Simon, C. et al., Embryonic regulation in implantation. *Semin. Reprod. Endocr.* 17, 267, 1999.
18. Bowen, J.A., Fullarbazer, W., and Burghardt, R.C. Spatial and temporal analyses of integrin and MUC-1 expression in porcine uterine epithelium and trophectoderm *in vivo*. *Biol. Reprod.* 55, 1098, 1996.
19. Kliman, H.J. et al. A mucin like glycoprotein identified by MAG (mouse ascite Golgi antibodies. Menstrual cycle dependent localization in human endometrium. *Am. J. Pathol.* 146, 166, 1995.
20. Lessey, B.A. Endometrial integrins and the establishment of uterine receptivity. *Hum. Reprod.* 13, 247, 1998.

21. Tamkun, J.W., Desimone, D.W. et al. Structure of integrins, glycoprotein involved in the transmembrane linkage between fibronectin and actin. *Cell* 46, 271, 1986.
22. Ruoslahti, E. and Pierschbacher, M.D. New prospectives in cell adhesion: RGD and integrins. *Science* 238, 491, 1987.
23. Vinatier, D. Integrins and reproduction. *Eur. J. Obst. Gynecol. Reprod. Biol.* 59, 71, 1995.
24. Tabibzadeh, S. and Babaknia, A. The signal and molecular pathways involved in implantation: symbiotic interaction between blastocyst and endometrium involving adhesion and tissue invasion. *Hum. Reprod.* 10, 1579, 1995.
25. Lessey B.A. et al. Luminal and glandular endometrial epithelium express integrins differentially throughout the menstrual cycle: implications for implantation, contraception and infertility. *Am. J. Reprod. Immunol.* 35, 195, 1996.
26. Klentzeris, L.D. et al. 1 integrin cell adhesion molecules in the endometrium of fertile and infertile women. *Hum. Reprod.* 8, 1223, 1993.
27. Bischof, P. et al. Localization of 2, 5 and 6 integrin subunits in human endometrium, deciduas and trophoblast. *Eur. J. Obst. Gynecol.* 51, 217, 1993.
28. Fazleabas, A.T. et al. Distribution of integrins and the extracellular matrix proteins in the baboon endometrium during the menstrual cycle and early pregnancy. *Biol. Reprod.* 56, 348, 1997.
29. Sueoka, K. et al. Integrins and reproductive physiology; expression and modulation in fertilization, embryogenesis and implantation. *Fertil. Steril.* 67, 799, 1997.
30. Vander Linden, P.J.Q. et al. Expression of integrins and E-cadherin in cells from menstrual effluent, endometrium, peritonial fluid, peritonium and endometriosis. *Fertil. Steril.* 61, 85, 1994.
31. Reddy, K.V.R. and Meherji P.K. Integrin cell adhesion molecules in endometrium of fertile and infertile women throughout menstrual cycle. *Ind. J. Exp. Biol.* 37, 323, 1999.
32. Reddy, K.V.R., Sadhana, G., and Meherji, P.K. Expression of integrin receptors on peripheral lymphocytes: correlation with endometrial receptivity. *Am. J. Reprod. Immunol.* 46, 188, 2001.
33. Fleming T.P. et al. Molecular maturation of cell adhesion systems during mouse early development. *Histochemistry* 101**,** 1, 1994.
34. Simon, C. et al. Embryonic regulation of integrin 3, 4 and 1 in human endometrial epithelial cells *in vitro*. *J. Clin. Endocrinol. Metab.* 82, 2607, 1997.
35. Almeida, E.A. et al. Mouse egg integrin $\alpha6\beta1$ function as a sperm receptor. *Cell* 81, 1095, 1995
36. Reddy, K.V.R., Rajeev, S.K., and Vijayalaxmi, G. 61 integrin is a potential marker for evaluating sperm quality in men. *Fertil. Steril.* 79, 1590, 2003.
37. Kimber S.J. Glycoconjugates and cell surface interactions in a pre- and peri-implantation development. *Int. Rev. Cytol.* 120, 153, 1990.
38. Stewart-Akers, A.M. et al. Endometrial leucocytes are altered numerically and functionally in women with implantation defects. *Am. J. Reprod. Immunol.* 39**,** 111, 1998.
39. Enders, A.C. Anatomical aspects of implantation. *J. Reprod. Fertil.* 25, 1, 1976.
40. Lessey, B.A. et al. Aberrant integrin expression in the endometrium of women with endometriosis. *J. Clin. Endocrinol. Metab.* 79**,** 643, 1994.
41. Lessey, B.A. and Young, S.L. Integrins and other cell adhesion molecules in endometrium and endometriosis. *Reprod. Endocrinol.* 15**,** 291, 1997.
42. Meyer, W.R. et al. Hydrosalphinges adversely affect markers of endometrial receptivity. *Hum. Reprod.* 12, 1393, 1997.

43. Reddy, K.V.R. and Mangale, S.S. Integrin receptors, the dynamic modulators of endometrial function. *Tissue Cell.* 35, 260, 2003.
44. Apparao, K.B.A. et al. Osteopontin and its receptor alpha V beta 3 integrin are coexpressed in the human endometrium during the menstrual cycle but regulated differentially. *J. Clin. Endocrinol. Metab.* 10, 4991, 2001.
45. Noyes, R.W., Hertig, A.T., and Rock, J. Dating the endometrial biopsy. *Fertil. Steril.* 1, 3–25, 1950.
46. Li, T.C. and Cook, I.D. Evaluation of the luteal phase. *Hum. Reprod.* 6, 484, 1991.
47. Sharpe-Timms, K.L. and Glasser, S.R. Models for the study of uterine receptivity for blastocyst implantation. *Semin. Reprod. Biol.* 17, 107, 1999.
48. Kamat, B.R. and Isaacson, P.G. The immunocytochemical distribution of leucocytic subpopulations in human endometrium. *Am. J. Pathol.* 127, 66, 1987.
49. Grudzinskar, J.G. and Nysenbaum, A.M. Failure of human pregnancy after implantation. *Ann. N.Y. Acad. Sci.* 442, 38, 1985.
50. Fujita, K et al. Administration of thymocytes derived from non pregnant mice induces an endometrial receptive stage and leukaemia inhibitory factor expression in the uterus. *Hum. Reprod.* 13, 2888, 1998.
51. Takabatake, K. et al. Splenocytes in early pregnancy promotes embryo implantation by regulating endometrial differentiation in mice. *Hum. Reprod.* 12, 2102, 1997.
52. Klentzeris, L.D. et al. Lymphoid tissue in the endometrium of women with unexplained infertility: morphometric and immunohistochemical aspects. *Hum. Reprod.* 9, 646, 1994.
53. Coulam, C.B. Immunotherapy for recurrent spontaneous abortion. Early pregnancy. *Biol. Med.* 1, 13, 1995.
54. Serle, E. et al. Endometrial differentiation in the peri-implantation phase of women with recurrent miscarriage: a morphological and immunohistochemical study. *Fertil. Steril.* 62, 989, 1994.
55. Figdor, C.G., Van Kooyk, K., and Keizer, G.D. On the mode of action of LFA (Leukocyte function antigen). *Immunol. Today* 11, 277, 1990.
56. King, A. et al. Immunocytochemical characterization of the unusual large granular lymphocytes in human endometrium throughout the menstrual cycle. *Hum. Immunol.* 24, 195, 1989.
57. Shahani, S.K., Gupta, S.M., and Meherji, P.K. Lymphocytes — their possible endocrine role in the regulation of fertility. *Am. J. Reprod. Immunol.* 35, 1, 1996.
58. Imura, H., Fukata, I., and Mori, T. Cytokines and endocrine function: an interaction between the immune and neuroendocrine systems. *Clin. Endocrinol.* 35, 107, 1991.

CHAPTER 9

Nipple Aspirate Fluid to Diagnose Breast Cancer and Monitor Response to Treatment

Edward Sauter

CONTENTS

9.1 INTRODUCTION

Anatomically the breast comprises ducts and lobules, surrounded by supporting adipose and connective tissue. During the immediate postpartum lactation period, the breast glands actively secrete milk into the ducts for the nurture of the newborn infant, but it has been long recognized from histologic studies that the nonpregnant breast also secretes small amounts of fluid containing sloughed epithelial and other cells. The epithelial cells that line the ducts and lobules are at risk for malignant degeneration and are the origin of 99% of breast cancers (Young et al., 2004). The diagnosis of breast cancer requires the presence of malignant cells in a cytologic or histologic preparation of breast cells. Obtaining these cells generally requires an invasive needle or surgical biopsy based on an abnormality that is palpated or detected on an imaging study.

Early detection is a major factor contributing to the steady decline in breast cancer death rates, with a 3.2% annual decline over the past 5 years (Weir et al., 2003). Unfortunately, currently available breast cancer screening tools such as mammography and breast examination miss up to 40% of early breast cancers and are least effective in detecting cancer in young women, whose tumors are often more aggressive. Thus, there has long been interest in developing a noninvasive method to determine if a woman has breast cancer.

Indeed, collecting samples from the breast noninvasively has been conducted for at least 90 years. The adult nonpregnant, nonlactating breast secretes fluid into the breast ductal system (Keynes, 1923). This fluid normally does not escape because the nipple ducts are occluded by smooth muscle contraction, dried secretions, and keratinized epithelium. Initial studies to evaluate the breast noninvasively assessed spontaneous nipple discharge (SND), fluid which comes spontaneously from the breast ducts through the nipple without compression of the breast. While bilateral spontaneous discharge is generally physiologic, unilateral single duct discharge, whether bloody or nonbloody, is generally pathologic. In 1914, a case report documented the detection of breast cancer through the evaluation of SND (Nathan, 1914). Additional studies were performed to evaluate the cells in SND (Cheatle and Cutler, 1931; Deaver and McFarland, 1917) for the presence of disease. Although of potential use in disease diagnosis, evaluating SND did not address the assessment of women who did not have spontaneous discharge.

9.2 INITIAL STUDIES OF NAF FOCUS ON FEASIBILITY

George Papanicolaou was the first to design a large study evaluating fluid aspirated from the nipple rather than collecting fluid that came forth spontaneously. In his 1958 report evaluating NAF, he stated, “The practicability of utilizing breast secretion smears in screening for mammary carcinoma is in great measure dependent upon obtaining secretion in a relatively large proportion of the female population” (Papanicolaou et al., 1958). He cleansed the nipple and applied gentle massage toward the areola. If NAF did not come forth, he used a breast pump to create mild suction. He reported a series of 917 women without breast complaints in whom he attempted to collect NAF from one or both breasts (Papanicolaou et al., 1958). He was able to obtain a sample in 18.5% of subjects.

In order for NAF to be useful as a screening tool, it is essential to collect a sample in the vast majority of women. As a result, increasing the success rate continued to be an important area of investigation for the next 30 years. Early studies indicated that the ease of collecting NAF was related to the ethnicity of the individual, with NAF being more difficult to collect from Asians than African Americans or Caucasians (Petrakis et al., 1975). This was presumed to be due to the physiology of the breast, a modified ceruminous gland and is probably related to the secretory pattern in the breast and other ceruminous glands, which provide less secretions in most Asians (Petrakis, 1971) and American Indians (Petrakis, 1969) who are thought to have come from Asia than in Caucasians and African Americans. Other variables (Petrakis et al., 1975) found linked to success in NAF collection included age (late premenopause had the highest yield) and menopausal status (premenopausal subjects more often provided NAF). Various nipple aspiration devices were created, notably one by Otto Sartorius (Sartorius et al., 1977), which provided NAF on average in 50 to 60% of subjects (Petrakis et al., 1975; Sartorius et al., 1977) and in up to 80% in the highest-yielding subset of subjects (Sartorius et al., 1977).

The ability to collect NAF was linked not only to age, race, and menopausal status, but also to body habitus. In a large sample of white and black women between the ages of 20 and 59 years old who did not have a history of breast cancer, the proportion of women from whom NAF was collected increased with increasing dietary fat consumption (Lee et al., 1992b). This association of NAF yield with fat consumption was especially strong among black women, and was most pronounced in women aged 30 to 44 years.

In the 1990s the aspiration technique was modified to emphasize warming the breast, breast massage, and multiple aspiration attempts after clearing the nipple of dried secretions (Sauter et al., 1996). Each of these techniques had been heretofore practiced, but the emphasis on persistence seemed to increase successful NAF collection, as did having the subject return for a second or third visit, if necessary, to collect NAF. This increased yield to 99% of subjects who had not undergone prior breast surgery in the subareolar region, and who had not received breast irradiation (Sauter et al., 1996). Others have reported success rates near 90% without repeat visits (Mitchell et al., 2002), and investigators with yields after one visit of 66% increased their yield to 78% with multiple visits (King et al., in press).

9.3 STUDIES EVALUATING CELLS IN NAF

These early studies focused on the evaluation of morphologic changes in the shed duct epithelial cells to diagnose cancer, determination if NAF volume and color were predictors of breast cancer risk, and assessment of chemicals in NAF in different subject populations.

9.3.1 Evaluation of Cell Morphology

As previously mentioned, Papanicolaou was the first to report the presence of breast epithelial cells in NAF, and found malignant cells in 1 of 438 asymptomatic women (Papanicolaou et al., 1958). NAF was found to contain not only epithelial cells, but also foam cells, a term used to describe the "foamy" appearance of the cytoplasm. He speculated, "It thus appears possible that under the term foam cell we are dealing with a variety of cell types that, although morphologically indistinguishable…, may vary in origin." Almost 50 years later, after numerous studies using panels of epithelial and macrophage markers, the origin of foam cells remains an area of debate (King et al., 1984; Krishnamurthy et al., 2002; Mitchell et al., 2001). In the report, Papanicolaou also evaluated breast cyst fluid collected from 100 subjects and contrasted cytologic findings in NAF with those in breast cyst fluid. He noted a relative scarcity of foam cells in breast cyst fluid, which are generally the most frequent cellular component of NAF. Leukocytes and macrophages were also scarce in cyst fluid but relatively common in NAF.

The number of epithelial and foam cells and ratio of epithelial to foam cells have been assessed in different breast cancer risk populations (King et al., 1984; Papanicolaou et al., 1958; Sauter et al., 1997). It was found that as breast cancer risk increased, the number of epithelial cells, as well as the ratio of epithelial to foam cells, increased.

Increased breast density suggests more proliferative activity. Increased breast density as seen on mammography has been linked to increased breast cancer risk (Wolfe, 1976). Among a population of women in whom NAF cytology was collected, those with the greatest mammographic density were found to have a fourfold increased risk of atypical hyperplasia (Lee et al., 1992a).

Longitudinal studies have demonstrated the usefulness of abnormal NAF cytology in predicting future breast cancer risk. A prospective study which enrolled 2071 Caucasian women found that, after an average of 12.7 years of follow-up, the relative risk (RR) for women who yielded various cytologic categories of NAF vs. women who yielded no NAF (RR = 1) were as follows: unsatisfactory specimen, 1.4; normal cytology, 1.8; epithelial hyperplasia, 2.5; and atypical hyperplasia, 4.9 (Wrensch et al., 2001). A follow-up study involving 4046 women who were followed for a median of 21 years found that, compared with women from whom no fluid was obtained, whose incidence of breast cancer was 4.7%, the adjusted RRs for women with various NAF cytologic findings were 1.4 for those with unsatisfactory aspirate specimens, 1.6 for those with normal cytology in the aspirates, 2.4 for epithelial hyperplasia, and 2.8 for atypical hyperplasia. Thus, longer follow-up demonstrated a consistent, albeit somewhat lower, increased risk related to worsening NAF cytology, and is consistent with the implications of a fine needle aspiration or excisional biopsy demonstrating atypical hyperplasia (Wrensch et al., 2001).

Multiple aspiration visits have been demonstrated to increase the detection of abnormal epithelial cells in NAF (King et al., in press). Two hundred seventy-six women without known breast cancer underwent nipple aspiration. Among women in whom NAF was collected, hyperplastic cells were found in 34/178 (19.1%) at visit 1, which increased to 73/209 (34.9%) by visit 5. Atypical cells were found in 6.7% at the initial visit, and in 18.2% of NAF specimens in at least one of five visits.

The presence of tumor at the margin of a surgical biopsy presents a treatment dilemma, since approximately half of the time re-excision fails to find residual tumor. On the other hand, tumor recurrence rates are significantly higher if margins are not resected until they are tumor free (Sauter et al., 1999). NAF cytology has been used to evaluate the presence of residual breast cancer. Atypical and malignant cytology observed in NAF samples collected after excisional breast biopsy but before or concurrent with definitive surgery (Sauter et al., 1999) were significantly associated with residual ductal carcinoma *in situ* (DCIS) or invasive cancer. It was felt that pathologic factors such as tumor distance from the biopsy margin, multifocal/multicentric disease, subtype and grade of DCIS or invasive cancer (IC), tumor and specimen size, tumor and biopsy cavity location, presence or absence of extensive DCIS, and biopsy scar distance from the nipple would optimize a model to predict the presence of residual breast cancer among women with a biopsy with an involved or close tumor margin. The model (Sauter et al., 2001), which included both NAF cytology and pathologic parameters, was superior in predicting residual breast cancer (94%) to models using NAF cytology (36%) or pathologic parameters (75%) alone. NAF cytology also was useful in predicting which patients had one or more lymph nodes involved with tumor, which could prove useful in determining which subjects should receive chemotherapy.

While numerous studies point to the high specificity of NAF cytology in breast cancer diagnosis (King et al., 1975; Papanicolaou et al., 1958; Sauter et al., 1997), cytologic findings are occasionally difficult to interpret. Perhaps the chief difficulty is in the differentiation of benign from malignant papillary growths. This dilemma is found primarily in the cytologic evaluation of SND, which is often the result of a benign papilloma on histopathologic review which can appear suspicious for carcinoma to the cytopathologist not highly familiar with NAF and SND cytologic evaluation (Papanicolaou et al., 1958; Sauter et al., in press-b).

9.3.2 Image Analysis

While NAF cytologic evaluation is very specific in the diagnosis of breast cancer, it is not very sensitive (Krishnamurthy et al., 2003; Papanicolaou et al., 1958; Sauter et al., 1997). One approach that has been used to increase the sensitivity of NAF is to evaluate the DNA content of the cells. Normal cells contain 46 chromosomes, are called diploid, and have a DNA index (DI) of 1.0. An abnormal amount of cellular DNA is called aneuploidy and is associated with a high nuclear grade. Hypertetraploidy is used to describe a cell that contains more than twice the normal DNA content, and has a DI > 2.0.

Since NAF samples have limited and mixed cellularity (epithelial, foam, and occasionally white or red blood cells), evaluating DNA content requires image analysis, where the cells of interest (epithelial cells) but not other cells can be evaluated for their DNA content and the percentage of cells in various stages of the cell cycle. Aneuploidy in NAF is associated with atypical and malignant NAF cytology and is associated with the presence of breast cancer (Sauter et al., 1997). Abnormal DNA ploidy is highly predictive of the presence of residual breast cancer after diagnostic biopsy (Sauter et al., 1999).

9.3.3 Nuclear DNA Alterations

Both deletions in DNA, evidenced by loss of heterozygosity (LOH), and changes (either gains or losses) in the number of repeat units of DNA (de la Chapelle, 2003), termed microsatellite instability (MSI), had been identified in a variety of human physiological fluids from subjects with cancer, including sputum (Arvanitis et al., 2003), urine (Neves et al., 2002), stool (Koshiji et al., 2002), blood (Schwarzenbach et al., 2004), and SND (Miyazaki et al., 2000). To determine if LOH and/or MSI could be identified in NAF from subjects with breast cancer, DNA from matched NAF and breast tissue samples was extracted and 11 microsatellite markers evaluated (Zhu et al., 2003). An identical LOH/MSI alteration was detected in NAF from 33% of proliferative and 43% of cancerous breasts which harbored the change in matched tissue.

9.3.4 DNA Methylation

In cancer cells, several tumor suppressor genes such as *p16*^{INK4a}, *VHL*, *hMLH1*, and *BRCA1* have been found to have hypermethylation of normally unmethylated CpG islands within the promoter regions. The hypermethylation is associated with transcriptional silencing of the gene (Baylin et al., 1998). Hypermethylation can be analyzed by the sensitive methylation specific-PCR (MSP) technique, which can identify up to one methylated allele in 1000 unmethylated alleles, appropriate for the detection of neoplastic cells in a background of normal cells (Herman et al., 1996). MSP has been used in recent studies for the successful detection of cancer cell DNA in bodily fluids; these have included the detection of liver (Wong et al., 1999), lung (Esteller et al., 1999) and head and neck cancer DNA in serum (Sanchez-Cespedes et al., 2000), lung cancer DNA in both sputum (Belinsky et al., 1998) and bronchial lavage (Ahrendt et al., 1999), and prostate cancer DNA in urine (Cairns et al., 2001). Using a panel of six normally unmethylated genes: *glutathione S-transferase* π *1* (*GSTP1*); *retinoic acid receptor-β2* (*RARβ2*); *p16*^{INk4a}; *p14*^{ARF}; *RAS association domain family protein 1A* (*RASSF1A*); and *death-associated protein kinase* (*DAP-kinase*) in 22 matched specimens of breast cancer tissue, normal tissue, and nipple aspirate fluid collected from breast cancer patients, hypermethylation of one or more genes was found in all 22 malignant tissues and identical gene hypermethylation detected in DNA from 18 of 22 (82%) matched NAF samples (Krassenstein et al., 2004). In contrast, hypermethylation was absent in benign and normal breast tissue and nipple aspirate DNA from healthy women.

9.3.5 Mutations in Mitochondrial DNA

While each cell contains one matched pair of nuclear DNA (nDNA), the same cell contains several hundred to thousands of mitochondria and each mitochondrion contains 1 to 10 mitochondrial genomes (Chen et al., 2002). Both because of the sheer abundance of mitochondrial DNA (mtDNA) per cell and the tendency for mtDNA mutations to be homoplastic, mtDNA may provide a distinct advantage in terms of feasibility and sensitivity over nDNA-based methods for cancer detection, especially when one is dealing with samples of low cellularity such as NAF. A recent report documents the feasibility of detecting mtDNA mutations in NAF (Cavalli et al., 2004). The authors collected six NAF samples from four women, two BRCA1 carriers and two noncarriers. mtDNA analysis was successful in 4/6 samples, and one mutation was found in a carrier. It is unclear if the other three samples lacking a mutation were from carriers or noncarriers.

A second report collected matched tumor and benign tissue and NAF from 15 women with breast cancer (Zhu et al., 2005). Fourteen of the 15 (93%) cancer samples had one or more somatic mtDHA mutations. Four of nineteen mtDNA mutations in the cancer samples were found in matched NAF. No mutations were found in five matched NAF samples from women whose cancers lacked a mutation in the same region.

9.4 STUDIES EVALUATING EXTRACELLULAR FLUID IN NAF

NAF contains a variety of chemical substances either secreted from or which passively diffuse through the epithelial cells into the ductal lumen. These include substances of endogenous origin, including α-lactalbumin, immunoglobulins, lipids, fatty acids, proteins, cholesterol and cholesterol oxidation products, and hormones (Petrakis, 1986), as well as exogenous substances including nicotine and cotinine from cigarette smoking (Petrakis et al., 1978) and mutagenic agents of undetermined origin (Scott and Miller, 1990). Many of these substances are concentrated in NAF relative to corresponding serum.

9.4.1 Endogenous Substances: Single Protein Analysis

9.4.1.1 Hormones and Growth Factors

A variety of hormones have been measured in NAF, including estrogens, androgens, progesterone, dehydroepiandrosterone sulfate, prolactin, growth hormone, and the growth factors epidermal growth factor, transforming growth factor-α, vascular endothelial growth factor, and basic fibroblast growth factor (Chatterton et al., 2004; Hsiung et al., 2002; Petrakis, 1989; Sauter et al., 2002b). Elevated levels of estrogens, cholesterol, and cholesterol epoxides have been suggested to have etiologic significance in breast disease (Petrakis, 1993).

Levels of a number of these factors have been compared to disease risk. With the exception of recent parity, no relation was found between levels of estrogen in NAF and breast cancer risk. Higher levels of estradiol and estrone were found in the NAF of women with benign breast disease than in controls (Ernster et al., 1987). There is a decrease in estradiol and estrone levels in NAF following pregnancy or lactation that persists for several years before returning to prepregnancy levels

(Petrakis et al., 1987). This period of decreased estrogen exposure of the breast epithelium of postpartum women has been suggested to partially explain the protective effect of early pregnancy.

Basic fibroblast growth factor (bFGF) and vascular endothelial growth factor (VEGF) are two of the most important angiogenic factors that stimulate tumor growth (Folkman and Klagsbrun, 1987; Folkman and Shing, 1992). A preliminary report that analyzed 10 patients with breast cancer and 10 controls found that bFGF levels in NAF were higher in women with breast cancer than in normal subjects (Liu et al., 2000). A larger study, which evaluated 143 NAF specimens (Hsiung et al., 2002), also found that mean NAF bFGF levels were significantly higher in women with breast cancer than in those without. VEGF levels in NAF were not associated with breast cancer. A logistic regression model including NAF levels of bFGF and clinical variables was 90% sensitive and 69% specific in predicting which women had breast cancer. Adding another biomarker linked to breast cancer, prostate-specific antigen (PSA), increased the sensitivity to 91% and the specificity to 83%.

Leptin is a hormone that plays a central role in food intake and energy expenditure (Macajova et al., 2004). Systemic levels of leptin are increased in obese individuals, and have been found to stimulate the growth of breast cancer cells *in vitro*. Leptin levels in NAF were more readily measured in post- than in premenopausal women and were significantly higher in postmenopausal women with a body mass index (BMI) < 25 (Sauter et al., 2004a). While NAF leptin levels were not associated with pre- or postmenopausal breast cancer, they were associated with premenopausal BMI.

9.4.1.2 Tumor Antigens

A number of proteins present in NAF have previously been associated with cancer in the blood. Two of these are PSA and carcinoembryonic antigen (CEA). PSA, a chymotrypsin-like protease first found in seminal fluid and associated with prostate cancer (Soderdahl and Hernandez, 2002), is also found in breast tissue (Howarth et al., 1997; Sauter et al., 2002b) and in NAF. PSA levels in cancerous breast tissue are lower than in benign breast tissue (Sauter et al., 2002b). PSA is thought to cleave insulin-like growth factor binding protein-3 (IGFBP-3), the major binding protein of IGF-I. Most (Sauter et al., 1996, 2002a, 2002b) but not all (Zhao et al., 2001) studies indicate that low NAF PSA levels are associated with the presence and progression (Sauter et al., 2004b) of breast cancer, whereas high levels of NAF IGFBP-3 have been linked to breast cancer (Sauter et al., 2002a). One explanation for the discrepancy in PSA results may be the difference in NAF yield, which was 97% of subjects in the studies finding an association, and 34% in the study where an association between NAF PSA and breast cancer was not found (Sauter and Diamandis, 2001).

Another protein that is concentrated in NAF is CEA, which was identified in 1965 as the first human cancer-associated antigen (Gold and Freedman, 1965). Serum CEA levels have been used clinically to assess and monitor tumor burden in patients with breast cancer (Ebeling et al., 2002). CEA titers in NAF samples from normal breasts are typically more than 100-fold higher than in corresponding serum (Foretova et al., 1998). CEA levels in NAF from 388 women, including 44 women with newly diagnosed invasive breast cancer, were analyzed. CEA levels were significantly higher in breasts with cancer, but the sensitivity of CEA for cancer detection was only 32% (Zhao et al., 2001).

9.4.2 Endogenous Substances: Proteomic Analysis

Recent advances in comprehensive molecular technologies have allowed the analysis of global gene expression or protein profiles in cancerous vs. normal tissues with the goal of identifying markers that are differentially expressed between benign and malignant tissue. One such study (Porter et al., 2001) used serial analysis of gene expression to identify molecular alterations involved in breast cancer progression. The authors concluded that many of the highly expressed genes encoded secreted proteins, which in theory would be present in NAF.

Breast tissue contains thousands of intracellular proteins. NAF contains a limited number of cells and extracellular fluid, the composition of which includes a relatively small set of secreted breast specific proteins. The few cells in NAF can be separated from the extracellular fluid. The remaining proteins are secreted and therefore represent their final processed form, which makes proteomic analyses less ambiguous and can provide clues to changes in protein translational rates, post-translational modification, sequestration, and degradation, which lead to disease.

9.4.2.1 Two-Dimensional Polyacrylamide Gel Electrophoresis

The traditional method of proteomic analysis is one- or two-dimensional polyacrylamide gel electrophoresis (2-D-PAGE). Using two-dimensional rather than one-dimensional PAGE allows better separation of proteins of equal molecular weight based on charge. Once a protein of interest is found, it can be cut from the gel and identified. 2-D-PAGE has been used to screen NAF because it provides a convenient and rapid method for protein identification based on matrix-assisted laser desorption-time-of-flight mass spectrometry (MALDI-TOF MS). At least two studies have analyzed the NAF proteome. One (Varnum et al., 2003) used liquid chromatography, while the second used 2-D-PAGE (Alexander et al., 2004). More than 60 proteins were identified in the first and 41 in the second study. Many of the proteins were the same, but a significant subset of proteins (35 in the first, 21 in the second) were unique to each study. Both studies should be considered when assessing the NAF proteome.

2-D-PAGE may serve as a screening platform to identify proteins in NAF that are differentially expressed in cancerous and benign breasts. These proteins can then be validated using one or more high-throughput proteomic approaches (Alexander et al., 2004). Three protein spots were detected using 2-D-PAGE that were upregulated in three or more NAF samples from breasts with cancer. These spots were identified to be gross cystic disease fluid protein (GCDFP)-15, apolipoprotein (apo)D, and alpha-1 acid glycoprotein (AAG). To validate these three potential biomarkers, 105 samples (53 from benign breasts and 52 from breasts with cancer) were analyzed using enzyme-linked immunosorbent assay (ELISA), a high-throughput method of evaluating protein concentration. Considering all subjects, GCDFP-15 levels were significantly lower and AAG levels significantly higher in breasts with cancer. This was also true in pre- but not postmenopausal women. GCDFP-15 levels were lowest and AAG levels highest in women with DCIS. Menopausal status influenced GCDFP-15 and AAG more in women without than with breast cancer. ApoD levels did not correlate significantly with breast cancer.

9.4.2.2 Surface-Enhanced Laser Desorption Ionization Time-of-Flight Mass Spectrometry (SELDI-TOF-MS)

Although 2-D-PAGE is quite powerful, it has limitations in protein separation and sensitivity. Recent advances in comprehensive molecular technologies allow the simultaneous analysis of multiple protein expression targets. The SELDI-TOF technique can be performed with 1 μl of NAF, can detect components in the high femtomole range, and the chip surface, which allows the rapid evaluation of 8 to 24 samples, has high-throughput potential. Candidate breast cancer biomarkers can be identified using mass spectrometric techniques or an immunoassay to the suspected protein can be used to confirm its identity.

A wide array of proteins are secreted into and highly concentrated in NAF and have been associated with breast cancer. We are aware of three pilot studies (Coombes et al., 2003; Paweletz et al., 2001; Sauter et al., 2002c) that demonstrate the feasibility of SELDI-TOF analysis of NAF in a limited number of subject samples, and that identified one or more protein mass peaks associated with breast cancer. A potential limitation of all three studies is that specific protein identification of the protein mass peak was not obtained. Although it has been proposed (Petricoin et al., 2002) that this is not necessary, validation studies to confirm that these protein masses are linked to breast cancer are easiest after the identification of the specific proteins, eliminating the confounder of multiple proteins of similar mass.

9.5 NAF AS A TOOL TO INVESTIGATE THE PRESENCE OF MUTAGENS IN THE BREAST

It is thought that environmental mutagens stored in the adipose tissue of the breast could affect carcinogenesis through direct exposure to the adjacent ductal epithelial cells, and that evaluating NAF would provide information on carcinogen exposure (Petrakis et al., 1980). A standard assay for the presence of mutagens is the Ames test using one of a variety of *Salmonella* strains to detect the mutagen. A number of studies using different *Salmonella* strains have been conducted (Klein et al., 2001; Petrakis et al., 1980; Scott and Miller, 1990). One limitation of the assays performed to date is the need for approximately 10 μl (microliters) of NAF, which is more than is obtained from some subjects. No association was found in the studies between mutagenic activity in NAF and breast cancer.

9.6 EFFECT OF BOTANICALS ON THE BREAST

The role of food in health and disease is of immense and ongoing interest. One of the most studied botanicals is soy. Soy has been reported to have protective effects against breast cancer in Asian women. At least two studies have evaluated the effect of soy isoflavones on the breast using NAF, one (Hargreaves et al., 1999) short term (2 weeks) and the other (Petrakis et al., 1996) for a longer duration (6 months). The short-term study administered 45 mg soy isoflavones to 84 healthy premenopausal women. They found that the isoflavones genistein and daidzein were concentrated in NAF compared to matched serum, both before and after soy supplementation, and that apolipoprotein D (apoD) levels were significantly lowered and pS2 levels were raised in response to soy ingestion (pS2 levels rise and apoD levels go down in response to estrogen; Harding et al., 2000). NAF cytology did not significantly change. In the longer-term study, which evaluated both pre- and postmenopausal white subjects, the effect of soy protein isolate containing 38 mg of genistein was assessed by NAF volume, cytology, and gross cystic disease fluid protein (GCDFP-15) levels (Petrakis et al., 1996) before and after taking soy protein isolate. There was little effect of soy on the NAF parameters in postmenopausal women. In premenopausal women, there was a two- to sixfold increase in NAF volume, a moderate decrease in GCDFP-15 levels, and evidence of epithelial hyperplasia, which was not seen before soy ingestion, as well as increased levels of plasma estradiol, suggesting that isoflavones in soy provided an estrogenic stimulus.

9.7 ASSESSING RESPONSE TO CHEMOPREVENTIVE AGENTS

Cyclooxygenase (COX) converts arachidonic acid to prostaglandins (PGs), including PGE_2. There are two forms of COX: COX-1 and COX-2. COX-2 is inducible and upregulation is associated with breast and other cancers. Most COX inhibitors such as aspirin and nonsteroidals block both forms of the COX enzyme. The COX-2 inhibitor celecoxib (celebrex), a medication approved by the FDA to treat osteoarthritis and to reduce the number of intestinal polyps in patients with familial adenomatous polyposis (Steinbach et al., 2000), had a breast cancer preventive effect in preclinical models (Abou-Issa et al., 2001; Howe et al., 2002), lowering PGE_2 levels.

To assess the ability of celecoxib to lower systemic (plasma) and organ specific (NAF) PGE_2 levels, women at increased breast cancer risk were administered 200 mg twice daily. PGE_2 levels were 81-fold higher in NAF than in matched plasma. There was not a significant decrease in PGE_2 NAF or plasma levels after celecoxib administration. While PGE_2 levels did not change, the findings demonstrate the feasibility of measuring biomarkers in NAF before and after treatment with a chemopreventive agent (Sauter et al., in press).

The effect of the chemopreventive agent tamoxifen was evaluated for its potential antiestrogenic effect on estrogenic biomarkers in NAF (Harding et al., 2000). Two estrogen-stimulated proteins (pS2 and cathepsin D) and two estrogen-inhibited proteins (GCDFP-15 and apoD) were measured in NAF from women on or off antiestrogen therapy. Following treatment with tamoxifen, NAF levels of pS2 fell, and apoD and GCDFP-15 rose significantly (Harding et al., 2000). Treatment with hormone replacement therapy resulted in a significant rise in NAF pS2 and decrease in apoD.

9.8 SUMMARY

Initial studies of NAF focused both on the ability to collect a sample and the analysis of cell morphology. Cytology remains the single most reliable marker to evaluate in NAF, for if malignant cells are found, the likelihood of the breast containing cancer is almost certain. Great strides have been made since the initial report of Papanicolaou in increasing our ability to collect NAF in all subjects, although the variability of success rates suggests that there is still room for improvement. A limitation of NAF is that the samples are of mixed and limited cellularity, and approximately 40% contain no or scant epithelial cells. The best way to minimize the number of the samples without epithelial cells is to collect more NAF, either at the same or a second visit.

Despite the fact that not all subjects will provide a NAF sample containing epithelial cells, samples that lack epithelial cells are more likely to come from a breast without cancer. More sensitive molecular techniques are demonstrating that NAF samples can be used to assess alterations in the methylation of nuclear DNA and to search for evidence of mtDNA mutations.

A great strength of NAF is the high concentration of proteins it contains. The protein concentration is such that often 1 μl is sufficient to perform ELISA and SELDI-TOF studies, and 3 to 5 μl is sufficient for 2-D-PAGE analyses. Mutagenesis studies require somewhat more NAF, limiting their usefulness as a method to screen for disease.

It is likely that a panel of biomarkers will be required to optimally harness the information present in NAF. Preliminary reports suggest that combining protein markers such as PSA and IGFBP-3, or bFGF and PSA, provides a more predictive model of breast cancer than does either marker alone. Using clinical and pathologic information available to a physician after tumor resection, along with NAF cytology, may assist in determining if re-excision is required to ensure complete removal of a subject's breast cancer. Studies are ongoing to determine the optimal mix of cellular and extracellular markers, in combination with clinical and pathologic factors, to improve the usefulness of NAF in predicting who has or will develop breast cancer.

NAF is likely not only to be increasingly useful in breast cancer prediction, but also in determining response to the ingestion of a food or chemical. Preliminary studies with NAF analysis before and after soy ingestion demonstrate the ability of NAF to assess response to treatment. Further evidence of this comes from the ability to evaluate the effect of PGE_2 levels in NAF before and after ingestion of celecoxib, and estrogenic markers in NAF before and after taking tamoxifen.

REFERENCES

Abou-Issa, H.M., Alshafie, G.A., Seibert, K., Koki, A.T., Masferrer, J.L., and Harris, R.E. (2001). Dose-response effects of the COX-2 inhibitor, celecoxib, on the chemoprevention of mammary carcinogenesis. *Anticancer Res.*, 21, 3425–3432.

Ahrendt, S.A., Chow, J.T., Xu, L.H., Yang, S.C., Eisenberger, C.F., Esteller, M., Herman, J.G., Wu, L., Decker, P.A., Jen, J., and Sidransky, D. (1999). Molecular detection of tumor cells in bronchoalveolar lavage fluid from patients with early stage lung cancer. *J. Natl. Cancer Inst.*, 91, 332–339.

Alexander, H., Stegner, A.L., Wagner-Mann, C., Du Bois, G.C., Alexander, S., and Sauter, E.R. (2004). Proteomic analysis to identify breast cancer biomarkers in nipple aspirate. *Clin. Cancer Res.*, 10, 7500–7510.

Arvanitis, D.A., Papadakis, E., Zafiropoulos, A., and Spandidos, D.A. (2003). Fractional allele loss is a valuable marker for human lung cancer detection in sputum. *Lung Cancer*, 40, 55–66.

Baylin, S.B., Herman, J.G., Graff, J.R., Vertino, P.M., and Issa, J.P. (1998). Alterations in DNA methylation: a fundamental aspect of neoplasia. *Adv. Cancer Res.*, 72, 141–196.

Belinsky, S.A., Nikula, K.J., Palmisano, W.A., Michels, R., Saccomanno, G., Gabrielson, E., Baylin, S.B., and Herman, J.G. (1998). Aberrant methylation of p16(INK4a) is an early event in lung cancer and a potential biomarker for early diagnosis. *Proc. Natl. Acad. Sci. U.S.A.*, 95, 11891–11896.

Cairns, P., Esteller, M., Herman, J.G., Schoenberg, M., Jeronimo, C., Sanchez-Cespedes, M., Chow, N.H., Grasso, M., Wu, L., Westra, W.B., and Sidransky, D. (2001). Molecular detection of prostate cancer in urine by GSTP1 hypermethylation. *Clin. Cancer Res.*, 7, 2727–2730.

Cavalli, L.R., Singh, B., Isaacs, C., Dickson, R.B., and Haddad, B.R. (2004). Loss of heterozygosity in normal breast epithelial tissue and benign breast lesions in BRCA1/2 carriers with breast cancer. *Cancer Genet. Cytogenet.*, 149, 38–43.

Chatterton, R.T., Jr., Geiger, A.S., Khan, S.A., Helenowski, I.B., Jovanovic, B.D., and Gann, P.H. (2004). Variation in estradiol, estradiol precursors, and estrogen-related products in nipple aspirate fluid from normal premenopausal women. *Cancer Epidemiol. Biomarkers Prev.*, 13, 928–935.

Cheatle, G.L. and Cutler, M. (1931). *Tumours of the Breast: Their Pathology, Symptoms, Diagnosis and Treatment*. E. Arnold & Co., London.

Chen, J.Z., Gokden, N., Greene, G.F., Mukunyadzi, P., and Kadlubar, F.F. (2002). Extensive somatic mitochondrial mutations in primary prostate cancer using laser capture microdissection. *Cancer Res.*, 62, 6470–6474.

Coombes, K.R., Fritsche, H.A., Jr., Clarke, C., Chen, J.N., Baggerly, K.A., Morris, J.S., Xiao, L.C., Hung, M.C., and Kuerer, H.M. (2003). Quality control and peak finding for proteomics data collected from nipple aspirate fluid by surface-enhanced laser desorption and ionization. *Clin. Chem.*, 49, 1615–1623.

de la Chapelle, A. (2003). Microsatellite instability. *N. Engl. J. Med.*, 349, 209–210.

Deaver, J.B. and McFarland, J. (1917). *The Breast: Its Anomalies, Its Diseases, and Their Treatment*. Blakiston & Son, Philadelphia.

Ebeling, F.G., Stieber, P., Untch, M., Nagel, D., Konecny, G.E., Schmitt, U.M., Fateh-Moghadam, A., and Seidel, D. (2002). Serum CEA and CA 15-3 as prognostic factors in primary breast cancer. *Br. J. Cancer*, 86, 1217–1222.

Ernster, V.L., Wrensch, M.R., Petrakis, N.L., King, E.B., Miike, R., Murai, J., Goodson, W.H., 3rd, and Siiteri, P.K. (1987). Benign and malignant breast disease: initial study results of serum and breast fluid analyses of endogenous estrogens. *J. Natl. Cancer Inst.*, 79, 949–960.

Esteller, M., Sanchez-Cespedes, M., Rosell, R., Sidransky, D., Baylin, S.B., and Herman, J.G. (1999). Detection of aberrant promoter hypermethylation of tumor suppressor genes in serum DNA from non-small cell lung cancer patients. *Cancer Res.*, 59, 67–70.

Folkman, J. and Klagsbrun, M. (1987). Angiogenic factors. *Science*, 235, 442–447.

Folkman, J. and Shing, Y. (1992). Angiogenesis. *J. Biol. Chem.*, 267, 10931–10934.

Foretova, L., Garber, J.E., Sadowsky, N.L., Verselis, S.J., Joseph, D.M., Andrade, A.F., Gudrais, P.G., Fairclough, D., and Li, F.P. (1998). Carcinoembryonic antigen in breast nipple aspirate fluid. *Cancer Epidemiol. Biomarkers Prev.*, 7, 195–198.

Gold, P. and Freedman, S.O. (1965). Demonstration of tumor-specific antigens in human colonic carcinomata by immunological tolerance and absorption techniques. *J. Exp. Med.*, 121, 439–462.

Harding, C., Osundeko, O., Tetlow, L., Faragher, E.B., Howell, A., and Bundred, N.J. (2000). Hormonally-regulated proteins in breast secretions are markers of target organ sensitivity. *Br. J. Cancer*, 82, 354–360.

Hargreaves, D.F., Potten, C.S., Harding, C., Shaw, L.E., Morton, M.S., Roberts, S.A., Howell, A., and Bundred, N.J. (1999). Two-week dietary soy supplementation has an estrogenic effect on normal premenopausal breast. *J. Clin. Endocrinol. Metab.*, 84, 4017–4024.

Herman, J.G., Graff, J.R., Myohanen, S., Nelkin, B.D., and Baylin, S.B. (1996). Methylation-specific PCR: a novel PCR assay for methylation status of CpG islands. *Proc. Natl. Acad. Sci. U.S.A.*, 93, 9821–9826.

Howarth, D.J., Aronson, I.B., and Diamandis, E.P. (1997). Immunohistochemical localization of prostate-specific antigen in benign and malignant breast tissues. *Br. J. Cancer*, 75, 1646–1651.

Howe, L.R., Subbaramaiah, K., Patel, J., Masferrer, J.L., Deora, A., Hudis, C., Thaler, H.T., Muller, W.J., Du, B., Brown, A.M., and Dannenberg, A.J. (2002). Celecoxib, a selective cyclooxygenase 2 inhibitor, protects against human epidermal growth factor receptor 2 (HER-2)/neu-induced breast cancer. *Cancer Res.*, 62, 5405–5407.

Hsiung, R., Zhu, W., Klein, G., Qin, W., Rosenberg, A., Park, P., Rosato, E., and Sauter, E. (2002). High basic fibroblast growth factor levels in nipple aspirate fluid are correlated with breast cancer. *Cancer J.*, 8, 303–310.

Keynes, G. (1923). Chronic mastitis. *Br. J. Surg.*, 11, 89–121.

King, E.B., Barrett, D., and Petrakis, N.L. (1975). Cellular composition of the nipple aspirate specimen of breast fluid. II. Abnormal findings. *Am. J. Clin. Pathol.*, 64, 739–748.

King, E.B., Kromhout, L.K., Chew, K.L., Mayall, B.H., Petrakis, N.L., Jensen, R.H., and Young, I.T. (1984). Analytic studies of foam cells from breast cancer precursors. *Cytometry*, 5, 124–130.

King, E.B., Chew, K.L., Hom, J.D., Miike, R., Wrensch, M.R., and Petrakis, N.L. (2004). Multiple sampling for increasing the diagnostic sensitivity of nipple aspirate fluid for atypical cytology. *Acta Cytol.* 48, 813–817.

Klein, P., Glaser, E., Grogan, L., Keane, M., Lipkowitz, S., Soballe, P., Brooks, L., Jenkins, J., Steinberg, S.M., DeMarini, D.M., and Kirsch, I. (2001). Biomarker assays in nipple aspirate fluid. *Breast J.*, 7, 378–387.

Koshiji, M., Yonekura, Y., Saito, T., and Yoshioka, K. (2002). Microsatellite analysis of fecal DNA for colorectal cancer detection. *J. Surg. Oncol.*, 80, 34–40.

Krassenstein, R., Sauter, E., Dulaimi, E., Battagli, C., Ehya, H., Klein-Szanto, A., and Cairns, P. (2004). Detection of breast cancer in nipple aspirate fluid by CpG island hypermethylation. *Clin. Cancer Res.*, 10, 28–32.

Krishnamurthy, S., Sneige, N., Ordonez, N.G., Hunt, K.K., and Kuerer, H.M. (2002). Characterization of foam cells in nipple aspirate fluid. *Diagn. Cytopathol.*, 27, 261–264; discussion 265.

Krishnamurthy, S., Sneige, N., Thompson, P.A., Marcy, S.M., Singletary, S.E., Cristofanilli, M., Hunt, K.K., and Kuerer, H.M. (2003). Nipple aspirate fluid cytology in breast carcinoma. *Cancer*, 99, 97–104.

Lee, M.M., Petrakis, N.L., and Wrensch, M.R. (1992a). Association of abnormal nipple aspirate cytology and mammographic pattern and density, presented at Annual Meeting American Society of Preventive Oncology.

Lee, M.M., Wrensch, M.R., Miike, R., and Petrakis, N.L. (1992b). The association of dietary fat with ability to obtain breast fluid by nipple aspiration. *Cancer Epidemiol. Biomarkers Prev.*, 1, 277–280.

Liu, Y., Wang, J.L., Chang, H., Barsky, S.H., and Nguyen, M. (2000). Breast-cancer diagnosis with nipple fluid bFGF. *Lancet*, 356, 567.

Macajova, M., Lamosova, D., and Zeman, M. (2004). Role of leptin in farm animals: a review. *J. Vet. Med. A Physiol. Pathol. Clin. Med.*, 51, 157–166.

Mitchell, G., Trott, P.A., Morris, L., Coleman, N., Sauter, E., and Eeles, R.A. (2001). Cellular characteristics of nipple aspiration fluid during the menstrual cycle in healthy premenopausal women. *Cytopathology*, 12, 184–196.

Mitchell, G., Sibley, P.E., Wilson, A.P., Sauter, E., A'Hern, R., and Eeles, R.A. (2002). Prostate-specific antigen in nipple aspiration fluid: menstrual cycle variability and correlation with serum prostate-specific antigen. *Tumour Biol.*, 23, 287–297.

Miyazaki, M., Tamaki, Y., Sakita, I., Fujiwara, Y., Kadota, M., Masuda, N., Ooka, M., Ohnishi, T., Ohue, M., Sekimoto, M., Tomita, N., Furukawa, J., Matsuura, N., and Monden, M. (2000). Detection of microsatellite alterations in nipple discharge accompanied by breast cancer. *Breast Cancer Res. Treat.*, 60, 35–41.

Nathan, M. (1914). Diagnostic precoce d'un neoplasme du sein par l'examen histologique de son suintement hemorragique. *Clinique* (Paris), 60, 38–39.

Neves, M., Ciofu, C., Larousserie, F., Fleury, J., Sibony, M., Flahault, A., Soubrier, F., and Gattegno, B. (2002). Prospective evaluation of genetic abnormalities and telomerase expression in exfoliated urinary cells for bladder cancer detection. *J. Urol.*, 167, 1276–1281.

Papanicolaou, G.N., Holmquist, D.G., Bader, G.M., and Falk, E.A. (1958). Exfoliative cytology of the human mammary gland and its value in the diagnosis of cancer and other diseases of the breast. *Cancer*, 11, 377–409.

Paweletz, C.P., Trock, B., Pennanen, M., Tsangaris, T., Magnant, C., Liotta, L.A., and Petricoin, E.F., III (2001). Proteomic patterns of nipple aspirate fluids obtained by SELDI-TOF: potential for new biomarkers to aid in the diagnosis of breast cancer. *Dis. Markers*, 17, 301–307.

Petrakis, N.L. (1969). Dry cerumen — a prevalent genetic trait among American Indians. *Nature*, 222, 1080–1081.

Petrakis, N.L. (1971). Cerumen genetics and human breast cancer. *Science*, 173, 347–349.

Petrakis, N.L. (1986). Physiologic, biochemical, and cytologic aspects of nipple aspirate fluid. *Breast Cancer Res. Treat.*, 8, 7–19.

Petrakis, N.L. (1989). Oestrogens and other biochemical and cytological components in nipple aspirates of breast fluid: relationship to risk factors for breast cancer. *Proc. R. Soc. Edinburgh*, 95B, 169–181.

Petrakis, N.L. (1993). Nipple aspirate fluid in epidemiologic studies of breast disease. *Epidemiol. Rev.*, 15, 188–195.

Petrakis, N.L., Mason, L., Lee, R., Sugimoto, B., Pawson, S., and Catchpool, F. (1975). Association of race, age, menopausal status, and cerumen type with breast fluid secretion in nonlactating women, as determined by nipple aspiration. *J. Natl. Cancer Inst.*, 54, 829–834.

Petrakis, N.L., Gruenke, L.D., Beelen, T.C., Castagnoli, N., Jr., and Craig, J.C. (1978). Nicotine in breast fluid of nonlactating women. *Science*, 199, 303–305.

Petrakis, N.L., Maack, C.A., Lee, R.E., and Lyon, M. (1980). Mutagenic activity in nipple aspirates of human breast fluid. *Cancer Res.*, 40, 188–189.

Petrakis, N.L., Wrensch, M.R., Ernster, V.L., Miike, R., Murai, J., Simberg, N., and Siiteri, P.K. (1987). Influence of pregnancy and lactation on serum and breast fluid estrogen levels: implications for breast cancer risk. *Int. J. Cancer*, 40, 587–591.

Petrakis, N.L., Barnes, S., King, E.B., Lowenstein, J., Wiencke, J., Lee, M.M., Miike, R., Kirk, M., and Coward, L. (1996). Stimulatory influence of soy protein isolate on breast secretion in pre- and postmenopausal women. *Cancer Epidemiol. Biomarkers Prev.*, 5, 785–794.

Petricoin, E.F., Ardekani, A.M., Hitt, B.A., Levine, P.J., Fusaro, V.A., Steinberg, S.M., Mills, G.B., Simone, C., Fishman, D.A., Kohn, E.C., and Liotta, L.A. (2002). Use of proteomic patterns in serum to identify ovarian cancer. *Lancet*, 359, 572–577.

Porter, D.A., Krop, I.E., Nasser, S., Sgroi, D., Kaelin, C.M., Marks, J.R., Riggins, G., and Polyak, K. (2001). A SAGE (serial analysis of gene expression) view of breast tumor progression. *Cancer Res.*, 61, 5697–5702.

Sanchez-Cespedes, M., Esteller, M., Wu, L., Nawroz-Danish, H., Yoo, G.H., Koch, W.M., Jen, J., Herman, J.G., and Sidransky, D. (2000). Gene promoter hypermethylation in tumors and serum of head and neck cancer patients. *Cancer Res.*, 60, 892–895.

Sartorius, O.W., Smith, H.S., Morris, P., Benedict, D., and Friesen, L. (1977). Cytologic evaluation of breast fluid in the detection of breast disease. *J. Natl. Cancer Inst.*, 59, 1073–1080.

Sauter, E.R. and Diamandis, E.P. (2001). Prostate-specific antigen levels in nipple aspirate fluid. *J. Clin. Oncol.*, 19, 3160.

Sauter, E.R., Daly, M., Linahan, K., Ehya, H., Engstrom, P.F., Bonney, G., Ross, E.A., Yu, H., and Diamandis, E. (1996). Prostate-specific antigen levels in nipple aspirate fluid correlate with breast cancer risk. *Cancer Epidemiol. Biomarkers Prev.*, 5, 967–970.

Sauter, E.R., Ross, E., Daly, M., Klein-Szanto, A., Engstrom, P.F., Sorling, A., Malick, J., and Ehya, H. (1997). Nipple aspirate fluid: a promising non-invasive method to identify cellular markers of breast cancer risk. *Br. J. Cancer*, 76, 494–501.

Sauter, E.R., Ehya, H., Babb, J., Diamandis, E., Daly, M., Klein-Szanto, A., Sigurdson, E., Hoffman, J., Malick, J., and Engstrom, P.F. (1999). Biological markers of risk in nipple aspirate fluid are associated with residual cancer and tumour size. *Br. J. Cancer*, 81, 1222–1227.

Sauter, E.R., Ehya, H., Mammen, A., and Klein, G. (2001). Nipple aspirate cytology and pathologic parameters predict residual cancer and nodal involvement after excisional breast biopsy. *Br. J. Cancer*, 85, 1952–1957.

Sauter, E.R., Chervoneva, I., Diamandis, A., Khosravi, J.M., Litwin, S., and Diamandis, E.P. (2002a). Prostate-specific antigen and insulin-like growth factor binding protein-3 in nipple aspirate fluid are associated with breast cancer. *Cancer Detect. Prev.*, 26, 149–157.

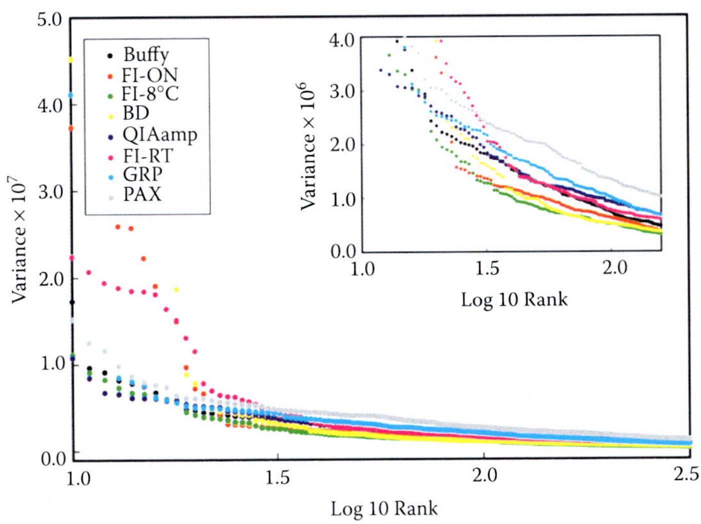

Figure 2.1 Overall variance for each probe set and technique. The variance for all genes was calculated within the respective groups, ordered by rank and plotted against the decade logarithm of the rank. Highlighted are the ranks with the highest variability. Abbreviations according to those described in Table 2.1.

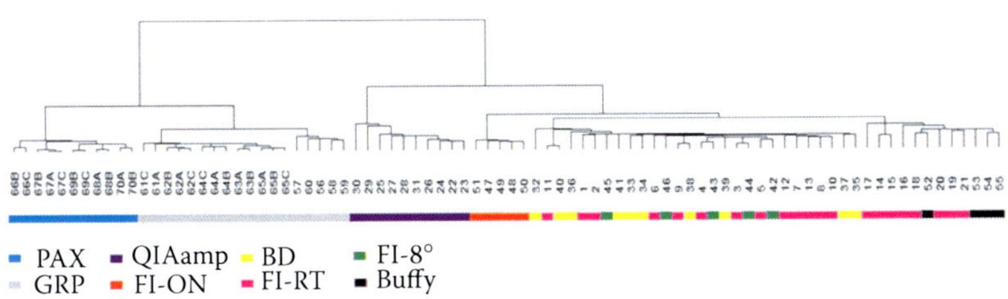

Figure 2.2 Hierarchical clustering of peripheral blood samples. Gene clusters associated with RNA isolation methods from whole blood (PAX, GRP, QIAamp) and isolation methods/conditions of PBMCs (FI-RT, FI-8°, BD, FI-ON, Buffy). Hierarchical clustering of samples was performed with Pearson's correlation algorithm and precalculated distances using dChip software. Prior clustering analysis genes were filtered with a statistical filter (0.5 < SD/mean < 10). Replicates are indicated by alphabetical suffixes.

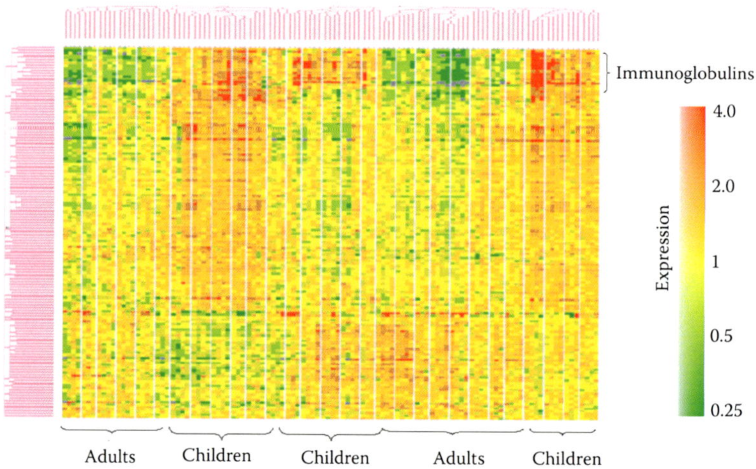

Figure 3.4 Age affects blood genomic expression. Hierarchical cluster analysis of 144 genes regulated between different age groups. Children and adults can be roughly separated although there are some misclassifications. The cluster of genes that correlates best with age relates to the immunoglobulins. (From Tang, Y. et al., *Mol. Brain Res.*, 2004. With permission.)

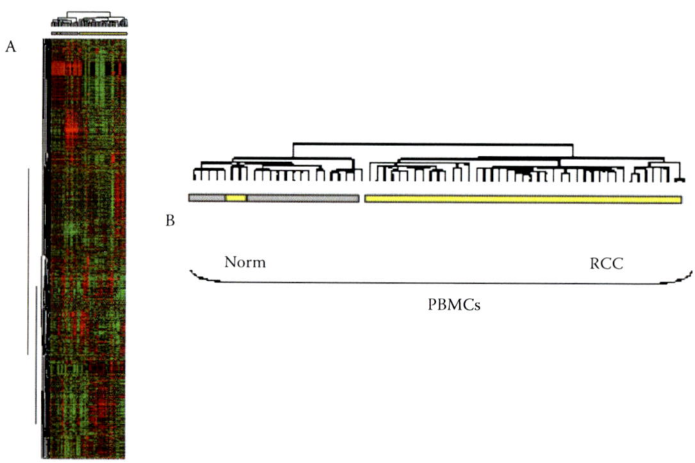

Figure 4.2 Global expression analysis of PBMCs from patients with RCC and normal volunteers. Total RNA obtained from PBMCs of 45 patients with RCC and PBMCs from 20 normal subjects were analyzed on oligonucleotide arrays containing more than 12,000 full-length human genes. In total, 65 samples were analyzed on individual arrays. In no case were samples pooled. (A) Unsupervised hierarchical cluster analysis of normal and RCC PBMCs using all expressed genes present in at least one sample and possessing a frequency of greater than 10 ppm in at least one sample (5249 genes total). Red indicates genes that are elevated relative to the average expression value across all experiments and green indicates genes that are decreased relative to the average expression value. (B) A dendrogram of sample relatedness using all expressed gene expression values is shown. RCC patient PBMC expression profiles are denoted by yellow bars and normal volunteer PBMC expression profiles are indicated by gray bars. (With permission from Twine et al., *Cancer Res.,* 63, 6069–6075, 2003.)

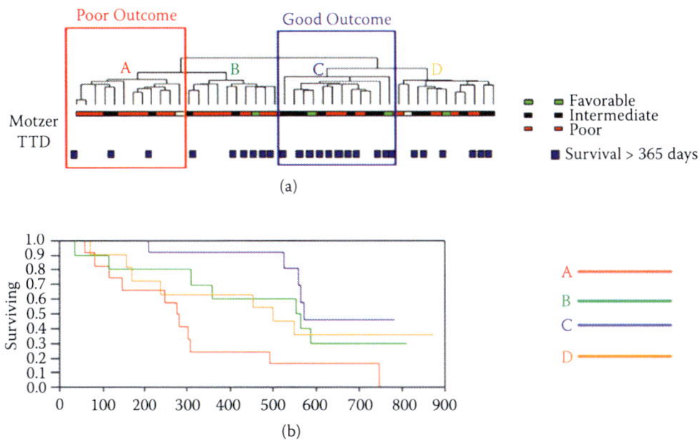

Figure 4.3 Unsupervised hierarchical clustering of RCC patient PBMC profiles and correlation with overall survival. (a) The dendrogram of sample relatedness using all 5424 genes' expression values is shown. Four distinct nodes were identified (nodes A, B, C, and D). Of the 12 patients with PBMC profiles in cluster A, 9 exhibited survival of less than 1 year (red outline), while 10 of the 12 patients with PBMC profiles in cluster C exhibited survival greater than 1 year (blue outline). The associated Motzer risk classifications (green = favorable, black = intermediate, red = poor, yellow = unassigned) are presented underneath the dendrogram. Year-long survival (blue squares indicate > 1 year survival) is also presented. (b) Kaplan–Meier survival curves for patients in the unsupervised analysis. Patients in cluster A possessed significantly shorter survival (median survival = 281 days) relative to patients in clusters B (median survival = 566 days), C (median survival = 573), and D (median survival = 502 days). (With permission from Burczynski et al., *Clin. Cancer Res.*, 11, 1181–1189, 2005.)

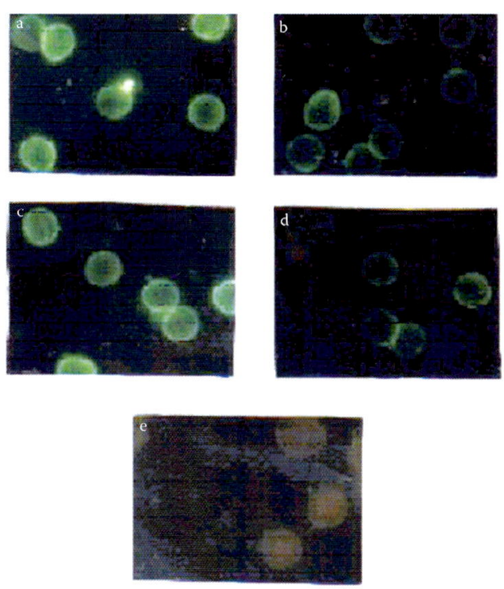

Figure 8.5 Immunofluorescence localization of integrins on PBL of fertile (a) and women with unexplained infertility (b) during midluteal phase. The expression of $\alpha 4\beta 1$ (a, b) and $\alpha v\beta 3$ (c, d) was significantly low in infertile cases (b, d). In negative controls (e) (without primary antibody), the antibody did not react with the cells.

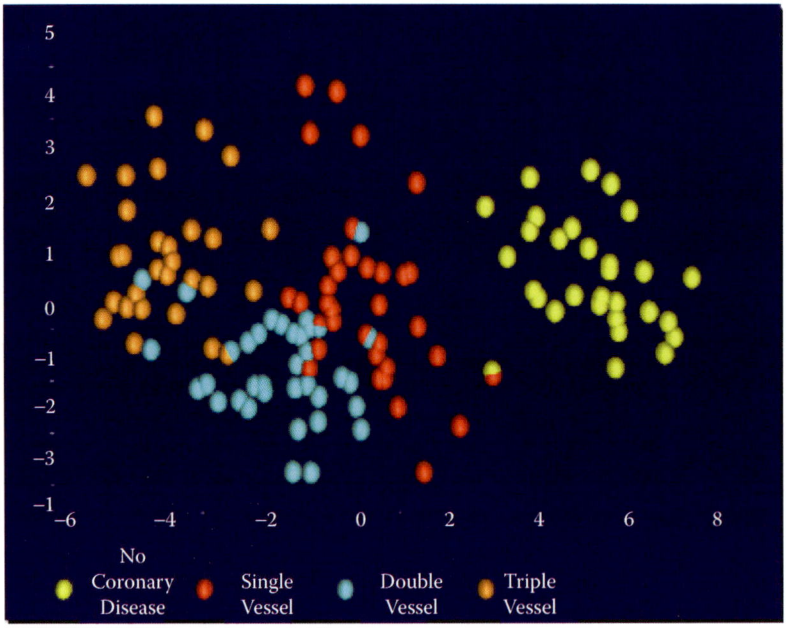

Figure 10.3 High-resolution ^{1}H NMR spectroscopy–based metabonomics has been used to screen patient blood serum samples for metabolic evidence of coronary artery disease. The plot shows a PLS-DA model following orthogonal signal correction that separates noncoronary artery disease from 1, 2, and 3 vessel disease. This work is taken from ref. 17.

Sauter, E.R., Tichansky, D.S., Chervoneva, I., and Diamandis, E.P. (2002b). Circulating testosterone and prostate-specific antigen in nipple aspirate fluid and tissue are associated with breast cancer. *Environ. Health Perspect.*, 110, 241–246.

Sauter, E.R., Zhu, W., Fan, X.J., Wassell, R.P., Chervoneva, I., and Du Bois, G.C. (2002c). Proteomic analysis of nipple aspirate fluid to detect biologic markers of breast cancer. *Br. J. Cancer*, 86, 1440–1443.

Sauter, E.R., Garofalo, C., Hewett, J., Hewett, J.E., Morelli, C., and Surmacz, E. (2004a). Leptin expression in breast nipple aspirate fluid (NAF) and serum is influenced by body mass index (BMI) but not by the presence of breast cancer. *Horm. Metab. Res.*, 36, 336–340.

Sauter, E.R., Klein, G., Wagner-Mann, C., and Diamandis, E.P. (2004b). Prostate-specific antigen expression in nipple aspirate fluid is associated with advanced breast cancer. *Cancer Detect. Prev.*, 28, 27–31.

Sauter, E.R., Schlatter, L., Hewett, J.E., Koivunen, D., and Flynn, J.T. (2004). Lack of effect of celecoxib on prostaglandin E2 concentrations in nipple aspirate fluid from women at increased risk of breast cancer. *Cancer Epidemiol. Biomarkers Prev.*, 13, 1745–1750.

Sauter, E.R., Schlatter, S., Lininger, J., and Hewett, J.E. (2004). The association of bloody nipple discharge with breast pathology. *Surgery.*

Schwarzenbach, H., Muller, V., Stahmann, N., and Pantel, K. (2004). Detection and characterization of circulating microsatellite-DNA in blood of patients with breast cancer. *Ann. N.Y. Acad. Sci.*, 1022, 25–32.

Scott, W.N. and Miller, W.R. (1990). The mutagenic activity of human breast secretions. *J. Cancer Res. Clin. Oncol.*, 116, 499–502.

Soderdahl, D.W. and Hernandez, J. (2002). Prostate cancer screening at an equal access tertiary care center: its impact 10 years after the introduction of PSA. *Prostate Cancer Prostatic Dis.*, 5, 32–35.

Steinbach, G., Lynch, P.M., Phillips, R.K., Wallace, M.H., Hawk, E., Gordon, G.B., Wakabayashi, N., Saunders, B., Shen, Y., Fujimura, T., Su, L.K., and Levin, B. (2000). The effect of celecoxib, a cyclooxygenase-2 inhibitor, in familial adenomatous polyposis. *N. Engl. J. Med.*, 342, 1946–1952.

Varnum, S.M., Covington, C.C., Woodbury, R.L., Petritis, K., Kangas, L.J., Abdullah, M.S., Pounds, J.G., Smith, R.D., and Zangar, R.C. (2003). Proteomic characterization of nipple aspirate fluid: identification of potential biomarkers of breast cancer. *Breast Cancer Res. Treat.*, 80, 87–97.

Weir, H.K., Thun, M.J., Hankey, B.F., Ries, L.A., Howe, H.L., Wingo, P.A., Jemal, A., Ward, E., Anderson, R.N., and Edwards, B.K. (2003). Annual report to the nation on the status of cancer, 1975–2000, featuring the uses of surveillance data for cancer prevention and control. *J. Natl. Cancer Inst.*, 95, 1276–1299.

Wolfe, J.N. (1976). Risk for breast cancer development determined by mammographic parenchymal pattern. *Cancer*, 37, 2486–2492.

Wong, I.H., Lo, Y.M., Zhang, J., Liew, C.T., Ng, M.H., Wong, N., Lai, P.B., Lau, W.Y., Hjelm, N.M., and Johnson, P.J. (1999). Detection of aberrant p16 methylation in the plasma and serum of liver cancer patients. *Cancer Res.*, 59, 71–73.

Wrensch, M.R., Petrakis, N.L., Miike, R., King, E.B., Chew, K., Neuhaus, J., Lee, M.M., and Rhys, M. (2001). Breast cancer risk in women with abnormal cytology in nipple aspirates of breast fluid. *J. Natl. Cancer Inst.*, 93, 1791–1798.

Young, J.L., Jr., Ward, K.C., Wingo, P.A., and Howe, H.L. (2004). The incidence of malignant non-carcinomas of the female breast. *Cancer Causes Control*, 15, 313–319.

Zhao, Y., Verselis, S.J., Klar, N., Sadowsky, N.L., Kaelin, C.M., Smith, B., Foretova, L., and Li, F.P. (2001). Nipple fluid carcinoembryonic antigen and prostate-specific antigen in cancer-bearing and tumor-free breasts. *J. Clin. Oncol.*, 19, 1462–1467.

Zhu, W., Qin, W., Ehya, H., Lininger, J., and Sauter, E. (2003). Microsatellite changes in nipple aspirate fluid and breast tissue from women with breast carcinoma or its precursors. *Clin. Cancer Res.*, 9, 3029–3033.

Zhu, W., Qin, W., Bradley, P., Wessel, A., Puckett, C.L., Sauter, E.R. (2005). Mitochondrial DNA mutations in breast cancer tissue and in matched nipple aspirate fluid. *Carcinogenesis*, 26, 145–152.

Section IV

Metabolomics and Other Approaches

CHAPTER 10

Metabonomics: Metabolic Profiling and Pattern Recognition Analysis of Body Fluids and Tissues for Characterization of Drug Toxicity and Disease Diagnosis

Julian L. Griffin and Nigel J. Waters

CONTENTS

10.1 OVERVIEW

To understand fully the impact of genetic modifications and toxicological interventions, global profiling tools are required to understand their consequences on the network of transcripts, proteins, and metabolites found within a cell, tissue, or organism. Metabonomics/metabolomics is one such technique used to globally profile the metabolite complement of a cell, tissue, or organism using either high-resolution ^{1}H nuclear magnetic resonance (NMR) spectroscopy or mass spectrometry (MS) in conjunction with statistical pattern recognition. Unlike other functional genomic tools, the approach is both high throughput and relatively cheap on a per sample basis. This chapter examines analytical advances in NMR spectroscopy, MS, and pattern recognition that have aided the development of this field, including high-resolution magic angle spinning NMR spectroscopy, cryogenically cooled NMR probes, high-throughput systems, and liquid chromatography MS. These advances have allowed metabonomic approaches to distinguish genetically modified yeast strains, distinguish both disease presence and severity in coronary heart disease, and build predictive models of drug toxicity. These techniques are also being used to data-mine other "-omic" technologies, including transcriptomics and proteomics.

10.2 INTRODUCTION

Since the completion of the Human Genome sequencing project, attention has focused on functional genomic tools to understand how a genetic modification or chemical manipulation results in a given phenotype. With the development of global transcriptional and proteomic profiling techniques such as DNA microarrays and two-dimensional gel electrophoresis, and the rapid increase in gene modification approaches for the production of genetically modified organisms, multivariate data sets are increasingly being produced in an attempt to understand a given pathology, genetic intervention, or drug effect/insult. However, to cross-compare results from different functional genomic investigations, it is necessary to have a description of the changing phenotype, and there is increasing recognition that the large-scale analysis of metabolites, such as by ^{1}H NMR spectroscopy or MS, provides such a process, bringing together differential mRNA and protein responses with specific metabolic pathways by defining a global metabolic phenotype.[1–3] This process of describing the phenotype of a cell, tissue, or organism through the global metabolites present has been referred to as "metabonomics" or "metabolomics." Figure 10.1 illustrates how metabonomics relates to the post-genomic organization of systems biology.

Proponents of both the words *metabolomics* and *metabonomics* have produced very similar definitions for the two words. For example, metabonomics has been defined as "the quantitative measurement of the multivariate metabolic responses of multicellular systems to pathophysiological stimuli or genetic modification,"[3] while metabolomics has been defined as "the complete set of metabolites/low molecular weight intermediates which is context dependent, varying according to the physiology, developmental or pathological state of the cell, tissue, organ or organism."[4] As well as the words metabolomics and metabonomics, some researchers have felt it necessary to distinguish the types of analytical techniques used in these approaches (for review, see Reference 5). Metabolic profiling has been proposed as a means of measuring the total complement of individual metabolites in a given biological sample, whereas metabolic fingerprinting refers to measuring a subclass, to create a "bar code" of metabolism.[2,6] In this latter approach only a limited number of metabolites are quantified and used to distinguish different disease and physiological

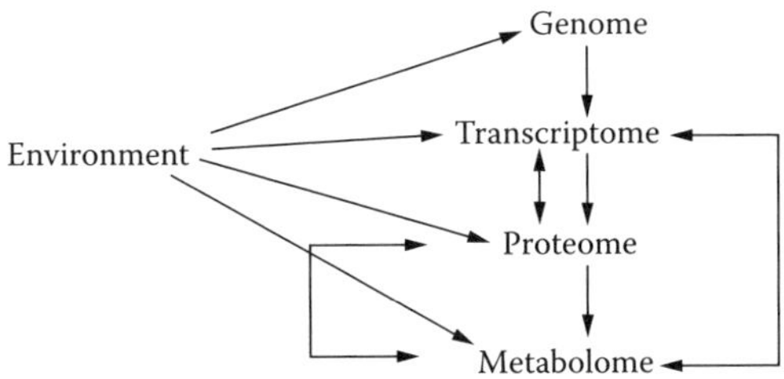

Figure 10.1 Metabonomics and how it fits into the tiered organization of systems biology.

states. However, there is significant overlap in the definitions and uses of these terms, and throughout this chapter the term metabonomics will be used for all approaches whereby a global analytical tool is used in conjunction with pattern recognition approaches to follow metabolic changes in a biofluid, tissue, or organism as this is currently the most widespread term used in the pharmaceutical industry.

Unlike other -omic technologies, both NMR- and MS-based metabonomics are inexpensive on a per sample basis and amenable to high sample throughput.[7–10] In addition to NMR spectroscopy and MS techniques, a range of other analytical tools has also been used (Table 10.1), although the former techniques currently dominate the literature. In terms of other global profiling tools the rapid generation of large metabonomic data matrices have two fundamental advantages. The first is that these techniques can be used as a "first-pass" screening tool to identify samples that should be analyzed using more costly -omic technologies. Alternatively, the analysis of many different samples can be used to circumvent one of the major statistical challenges of -omic technologies. Most approaches produce long, lean data sets consisting of a small number of experiments with many variables. For transcriptomics or proteomics it is sometimes too costly or difficult to obtain samples for a complete time course. Metabonomics can be used to produce more square data matrices, which are less prone to false positives during statistical analysis. Metabonomic analysis of biofluids can also highlight the key time points in a toxic insult, and hence direct the other functional genomic analyses.

The analytical approaches involved in metabonomics are also readily transferable between species, unlike technologies such as DNA microarrays based on sequence-specific hybridization or proteomic approaches based on immunochemistry. As a result, the approach has been applied to a number of environmental toxicology problems, such as examining cadmium and arsenic toxicity in the bank vole,[11–13] an animal with no sequenced genome. In terms of interfacing biofluid-based metabonomics with current toxicology approaches, the collection of urine and blood plasma is minimally invasive, and sample volumes are usually small enough to allow multiple sampling across time courses for rats and larger animals. Indeed, with recent advances in both NMR and MS techniques it is now possible to obtain reasonable data from as little as ~5 μl of blood plasma or cerebral spinal fluid (CSF), allowing multiple sampling even in mouse studies.[14]

In this chapter, advances in NMR- and MS-based global metabolic profiling technology and associated pattern recognition tools are reviewed for the fields of

Table 10.1 Different Spectroscopic Methods Used in Metabonomics for Analysis of Metabolites

Technique	Description	Advantages	Disadvantages
Fourier-transform infrared	Uses vibrational frequencies of metabolites to produce a fingerprint of metabolism	Cheap and good for high-throughput first screening; Oliver and colleagues[64] have used this to differentiate yeast respiratory mutants from wild-type strains	Very difficult to identify which metabolites are responsible for causing changes; very poor at distinguishing metabolites within a class of compounds
Gas chromatography mass spectrometry (GC-MS)	Method of choice for plant metabolomics; uses GC to separate metabolite mixtures, prior to MS to identify the different metabolites	A relatively cheap and robust method, which also has a high degree of sensitivity in terms of metabolite detection	Metabolites must be derivatized first (with different classes of compounds requiring different derivatizations), and this can be time-consuming; not all metabolites can be derivatized into volatile compounds suitable for GC
Liquid chromatography mass spectrometry (LC-MS)	A similar approach to GC-MS, except separation occurs during LC	This method is increasingly being used in place of GC-MS as it has the advantage that metabolites do not have to be derivatized to make them volatile for GC; also, similar to GC-MS, very sensitive	Relatively more costly than GC-MS and critically depends on the reproducibility of the LC (potentially more difficult to control than GC); also can suffer from ion suppression where metabolites are poorly ionized when in the presence of cations and anions
Metabolite arrays	These devices use a 96-well plate assay system for phenotyping; such arrays have been used to phenotype *E. coli* by 700 different assay mixtures ("assay-on-a-chip")[65]	Good as a screening tool when produced for a given situation	The number of metabolites that can be measured is limited by the number placed on the chip; difficult to screen for unknowns and follow metabolism of xenobiotics
NMR spectroscopy	This approach has been widely used by the pharmaceutical industry and in the screening of human patients through urinary and blood plasma metabolic profiles	A noninvasive technique — the use of magnetic resonance spectroscopy demonstrates that metabolomics analysis of tissues in human patients is possible; can be fully automated and has a high degree of reproducibility; relatively easy to identify metabolites from simple one-dimensional spectra	Lower sensitivity than MS; co-resonant metabolites can be difficult to quantify; drug metabolites may be co-resonant with metabolites of interest

Raman spectroscopy[66]	An extension of FT-IR and UV/visible spectroscopy; relies on light scattering following irradiation with a laser	This approach has the advantage over FT-IR in that water has only a weak Raman spectrum and many functional groups can be observed using Raman spectroscopy but not IR (e.g., better distinction of carbon–carbon bonds)	Very difficult to identify which metabolites are responsible for causing changes; very poor at distinguishing classes of compounds
Thin layer chromatography (TLC)	Tweeddale and co-workers[67] have used TLC to follow the metabolic fate of ^{14}C-glucose in *E. coli* under different culture conditions	A particularly cheap method	Open to inter-assay variation, and limited in terms of the metabolites that can be quantified

mammalian toxicology and pathology. The drive for analytical chemists engaged in metabonomics is to increase both the number of metabolites quantifiable and the ease with which these can be identified. These approaches are already being applied to validate animal models of disease,[15,16] diagnose disease and monitor treatment in human patients,[17,18] to assess toxic insults in model systems, and to cross-correlate other -omic technologies such as DNA microarrays and global proteomics.

10.3 HIGH-THROUGHPUT METABOLIC PROFILING IN DRUG TOXICOLOGY

To maximize the information obtainable from multivariate data sets, a high-throughput technology is desirable so that the data matrices produced can fully define both the variation associated with a disorder and the innate variation associated with the biological system, while minimizing false positives associated with such global multivariate analyses. Biofluid NMR spectroscopy in conjunction with pattern recognition techniques has proved a highly successful approach for monitoring changes in systemic metabolism during drug toxicity studies.[19] One of the major successes of this approach has been the prediction of organ specific toxicity from biofluid analysis, allowing the assessment of systemic metabolism through a minimally invasive process. Beckwith-Hall and co-workers[20] demonstrated that the technique can distinguish model liver and kidney toxins, while Nicholls and colleagues have used the approach to determine the mechanism of liver injury caused by phospholipidosis.[21,22] With improvements both in automation of NMR spectroscopy and liquid chromatography (LC)-MS, sample throughput for metabolite-rich fluids such as urine and blood plasma is as high as ~300 and ~60 samples per day, respectively, with no significant costs or time associated with sample preparation.

Using such an approach, the consortium for metabonomic toxicology (COMET) consisting of Imperial College London, U.K., Bristol-Myers Squibb, Eli Lilly and Company, Hoffman-LaRoche, NovoNordisk, Pfizer Incorporated, and the Pharmacia Corporation, has been investigating ~150 model liver and kidney toxins over a 3-year period through NMR-based analysis of urinary metabolites.[23] The final COMET database will comprise ~100,000 NMR spectra. To achieve this it has first been necessary to determine how reproducible such a database would be in terms of both the collection of samples and the resultant NMR analysis. This initial study was performed at two sites, using a 500-MHz spectrometer at one site and a 600-MHz system at the other and using two identical (split) sets of urine samples from rats administered with hydrazine. Lindon and colleagues[24] found that the variation in NMR-based metabonomics as a result of conducting a study across seven different laboratories was minor in comparison to the metabolic changes associated with the toxic lesion. Despite the difference in spectrometer operating frequency, a high degree of consistency was observed between both NMR data sets. Figure 10.2 shows a principal component plot of the data from the two different sites, showing how the two sets of spectra mapped closely to one another, demonstrating the most important variance in the data set was associated with time after dosing and not the spectrometer on which the data were acquired. It is hoped that such an approach

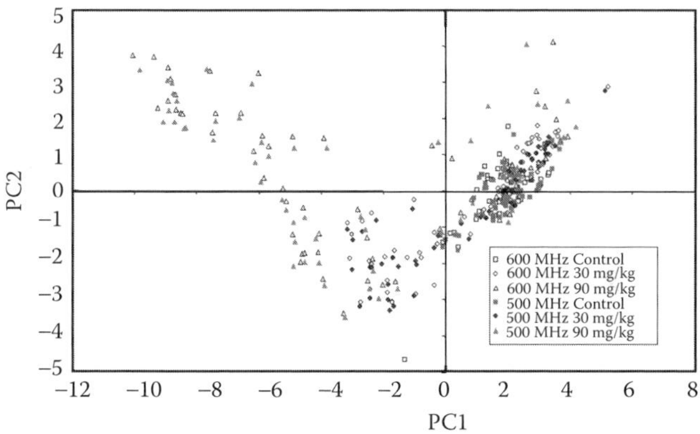

Figure 10.2 Scores plot of the first and second principal components from PCA of urinary NMR spectral data from a hydrazine toxicity study in the rat, illustrating the high degree of biochemical consistency between NMR spectra measured at two different field strengths and at two different sites.

will allow the generation of expert systems where liver and kidney toxicity can be predicted for model drug compounds, with the databases being easily transferable between laboratories. This reproducibility and robustness are enviable when compared with other -omic technologies used for toxicology and pathology.

To fully interrogate the large multivariate data sets that are rapidly produced by studies such as COMET, pattern recognition tools have become an integral part of these approaches.[24,25] Both unsupervised and supervised techniques can be used to derive metabolic profiles.[24] To investigate the innate variation in a data set, unsupervised techniques such as principal components analysis (PCA) or hierarchical cluster analysis (HCA) have been applied. However, where specific questions are being posed, supervised techniques such as prediction to latent structures through partial least squares (PLS),[26] genetic programming,[27] and neural networks may be more appropriate. PLS, the regression extension of PCA, can also be used as a means of data filtering, referred to as orthogonal signal correction (OSC).[28] Variation that is orthogonal to the trend of interest is removed using PLS. To assess which chemometric methods are best to process the data produced by NMR-based metabonomics, the toxicity of 19 compounds was classified according to the main organ of toxicity using density superposition, HCA, and *k*-nearest neighbor approaches.[29,30] Of these approaches the HCA approach fared the worst in terms of prediction, while the others produced highly predictive models of organ toxicity.

10.4 MASS SPECTROMETRY, METABONOMICS, AND TOXICOLOGY

Electrospray ionization (ESI)–MS coupled with LC is the analytical platform of choice for both quantitative and qualitative analysis in the great majority of drug metabolism departments in the pharmaceutical industry. This is in stark contrast to the high capital cost and limited availability of high-field NMR spectrometers. MS technology provides a robust and selective method which is inherently more sensitive than NMR spectroscopy (pg/ml range). It is for these reasons that LC-MS(-MS) has recently been employed in metabonomic studies. Typically, this has involved quadrupole time-of-flight (QTOF) instrumentation to enable sensitive LC-MS-MS together with exact mass measurement. Additionally, gas chromatography (GC)-MS has also been used in plant metabolomics.[31]

The complexity of biofluid ^{1}H NMR spectra often gives rise to a plethora of overlapping signals, which require deconvolution to a second dimension such as correlation spectroscopy (COSY), J-resolved, or diffusion-edited experiments. Pattern recognition (PR) analysis of the one-dimensional spectra and removal of xenobiotic-derived signals also removes endogenous components falling within the same spectral integral region, which can complicate interpretation. With an LC-MS approach, chromatographic separation and MS-MS selectivity of each metabolite can remove this obstacle, leading to generation of simpler spectra. Also unlike ^{1}H NMR spectroscopy, MS offers the ability to detect nonproton species. The ability to apply PCA and other similar PR analyses to MS spectra makes this approach amenable to metabonomic investigations. The availability of combined LC-MS processing and PR analysis software programs, such as Micromass MarkerLynx™ Application Manager, has also aided such studies.

LC is utilized in preference to direct infusion as this distinguishes isobaric species and negates the ion suppression issues as a consequence of competing analytes entering the ion source at any given time, which ultimately results in an improved limit of detection. Using short columns and rapid gradients one can achieve sample analysis times of about a minute giving reasonable LC time resolution, MS sensitivity, and inherent structural information. Employing a purge–wash–purge cycle with an aqueous-organic wash solvent helps to minimize carryover between injections. The high polarity of metabolites in biofluids such as urine means it is only necessary to employ a 0 to 30% organic gradient to get entire elution of all components. Reverse phase (RP) LC is not recommended, because it fails to provide adequate chromatographic separation of the highly polar endogenous metabolites, e.g., amino acids and sugars.

Lenz and co-workers[32,33] have advocated the complementary nature of NMR- and MS-based metabonomic analysis, highlighting the different metabolites detected by each technique, using cyclosporine A and mercuric chloride as model nephrotoxins. High-resolution NMR spectroscopy requires minimal sample preparation and no need to preselect analyte or analytical conditions. Despite the obvious advantages of MS, some metabolites will not be detected with this system, including certain volatile species, nonionizables, and those compounds susceptible to thermal degradation in the ion source. The requirement for ionization also means biofluid samples must be analyzed in both positive and negative modes to ensure optimum chance of detecting the greatest number of metabolites. Nevertheless, negative mode does tend to lead to richer data sets due to the high anionic content in biofluids like urine. MS combined with exact mass measurement (and thus elemental composition) provides a means not only to detect, but also to identify putative biomarkers. This MS-based strategy, either alone or in combination with NMR spectroscopy, has been demonstrated to be a viable option for metabonomics applications in drug discovery and development.[32,33]

The future work to fully integrate MS applications into metabonomic studies will revolve around a number of analytical issues. Many of the endogenous biofluid NMR-detected metabolites such as amino acids, organic acids, and sugars are not optimized for MS detection. Various polar compounds suffer poor retention by high-performance liquid chromatography (HPLC) or poor ionization or both, and so optimization of chromatography or the use of derivatizing agents may be required to improve detection of these species. Many studies to date have detected diagnostic ions in the absence of complete structure elucidation, reflecting the current "exploratory" nature of MS-based metabonomics. This type of metabonomic study was able to differentiate mouse urine samples based on strain, diurnal, and gender differences without identifying the structure of the ions responsible for categorization.[8] On completion of full structure elucidations (by NMR, MS, IR, etc.), databases can be constructed incorporating the toxicological or etiological significance of metabolite markers.

10.5 DISEASE DIAGNOSIS

The ease of automation for NMR-based metabonomics also makes it an ideal technique for screening human populations for common metabolic disorders. One notable example has used a PR-based expert system to predict both the occurrence and severity of coronary artery disease (CAD) through ^{1}H NMR spectroscopic analysis of blood plasma samples[17] (Figure 10.3). Brindle and co-workers[17] identified a number of metabolic patterns that could be used to distinguish whether 1, 2, or 3 coronary arteries were affected during the disease as well as identify which patients suffered from CAD. The researchers have also shown that such an analysis can be used to predict patients with high blood pressure, suggesting that a combination of ^{1}H NMR spectroscopy and PR may be used to diagnose a range of cardiovascular disorders. If such systems can be applied to the clinical situation for predicting CAD, significant financial savings could be made over invasive angiography, which is currently the gold standard for diagnosis. This study has since been extended to include microarray data in an attempt to improve the rate of prediction for CAD above the current >90% capability of the NMR process alone, in the hopes of approaching the >99% capability of angiography.

10.6 CORRELATION OF METABONOMICS WITH OTHER -OMIC TECHNOLOGIES

To maximize the use of DNA microarray and proteomic approaches it is often prudent to target the analysis to key time points, thus maximizing the repetition number for the large data sets produced. Metabonomics provides a lower-cost mechanism for identifying key time points and metabolic events to be further investigated and has been used in such a manner by a number of researchers.

Griffin and co-workers[15] have examined orotic acid–induced fatty liver disease in rats using metabonomics, transcriptomics, and proteomics. One of the benefits of using NMR spectroscopy as part of this global functional genomic approach was

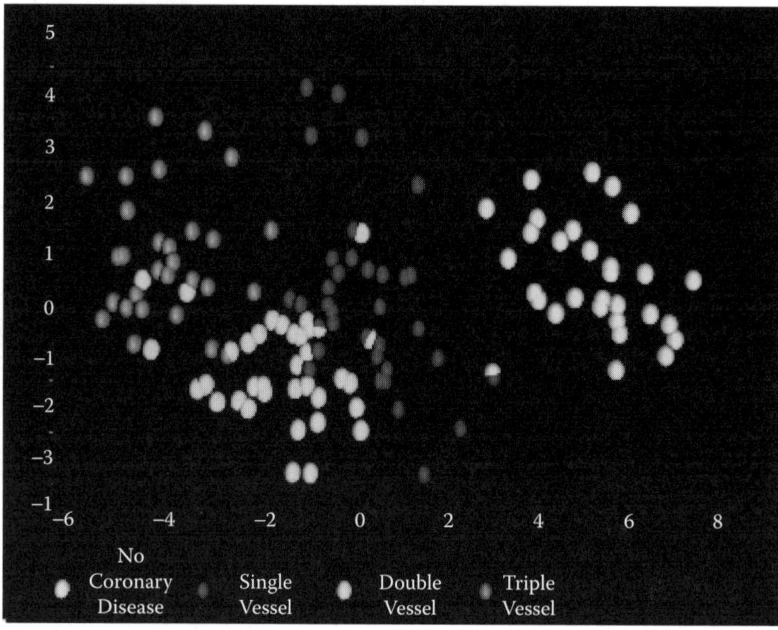

Figure 10.3 **(Color figure follows p. 138.)** High-resolution ^{1}H NMR spectroscopy–based metabonomics has been used to screen patient blood serum samples for metabolic evidence of coronary artery disease. The plot shows a PLS-DA model following orthogonal signal correction that separates noncoronary artery disease from 1, 2, and 3 vessel disease. This work is taken from ref. 17.

that a range of tissues could be examined including the liver, blood, and urine, placing changes in the liver in context with the overall global metabolism of the animal. Furthermore, by providing a metabolic phenotype, this approach allowed the comparison of the in-bred Kyoto strain and the out-bred Wistar strain of rat. Kyoto rats were particularly susceptible to fatty liver accumulation, and metabonomic analysis identified that this strain of rat had an increased cytosolic lipid triglyceride content compared with the out-bred Wistars.

Ringeissen and colleagues[35] have similarly used a joint metabonomic and transcriptomic approach to investigate the action of peroxisome proliferators-activated receptor (PPAR) ligands on systemic metabolism in the rat. They correlated changes in *N*-methylnicotinamide (NMN) and *N*-methyl-4-pyridone-3-carboxamide (4PY) concentrations with peroxisome proliferation, as measured by electron microscopy, and key enzymes in the tryptophan-NAD$^+$ pathway, measured using reverse transcription-polymerase chain reaction (RT-PCR). This elegant paper demonstrated how metabonomics could be used to go from a complex multivariate problem involving systemic metabolism changes to identifying two biomarkers that could be measured to monitor peroxisome proliferation. Given the explosion in applications of -omic technologies within the pharmaceutical industry it is likely that similar approaches will be used increasingly.[36] Chapter 17 by Pennie et al. in the final section of this textbook reviews the concept of pan-omic approaches in greater detail.

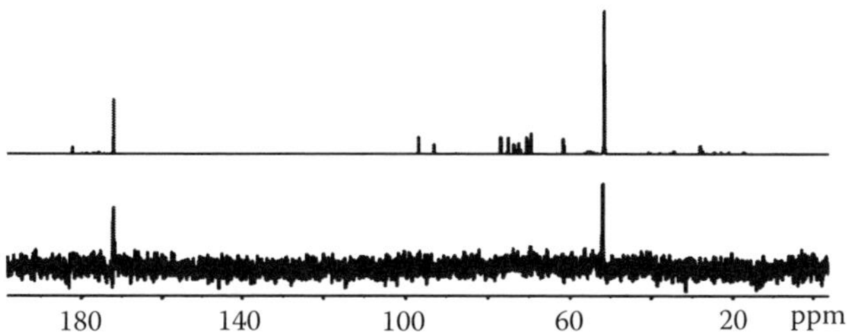

Figure 10.4 The improvements achievable in ^{13}C NMR spectroscopy using cryoprobe technology. Both ^{13}C spectra are acquired on the same sample at 500 MHz using a ^{13}C direct geometry probe and with 256 scans. The top spectrum uses a cryoprobe while the bottom uses a conventional probe.

10.7 CRYOPROBE TECHNOLOGY

To date, NMR-based metabolic profiling has centered on ^{1}H NMR spectroscopy. However, given the relatively small chemical shift range of the nucleus, there is significant overlap between many metabolites in conventional one-dimensional spectroscopy. These resonances can be separated into further dimensions using pulse sequences, such as COSY, Carr Purcell Meiboom and Gill (CPMG), and diffusion ordered spectroscopy (DOSY), which rely on physical properties including J-coupling, relaxation, and diffusion.[16]

An alternative is to examine metabolites through ^{13}C NMR spectroscopy, where resonances are spread over a ~200 ppm chemical shift range.[10] To compensate for the lower sensitivity of the ^{13}C nucleus and a natural abundance of only 1.1%, superconducting NMR probe technology ("cryoprobes") can be applied, significantly reducing NMR acquisition times and allowing natural abundance detection of metabolites (Figure 10.4). This approach relies on cooling the NMR radiofrequency detector and preamplifier to ~20K, or less.[37] As thermal noise is reduced by a factor equivalent to ~temperature$^{1/2}$, the thermal noise is reduced by ~4-fold, giving a ~16-fold reduction in acquisition time for the same signal to noise using a conventional probe. Using this approach, Keun and colleagues[10] readily detected hepatic toxicity using ^{13}C NMR spectroscopy of urine detecting metabolites via natural abundance labeling.[10]

10.8 HIGH-RESOLUTION MAGIC ANGLE SPINNING ^{1}H NMR SPECTROSCOPY

Direct observation of metabolites within tissues is impaired by a number of physical processes that serve to broaden spectral resonances. Relaxation times are often short, giving rise to broader lines, and anisotropic NMR parameters are not averaged completely to zero, also causing line broadening. For ^{1}H NMR spectroscopy, chemical shift anisotropies are small, quadrupolar couplings are not present, and J-coupling anisotropy is negligible. However, both dipolar coupling and dia-

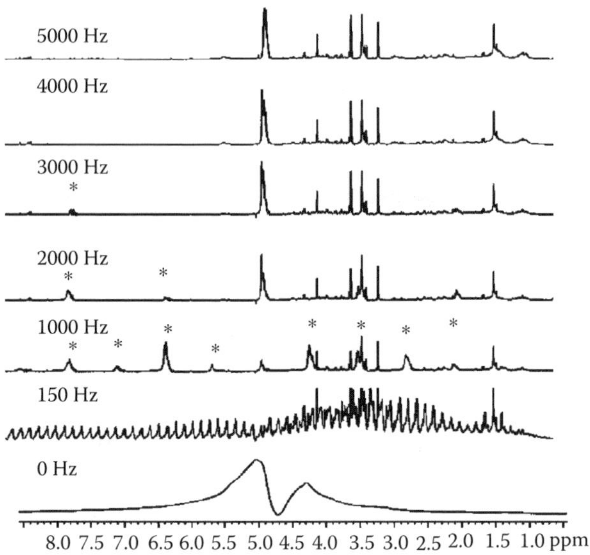

Figure 10.5 High-resolution magic angle spinning ^{1}H NMR spectra of cardiac tissue acquired at 700 MHz and at various spin rates (ν) demonstrating the effects of spin rate on spectral resolution. * Marks major spinning side bands arising in the spectra at slower spin speeds.

magnetic susceptibility anisotropy are significant, giving rise to broadened lines in spectra.[38] The dipolar Hamiltonian (in hertz) for two spin one-half nuclei in a rigid solid is

$$H_D/h = \Sigma\ (h/8\pi^2)\ \gamma_i\gamma_j r_{ij}^{-3}\ (3\ \cos^2\theta_{ij} - 1)(I_i I_j - 3I_{zi}I_{zj})$$

The value of the dipolar coupling depends on the angle (θ) that the internuclear vector makes with the field direction and the internuclear distance. If there is some molecular motion the angular term is partially averaged and for tissues this can be considerable, leaving line widths of the order of 1 kHz. However, if the sample is spun at a rate large compared to the partially averaged dipolar coupling, then the angular term is completely averaged to zero[39] (Figure 10.5).

With increasing spin rates to ensure spinning side bands are at the periphery of spectra the samples may suffer degradation, especially for softer tissues or cells, as centripetal force increases with the square of spin speed. This has led to the development of pulse sequences, such as TOSS and PASS, for use with ^{1}H magic angle spinning (MAS) NMR spectroscopy to minimize tissue degradation.[40] Even at speeds of 5000 to 6000 Hz, cultured adipocytes and neuronal cells appear to be viable with minimal damage to cell membranes.[41,42] The nondestructive nature of this technique enables histopathological assessment post-NMR analysis and therefore facilitates a direct tissue structure–function correlation.

For *in vivo* NMR spectroscopy, spectral quality is severely compromised by high heterogeneity leading to magnetic field inhomogeneity and poor resolution. Chemical extraction of *ex vivo* tissue samples results in a loss of tissue components, a loss of compartmentalization, and ultimately destruction of the structural and functional integrity of the tissue or cells under study. A comparative investigation of high-resolution MAS (HRMAS) NMR spectroscopy of intact liver tissue vs. conventional liquid-state NMR spectroscopy of aqueous and lipophilic liver extracts highlighted that no additional alpha-naphthylisothiocyanate (ANIT)-induced liver biomarker information was obtained by the laborious process of tissue extraction. HRMAS-NMR was able to detect a multitude of tissue metabolites with minimal disruption to the system and without metabolite discrimination, demonstrating it as a powerful tool for drug toxicology and disease etiology studies.[43] The minimal sample preparation required to study tissues by HRMAS allows the visualization of dynamic processes *ex vivo* (e.g., the selective deuteration of alanine by alanine aminotransferase), and by employing spectral editing methods one can observe the compartmentation within various cellular environments.

A whole array of tissues, cells, and organelles have been studied by HRMAS-NMR spectroscopy including liver, kidney, brain, testes, heart tissue and mitochondria, erythrocytes, endometrial cells, adipocytes, lymph nodes, breast, and prostate tissue. Garrod and colleagues[44,45] have used HRMAS ^{1}H NMR spectroscopy and pattern recognition to correlate histopathology and urinary biomarkers in 2-bromoethanamine toxicity, a known renal papillary toxin, with biochemical changes in the kidneys. The drug induces mitochondrial dysfunction and inhibits fatty acyl-CoA dehydrogenases. HRMAS ^{1}H NMR spectroscopy detected a transient rise in glutaric acid in the renal cortex, renal papilla, and the liver, indicating the metabolic events that produced the characteristic urinary metabolite changes.

Waters and co-workers[46] employed a HRMAS ^{1}H NMR spectroscopic and pattern recognition approach that enabled the detailed study of biochemical perturbations in intact liver tissue spectra following an ANIT-induced hepatotoxic insult, thereby allowing a direct correlation with biofluid NMR spectra, histopathological data, and clinical chemistry parameters. The use of NMR-based metabonomic techniques allowed the visualization of key time periods in the development of a toxic injury, enabling the identification of lesion-specific, matrix-specific biomarkers of cholestasis and hepatotoxicity. The variety and complexity of the biochemical changes arising from a single dose of the hepatotoxicant over time showed the importance of the use of multiparametric analytical approaches to the study of toxic episodes to relate biochemical changes to classically accepted pathological end points (Figure 10.6). Such a holistic approach to the study of time-related toxic effects in the intact system enabled the characterization of key metabolic effects during the development and recovery from a toxic lesion and has been termed "integrated metabonomics."[46]

Coen and colleagues[36] employed a similar strategy to the investigation of acetaminophen toxicity in the mouse. Metabolic effects in intact liver tissue and lipid soluble liver tissue extracts from animals treated with the high dose level of acetaminophen included an increase in lipid triglycerides and monounsaturated fatty acids together with a decrease in polyunsaturated fatty acids, indicating mitochon-

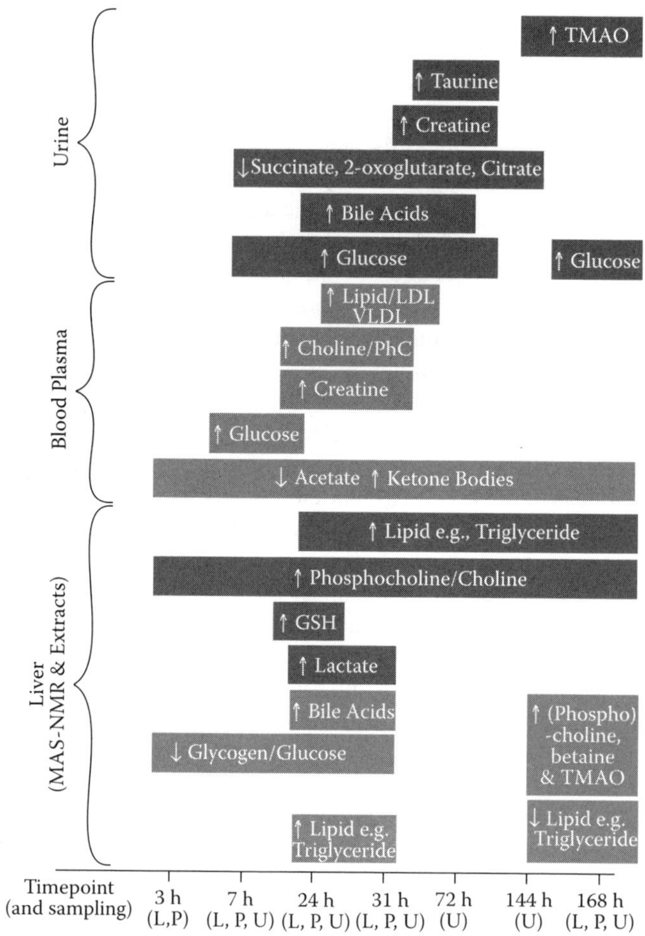

Figure 10.6 An integrated metabonome diagram describing the NMR-detected biochemical changes over time observed in rat urine, blood plasma, and liver following ANIT treatment. Such an approach enables a mapping of key metabolic perturbations and thus gives a more detailed insight into mechanisms of toxicity or disease progression. (Key: GSH = glutathione; LDL = low-density lipoprotein; PhC = phosphatidylcholine; TMAO = trimethylamine-*N*-oxide; VLDL = very low density lipoprotein. Metabolite changes glycogen and bile acids were observed in both extract and MAS-NMR spectra. Note: L, P, and U refer to sampling of liver, plasma and urine, respectively).

drial dysfunction with concomitant compensatory increase of peroxisomal activity. In addition, a depletion of phospholipids was observed in treated liver tissue, which suggested an inhibition of enzymes involved in phospholipid synthesis. There was also a depletion in the levels of liver glucose and glycogen. In addition, the aqueous soluble liver tissue extracts from high-dose animals also revealed an increase in lactate, alanine, and other amino acids, together with a decrease in glucose. Plasma spectra showed increases in glucose, acetate, pyruvate, and lactate. These observations all provided evidence for an increased rate of glycolysis, together with a mitochondrial inability to use pyruvate in the citric acid cycle, and also revealed the impairment of fatty acid beta-oxidation in liver mitochondria of such treated mice. Mortishire-Smith and colleagues[47] have used a combination of HRMAS ^{1}H NMR spectroscopy, biofluid NMR, and *in vitro* assays to assess impaired fatty acid metabolism as a mechanism of drug-induced toxicity. They identified decreases in tricarboxylic acid (TCA) cycle intermediates and increases in medium-chain dicarboxylic acids in urine as being correlated with increased lipid triglycerides in liver as identified by HRMAS spectroscopy, confirming the drug impaired lipid metabolism by *in vitro* experiments.

Furthermore, HRMAS ^{1}H NMR spectroscopy can demonstrate when a biofluid biomarker does not originate in a given organ. Nicholson and colleagues[48] demonstrated that acute exposure of male rats to cadmium chloride resulted in creatinuria following testes specific toxicity. Thus, it seemed reasonable that similar creatinuria detected in a chronic exposure study of male rats to cadmium chloride resulted from testicular damage.[12] However, using HRMAS ^{1}H NMR spectroscopy, no biochemical changes were detected in testicular tissue, and in particular there was no decrease in tissue creatine content or a change in redox potential in the tissue, known to precede cadmium-induced testicular toxicity. Instead, the most likely explanation for the creatinuria was breakdown of muscle tissue to supply glutamine to renal tissue and prevent renal tubular acidosis.

In addition, Waters et al.[49] were able to deconvolute the series of biochemical events in the onset and progression of toxicities in more than one tissue, exemplified by the model nephro- and hepatotoxin, thioacetamide. Utilizing HRMAS ^{1}H-NMR spectroscopy of intact tissues provides the essential link between the metabolite profiles obtained from biofluid NMR and the structural progression of the lesion observed by histopathological techniques. The thioacetamide-induced biochemical manifestations included a renal and hepatic steatosis accompanied by hypolipidemia; an increased urinary excretion of taurine and creatine concomitant with elevated creatine in liver, kidney, and plasma; a shift in energy metabolism characterized by depleted liver glucose and glycogen, reduced urinary excretion of tricarboxylic acid cycle intermediates, and raised plasma ketone bodies; increased levels of tissue and plasma amino acids leading to aminoaciduria verifying necrosis-enhanced protein degradation and renal dysfunction; and elevated hepatic and urinary bile acids indicating secondary damage to the biliary system. The ability of integrated metabonomic studies to delineate and define the tissue origin of biomarkers present in biofluids lends itself to novel drug candidate safety investigations in the pharmaceutical discovery setting, where embedded toxicity is not uncommon. Such an approach allows the deconvolution of embedded pathologies, identifying and locating sites of toxin-induced damage, and is thus able to direct histopathology.

These studies have demonstrated the strength of an integrated metabonomic approach in assessing drug toxicity.[46,49] For example, HRMAS ^{1}H NMR spectroscopy coupled with chemometric methods identified a toxin-induced hepatic steatosis with both ANIT and thioacetamide. In the absence of biofluid NMR-PCA data, this is all that can be inferred. However, on correlating the NMR-PCA liver data with that of blood plasma and urine, it is clear that the reduced low density lipoprotein (LDL) levels in plasma from thioacetamide-treated rats indicate reduced lipid transport and secretion by the liver. The blood plasma NMR-PCA data from ANIT-treated individuals showed elevated LDL and thus implied increased lipid synthesis or, as proposed, a bile acid-mediated micellar solubilization effect. Therefore, mapping changes by NMR-based metabonomics in more than one biological matrix allows inferences to be made concerning the mechanism of action and the redistribution and metabolism of endogenous low-molecular-weight metabolites during the progression of and recovery from drug-induced tissue damage. It is this mechanistic insight into drug toxicity and disease progression which makes HRMAS ^{1}H NMR spectroscopy such an invaluable exploratory tool.

10.9 DRUG DEVELOPMENT

Metabonomics has been used to monitor the effects of various anticancer drugs in tumor cells. Indeed, PR and NMR spectroscopy have been used for a number of years to follow metabolic changes that occur in tumors in response to therapy.[50–53] For example, "neural networks" — pattern recognition processes that iteratively search for the best solution using a network construction similar to neurons in the brain — have been used to identify metabolic profiles of chemotherapy-resistant gliomas in humans prior to treatment.[54] In this regard, metabolic profiles could be used to predict which tumors are most likely to respond, or become resistant, to a specific type of therapy.

In a similar manner, HRMAS ^{1}H NMR spectroscopy of intact Ishikawa cells was used to investigate the action of tamoxifen and other specific estrogen receptor modulators (SERMs).[55] Ishikawa human endometrial adenocarcinoma cells are hormone responsive, and are therefore ideal for investigating drugs that modulate the estrogen receptor. This study collected metabolic fingerprints, made up of about 20 metabolites, in intact cells and generated PR models that correlated metabolic changes with varying doses of different SERMs. The metabolites analyzed in this model included ethanolamine, myo-inositol, uridine, and adenosine, suggesting alterations in both membrane turnover and DNA transcription. Furthermore, the metabolic effects of other estrogen modulators could be monitored using this PR model. This identification of specific metabolomic fingerprints that are associated with various drug types and dosages will allow researchers to determine how well certain tumor cells respond to different doses of drugs such as tamoxifen.

Metabonomics has also been used to identify surrogate biomarkers for the pharmacodynamic monitoring of tumor response to drug intervention in human colorectal xenografts grown in mice.[56,57] 17-Allylamino-17-demethoxygeldanamycin (17-AAG) prevents tumor cell growth by inhibiting the action of heat shock protein-90, a molecular chaperone. Although this drug's exact *in vivo* mechanism of action is yet to be determined, ^{31}P NMR spectroscopy and *in vitro* metabonomic analysis have indicated that it functions by perturbing cell membrane metabolism. These characteristic metabolic perturbations also can be used to follow treatment efficacy *in vivo*.

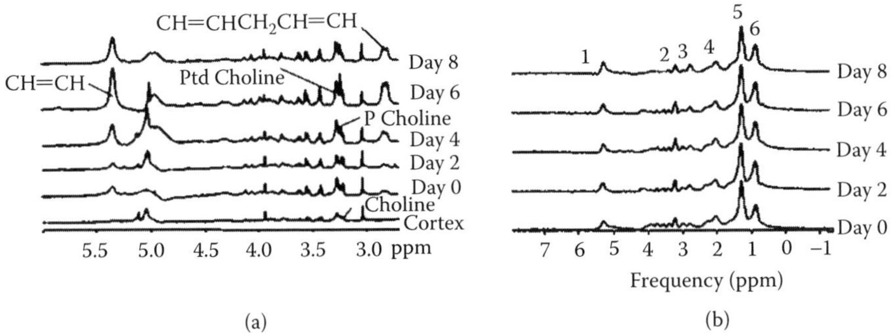

Figure 10.7 High-resolution MAS ^{1}H NMR spectrum (A) and *in vivo* ^{1}H NMR spectra of rat glioma during gene therapy–induced apoptosis (B). Key: 1. Ch=CH; 2. Choline containing metabolites; 3. $CH{=}CHCH_2CH{=}CH$; 4. $CH_2CH{=}CH$; 5. $–CH_2–$; 6. $–CH_3–$ (Adapted from ref. 58)

10.10 METABONOMICS *IN VIVO*

The ultimate aim of many studies involving NMR-based metabonomics of tissues is to quantify biomarker changes *in vivo* using clinical magnetic resonance spectroscopy (MRS) systems. One such study has used a combination of *in vivo*, *in vitro,* and HRMAS ^{1}H NMR spectroscopy in conjunction with PCA to examine polyunsaturated fatty acids (PUFA) that accumulate in BT4C glioma cells during gene therapy–induced apoptosis[58] (Figure 10.7). In the study, apoptosis was induced in rat gliomas by administration of ganciclovir and targeting tumor cells that carried a herpes simplex thymidine kinase (HSV-tk) expressing vector. Metabonomic analysis of glioma using both *in vivo* MRS and HRMAS ^{1}H NMR of the glioma removed at postmortem demonstrated that the concentration of PUFAs, detected as CH=CH and $CH{=}CHCH_2CH{=}CH$ resonances by ^{1}H NMR, increased threefold. These PUFA lipids are readily observable *in vivo* using MRS and could be used in the future to monitor the efficacy of gene therapy treatments. Furthermore, while histology and TUNEL staining could be used to follow the rate of apoptosis in excised tumors, the NMR observable changes indicated the metabolic pathways that accompany apoptosis. It remains to be seen whether this characteristic rise in polyunsaturated lipids in glioma undergoing apoptosis is a general feature of tumors during programmed cell death.

The analysis of intact tissue by HRMAS in conjunction with PR techniques has also been used to study cervical biopsies and correlate this information with histopathology, in particular correlating lactate concentration with metastatic spread.[59] PCA also classified the patients according to diagnosis, largely according to raised cholines, amino acids, and reduced glucose in the malignant tissue. Considering these and similar metabonomic studies both *in vivo* and *ex vivo,*[60–64] these metabolic biomarkers should be usable for the noninvasive monitoring of treatments based on events such as apoptosis. By also providing a noninvasive tool for monitoring tumor phenotype changes in animal models, they also offer a unique insight into the disease process not obtainable using histology, transcriptomics, or proteomics. However, one drawback is that the amount of metabolites that can currently be detected *in vivo* is relatively small, making it difficult to determine exactly which metabolic pathways are responsible for a given change.

10.11 CONCLUSIONS

Metabonomics is a relatively new addition to the list of tools that the toxicologist can use in the drug safety assessment process. These techniques are also being used to monitor treatment in animal models and in humans both via biofluid analysis and even MRS *in vivo*. Metabonomics has several benefits when compared with conventional techniques and other -omic approaches and, in particular, is amenable to high sample throughput. The future is likely to see the development of a number of databases centered on this technology, and aimed at producing expert predictive computer systems to determine organ toxicity and to assess the mode of action of certain drugs.

The increasing availability and technological advances in high-throughput, high-resolution NMR and MS analytical platforms coupled with statistical pattern recognition tools has enabled metabonomic studies in a whole array of biomedical fields including drug and environmental toxicology, functional genomics, disease diagnosis and etiology, drug efficacy, and pharmacodynamics. As such, with further validation and exploration, the role of metabonomics within pharmaceutical discovery and development will continue to expand as part of the postgenomic systems biology era.

REFERENCES

1. Raamsdonk, L.M., Teusink, B., Broadhurst, D., Zhang, N., Hayes, A., Walsh, M.C., Berden, J.A., Brindle, K.M., Kell, D.B., Rowland, J.J., Westerhoff, H.V., van Dam, K., and Oliver, S.G. A functional genomics strategy that uses metabolome data to reveal the phenotype of silent mutations. *Nat. Biotechnol.* 19(1), 45–50, 2001.
2. Fiehn, O. Metabolomics — the link between genotypes and phenotypes. *Plant Mol. Biol.* 48, 155–171, 2002.
3. Nicholson, J.K., Connelly, J., Lindon, J.C., and Holmes, E. Metabonomics: a platform for studying drug toxicity and gene function. *Nat. Rev. Drug Discovery* 1, 153 –161, 2002.
4. Oliver, S.G. Functional genomics: lessons from yeast. *Philos. Trans. R. Soc. Lond. B* 357, 17–23, 2002.
5. Harrigan, G.C. and Goodacre, R., Eds. *Metabolic Profiling, Its Role in Biomarker Discovery and Gene Function Analysis*. Kluwer Academic, Amsterdam, 2003, 83–94.
6. Fiehn, O. Combining genomics, metabolome analysis and biochemical modeling to understand metabolic networks. *Comp. Function Genomics* 2, 155–168, 2001.
7. Stitt, M. and Fernie, A.R. From measurements of metabolites to metabolomics: an "on the fly" perspective illustrated by recent studies of carbon-nitrogen interactions. *Curr. Opin. Biotechnol.* 14, 136–144, 2003.

8. Plumb, R., Granger, J., Stumpf, C., Wilson, I.D., Evans, J.A., and Lenz, E.M. Metabonomic analysis of mouse urine by liquid-chromatography-time of flight mass spectrometry (LC-TOFMS): detection of strain, diurnal and gender differences. *Analyst* 128, 819–823, 2003.
9. Allen, J., Davey, H.M., Broadhurst, D., Heald, J.K., Rowland, J.J., Oliver, S.G., and Kell, D.B. High-throughput classification of yeast mutants for functional genomics using metabolic footprinting. *Nat. Biotech*, 2003, Advance On-line Publication.
10. Keun, H.C., Beckonert, O., Griffin, J.L., Richter, C., Moskau, D., Lindon, J.C., and Nicholson, J.K. Crogenic probe 13C NMR spectroscopy of urine for metabonomic studies. *Anal. Chem.* 74, 4588–4593, 2002.
11. Griffin, J.L., Walker, L.A., Shore, R.F., and Nicholson, J.K. *Chem. Res. Toxicol.* 14(10), 1428–1434, 2001.
12. Griffin, J.L., Walker, L.A., Troke, J., Osborn, D., Shore, R.F., and Nicholson, J.K. *FEBS Lett.* 478(1–2), 147–50, 2000.
13. Griffin, J.L., Walker, L., Shore, R.F., and Nicholson, J.K. *Xenobiotica.* 31(6), 377–385, 2001.
14. Griffin, J.L., Nicholls, A.W., Keun, H.C., Mortishire-Smith, R.J., Nicholson, J.K., and Kuehn, T. Metabolic profiling of rodent biological fluids via 1H NMR spectroscopy using a 1mm Microlitre probe. *Analyst* 127(5), 582–584, 2002.
15. Griffin, J.L., Bonney, S.A., Mann, C., Hebbachi, A.M., Gibbons, G.F., Nicholson, J.K., Shoulders, C.C., and Scott, J. *Physiol Genomics,* Epub ahead of print, Jan 27, 2004.
16. Griffin, J.L., Williams, H.J., Sang, E., and Nicholson, J.K. Abnormal lipid profile of dystrophic cardiac tissue as demonstrated by one- and two-dimensional magic-angle spinning (1)H NMR spectroscopy. *Magn. Reson. Med.* 46, 249–255, 2001.
17. Brindle, J.T., Antti, H., Holmes, E., Tranter, G., Nicholson, J.K., Bethell, H.W., Clarke, S., Schofield, P.M., McKilligin, E., Mosedale, D.E., and Grainger, D.J. *Nat. Med.* 8(12), 1439–1444, 2002.
18. Brindle, J.T., Nicholson, J.K., Schofield, P.M., Grainger, D.J., and Holmes, E. *Analyst*, 128(1), 32–36, 2003.
19. Lindon, J.C., Holmes, E., and Nicholson, J.K. Pattern recognition methods and applications in biomedical magnetic resonance. *Prog. Nucl. Magn. Reson.* 39, 1–40, 2001.
20. Beckwith-Hall, B.M., Nicholson, J.K., Nicholls, A.W., Foxall, P.J., Lindon, J.C., Connor, S.C., Abdi, M., Connelly, J., and Holmes, E. *Chem. Res. Toxicol.* 11(4), 260–272, 1998.
21. Nicholls, A.W., Nicholson, J.K., Haselden, J.N., and Waterfield, C.J. *Biomarkers* 5, 410–423, 2000.
22. Espina, J.R., Shockcor, J.P., Herron, W.J., Car, B.D., Contel, N.R., Ciaccio, P.J., Lindon, J.C., Holmes, E., and Nicholson, J.K. *Magn. Reson. Chem.* 39(9), 559–565, 2001.
23. Lindon, J.C., Nicholson, J.K., Holmes, E., Antii, H., Bollard, M.E., Keun, H., Beckonert, O., Ebbels, T.M., Reily, M.D., Robertson, D. et al. The role of metabonomics in toxicology and its evaluation by the COMET project. 2003, *Toxicol. Appl. Pharmacol.* 187, 137–146, 2003.
24. Lindon, J.C., Holmes, E., and Nicholson, J.K. *Expert Rev. Mol. Diagn.* 4(2), 189–199, 2004.
25. Valafar, F. Pattern recognition techniques in microarray data analysis. *Ann. N.Y. Acad. Sci.* 980, 41–64, 2002.

26. Eriksson, L., Johansson, E., Kettaneh-Wold, N., and Wold, S. *Introduction to Multi- and Megavariate Data Analysis Using Projection Methods (PCA & PLS).* Umetrics, Umea, Sweden, 1999.
27. Kell, D.B. Metabolomics and machine learning: explanatory analysis of complex metabolome data using genetic programming to produce simple, robust rules. *Mol. Biol. Rep.* 2002, 29(1–2), 237–241.
28. Wold, S., Antti, H., Lingren, F., and Ohman, J. Orthogonal signal correction of near-infrared spectra. *Chemometrics Intell. Lab. Syst.* 44, 175–185, 1998.
29. Ebbels, T.M.D., Keun, H.C., Beckonert, O., Antti, H., Bollard, M.E., Holmes, E., Lindon, J.C., and Nicholson, J.K. *Anal. Chim. Acta* 490, 109–122, 2003.
30. Beckonert, O., Bollard, M.E., Ebbels, T.M.D., Keun, H.C., Antti, H., Holmes, E., Lindon, J.C., and Nicholson, J.K. *Anal. Chim. Acta* 490, 3–15, 2003.
31. Fiehn, O., Kopka, J., Dormann, P., Altmann, T., Trethewey, RN., and Willmitzer, L. Metabolite profiling for plant functional genomics. *Nat. Biotech.* 18, 1157–1161, 2000.
32. Lenz, E.M., Bright, J., Knight, R., Wilson, I.D., and Major, H. Cyclosporin A-induced changes in endogenous metabolites in rat urine: a metabonomic investigation using high field 1H NMR spectroscopy, HPLC-TOF/MS and chemometrics. *J. Pharm. Biomed. Anal.* 35, 599–608, 2004.
33. Lenz, E.M., Bright, J., Knight, R., Wilson, I.D., and Major, H. A metabonomic investigation of the biochemical effects of mercuric chloride in the rat using 1H NMR and HPLC-TOF/MS: time dependent changes in the urinary profile of endogenous metabolites as a result of nephrotoxicity. *Analyst* 129, 535–541, 2004.
34. Plumb, R.S., Stumpf, C.L., Gorenstein, M.V., Castro-Perez, J.M., Dear, G.J., Anthony, M., Sweatman, B.C., Connor, S.C., and Haselden, J.C. Metabonomics: the use of electrospray mass spectrometry coupled to reversed-phase liquid chromatography shows potential for the screening of rat urine in drug development. *Rapid Commun. Mass Spectrum.* 16(20), 1991–1996, 2002.
35. Ringeissen, S., Connor, S.C., Brown, H.R., Sweatman, B.C., Hodson, M., Kenny, S.P., Haworth, R.I., McGill, P., Price, M.A., Aylott, M.C., Nunez, D.J., Haselden, J.N., and Waterfield, C.J. *Biomarkers* 8, 240–271, 2004.
36. Coen, M., Lenz, E.M., Nicholson, J.K., Wilson, I.D., Pognan, F., and Lindon, J.C. An integrated metabonomic investigation of acetaminophen toxicity in the mouse using NMR spectroscopy. 2003 *Chem. Res. Toxicol.* 16, 295–303, 2003.
37. Styles, P., Soffe, N.F., Scott, C.A., Cragg, D.A., Row, F., White, D.J., and White, P.C.J. A high resolution NMR probe in which the coil and preamplifier are cooled with liquid helium. *J. Magn Reson.* 60, 397–404, 1984.
38. Andrew, E.R., Bradbury, A., and Eades, R.G. Removal of dipolar broadening of NMR spectra of solids by specimen rotation. *Nature* 183, 1802, 1959.
39. Cheng, L.L., Lean, C.L., Bogdanova, A., Wright, S.C., Ackerman, J.L., Brady, T.J., and Garrido, L. Enhanced resolution of proton NMR spectra of malignant lymph nodes using magic angle spinning. *Magn. Reson. Med.* 36, 653–658, 1996.
40. Wind, R.A., Zhi Hu, J., and Rommereim, D.N. High-resolution (1)H NMR spectroscopy in organs and tissues using slow magic angle spinning. *Magn. Reson. Med.* 46, 213–218, 2001.
41. Weybright, P., Millis, K., Campbell, N., Cory, D.G., and Singer, S. Gradient, high-resolution, magic angle spinning 1H nuclear magnetic resonance spectroscopy of intact cells. *Magn. Reson. Med.* 39, 337–344, 1998.

42. Griffin, J.L., Bollard, M.E., Nicholson, J.K., and Bhakoo, K. Metabolic profiles of intact cultured neuronal and glial cells derived from HRMAS ^{1}H NMR spectroscopy. *NMR Biomed.* 15, 375–384, 2002.
43. Waters, N.J., Holmes, E., Waterfield, C.J., Farrant, R.D., and Nicholson, J.K. *Biochem. Pharmacol.* 64(1), 67–77, 2002.
44. Garrod, S., Humpfer, E., Spraul, M., Connor, S.C., Polley, S., Connelly, J., Lindon, J.C., Nicholson, J.K., and Holmes, E. *Magn. Reson. Med.* 41(6), 1108–1118, 1999.
45. Garrod, S., Humpher, E., Connor, S.C., Connelly, J.C., Spraul, M., Nicholson, J.K., and Holmes, E. *Magn. Reson. Med.* 45(5), 781–790, 2001.
46. Waters, N.J., Holmes, E., Williams, A., Waterfield, C.J., Farrant, R.D., and Nicholson, J.K. *Chem. Res. Toxicol.* 2001 14(10), 1401–1412, 2001.
47. Mortishire-Smith, R.J., Skiles, G.L., Lawrence, J.W., Spence, S., Nicholls, A.W., Johnson, B.A., and Nicholson, J.K. *Chem. Res. Toxicol.* 17(2), 165–173, 2004.
48. Nicholson, J.K., Higham, D.P., Timbrell, J.A., and Sadler, P.J. *Mol. Pharmacol.* 36(3), 398–404, 1989.
49. Waters, N.J., Waterfield, C.J., Farrant, R.D., Holmes, E., and Nicholson, J.K., Metabonomic deconvolution of embedded toxicity: application to thioacetamide hepato- and nephro-toxicity. *Chem. Res. Toxicol.* 18(4), 639–54, 2005.
50. Preul, M.C., Caramanos, Z., Leblanc, R., Villemure, J.G., and Arnold, D.L. Using pattern analysis of in vivo proton MRSI data to improve the diagnosis and surgical management of patients with brain tumors. *NMR Biomed.* 11(4–5), 192–200, 1998.
51. Hagberg, G. From magnetic resonance spectroscopy to classification of tumors. A review of pattern recognition methods. *NMR Biomed.* 11(4–5), 148–156, 1998.
52. Gerstle, R.J., Aylward, S.R., Kromhout-Schiro, S., and Mukherji, S.K. The role of neural networks in improving the accuracy of MR spectroscopy for the diagnosis of head and neck squamous cell carcinoma. *Am. J. Neuroradiol.* 21(6), 1133–1138, 2000.
53. Gray, H.F., Maxwell, R.J., Martinez-Perez, I., Arus, C., and Cerdan, S. Genetic programming for classification and feature selection: analysis of 1H nuclear magnetic resonance spectra from human brain tumour biopsies. *NMR Biomed.* 11(4–5), 217–224, 1998.
54. Underwood, J., Tate, A.R., Luckin, R., Majos, C., Capdevila, A., Howe, F., Griffiths, J. and Arus, C. A prototype decision support system for MR spectroscopy-assisted diagnosis of brain tumours. *Medinfo* 10(Pt. 1), 561–565, 2001.
55. Griffin, J.L., Pole, J.C., Nicholson, J.K., and Carmichael, P.L. Cellular environment of metabolites and a metabonomic study of tamoxifen in endometrial cells using gradient high resolution magic angle spinning 1H NMR spectroscopy. *Biochim. Biophys. Acta* 1619(2), 151–158, 2003.
56. Chung, Y.-L, Troy, H., Banerji, U. et al. The pharmacodynamic effect of 17-AAG on HT29 xenografts in mice monitored by magnetic resonance spectroscopy. *Proc. Am. Assoc. Cancer Res.* 43, 73, 2002.
57. Chung, Y.-L, Troy, H., Banerji, U., Jackson, L.E., Walton, M.I., Stubbs, M., Griffiths, J.R., Judson, I.R., Leach, M.O., Workman, P., and Ronen, S.M. Magnetic resonance spectroscopic pharmacodynamic markers of Hsp90 inhibitor, 17-allylamino-17-demethoxygeldanamycin, in human colon cancer models. *J. Natl. Cancer Inst.* 95, 1624–1633, 2003.
58. Griffin, J.L., Lehtimaki, K.K., Valonen, P.K., Grohn, O.H., Kettunen, M.I., Yla-Herttuala, S., Pitkanen, A., Nicholson, J.K., and Kauppinen, R.A. Assignment of 1H nuclear magnetic resonance visible polyunsaturated fatty acids in BT4C gliomas undergoing ganciclovir-thymidine kinase gene therapy-induced programmed cell death. *Cancer Res.* 63(12), 3195–3201, 2003.

59. Sitter, B., Bathen, T., Hagen, B., Arentz, C., Skjeldestad, F.E., and Gribbestad, I.S. Cervical cancer tissue characterized by high-resolution magic angle spinning MR spectroscopy. *MAGMA* 16(4), 174–181, 2004.
60. Cheng, L.L., Chang, I.W., Smith, B.L., and Gonzalez, R.G. Evaluating human breast ductal carcinomas with high-resolution magic-angle spinning proton magnetic resonance spectroscopy. *J. Magn. Reson.* 135(1), 194–202, 1998.
61. Chen, J.-H., Enloe, B.M., Fletcher, C.D., Cory, D.G., and Singer, S. Biochemical analysis using high-resolution magic angle spinning NMR spectroscopy distinguishes lipoma-like well-differentiated liposarcoma from normal fat. *J. Am. Chem. Soc.* 123(37), 9200–9201, 2001.
62. Millis, K., Weybright, P., Cambell, N., Fletcher, J.A., Fletcher, C.D., Cory, D.G., and Singer, S. Classification of human liposarcoma and lipoma using ex vivo proton NMR spectroscopy. *Magn. Reson. Med.* 41, 257–267, 1999.
63. Tomlins, A., Foxall, P.J.D., Lindon, J.C., Lynch, M.J., Spraul, M., Everett, J., and Nicholson, J.K. High resolution magic angle spinning ^{1}H nuclear magnetic resonance analysis of intact prostatic hyperplastic and tumour tissues. *Anal. Commun.* 35, 113–115, 1998.
64. Oliver, S.G., Winson, M.K., Kell, D.B., and Baganz, F. Systematic functional analysis of the yeast genome. *Trends Biotechnol.* 16, 373–378, 1998.
65. Bochner, B.R., Gadzinski, P., and Panomitros, E. Phenotype microarrays for high-throughput phenotypic testing and assay of gene function. *Genome Res.* 11, 1246–1255, 2001.
66. Hanlon, E.B., Manoharan, R., Koo, T.-W. et al. Prospects for *in vivo* Raman spectroscopy. *Phys. Med. Biol.* 45, R1–R59, 2000.
67. Tweeddale, H., Notley-McRobb, L., and Ferenci, T. Effect of slow growth on metabolism of Escherichia coli, as revealed by global metabolite pool ("metabolome") analysis. *J. Bacteriol.* 180, 5109–5116, 1998.

CHAPTER 11

Comprehensive Metabolomic Profiling of Serum and Cerebrospinal Fluid: Understanding Disease, Human Variability, and Toxicity

Shawn Ritchie

CONTENTS

11.1 INTRODUCTION

Sequencing of the human genome represents a monumental achievement and has provided new insights into the underlying genetic mechanisms of many diseases.[1,2] Subsequently, this effort has spawned new scientific endeavors collectively referred to as functional genomics, intended to delineate gene function and ultimately expand our understanding of gene–activity relationships. Functional genomics subdisciplines can include transgenics, RNAi, proteomics, and metabolomics. Relative to the genome, the proteome and metabolome comprise a vast and diverse spectrum of molecular structures and events, influenced not only by genetic predisposition, but by factors such as environment, nutrition, and lifestyle. Studies in the field of proteomics aim to catalog and understand protein networks; research in "metabolomics" aims to do the same for the "metabolome," the comprehensive small molecule composition of a biological sample at any given state or time. Using a systems biology approach to understand the relationships between genes, proteins, and metabolites, and how their interactions contribute to human health is, and will continue to be, a major focal point for researchers in the 21st century.

Compared to genomics methods such as DNA sequencing and transcriptomics, characterizing a metabolome has inherent obstacles that stem from the fact that phenotype is not a finite entity and can be difficult to quantify objectively. In large part, this difficulty can be attributed to the chemical complexity of metabolites, technical hurdles for simultaneously capturing and assaying every metabolite present in a sample, and our inadequate understanding of the metabolic milieu. Metabolite pools are functionally dynamic entities, which represent a given state of enzyme activity within a cell. Enzymes are subject to regulation via a myriad of factors such as hereditary or sporadically acquired mutations, gene and protein expression level, and post-translational status. Regulation of enzyme activities through these mechanisms will result in new equilibrium states and altered abundances of metabolic substrates and products. Therefore, capturing the relative abundances of metabolite pools can represent a direct measurement of the end product of a biological event. Such events can be stimulated by environmental or endogenous biological phenomena such as drug exposure or cancer, resulting in a cascade of signaling events, e.g., activation of the MAP kinase cascade. These biological events can sometimes result in altered gene expression and consequently a change in enzyme regulation. The target gene upregulated by a signaling pathway, for example, may be the enzyme itself or a kinase that can post-translationally modify the enzyme responsible for the metabolite pool. In either case, the perturbation, at least of enzyme activities inside a cell, is expected to ultimately manifest as a change in endogenous metabolite abundances and establishment of new equilibriums. (*Note:* For clarification, the term "expression" should be solely reserved for either the production of mRNA from DNA, or protein from mRNA. Metabolites are not expressed, but rather produced enzymatically, and should therefore be referred to as having a given level of intensity or abundance.) The goal of metabolomics, therefore, is to capture such end points and use the knowledge for identifying metabolite biomarkers for disease indices, drug efficacy, toxicity, nutrition, patient stratification, etc.

Metabolomics, in theory, is capable of accurately describing a phenotype with a high degree of sensitivity and reproducibility. Conceptually, therefore, metabolomics is ideally suited for many of the same biomarker discovery applications as transcriptomics and proteomics. A small-molecule biomarker can have several advantages over transcripts or whole proteins, including sensitivity, reproducibility, deployment of high-throughput screening (HTS) strategies, and straightforward identification of discriminatory molecules. Regardless of the type of metabolomics strategy employed (expanded on in the following section), identification of metabolite structures is relatively straightforward, even for novel molecules. This is a significant advantage over proteomics-based profiling methods. For example, surface-enhanced laser desorption ionization (SELDI) mass spectrometry (MS) can generate disease-specific patterns but offers no identification of any of the resulting spectral peaks unless off-line methods are employed. A diagnostic biomarker that is not identified can be useful in a correlative sense but is superseded by markers with known identities because they point directly to the pathway involved in a particular disease etiology. The knowledge of such pathways can then be exploited for the selection of drugable targets, for example, by screening a combinatorial library of compounds in a high-throughput fashion for pathway-specific biological activity.

In addition to the identification of diagnostic markers, metabolomics can aid in drug development by identifying markers of drug efficacy and toxicity. As discussed in detail later in the chapter, metabolomics strategies can be employed early in the discovery and preclinical phases of drug discovery, where they can complement or even substitute for conventional clinical chemistry measurements. Identification of metabolic signatures correlating with adverse reactions and toxicity can be used to optimize strategies for lead selection and enrollment of patients into clinical trials. Comparable approaches are applicable to late-phase clinical trials as well; however, metabolomics can offer an added benefit by stratifying patients *a priori* into cohorts that are most likely to respond to a given therapy. Such methods represent the first steps toward the realization of personalized medical treatments.

Metabolomics also holds promise for improving human health by helping researchers and clinicians to understand how dietary factors, genetic predisposition data, disease, and miscellaneous lifestyle-associated variables affect metabolite networks. For example, it may be possible using knowledge gained from the integration of metabolic data with other lifestyle-associated meta-data to stratify individuals into various health state categories associated with higher susceptibilities to certain diseases. Depending on the disease and the susceptibility, protocols ranging from simple dietary or exercise interventions to more complex longitudinal screening programs could then be implemented to help individuals return to a more "healthy" or minimal susceptibility phenotype. The assimilation of metabolic, genetic, and lifestyle-associated data at a systems biology level will ultimately lead to an enhanced quality of life and improved longevity of the human population. The remainder of this chapter briefly reviews current metabolomics analytical methods and shows examples of typical serum and cerebrospinal fluid (CSF) comprehensive metabolomic data. The issue of human variability when using surrogate tissues is also addressed. The chapter concludes with a discussion on the utility of metabolomics for pharmaceutical and toxicological applications.

11.2 ANALYTICAL METHODOLOGIES

The study of metabolomics can be subclassified into two genres depending on the objectives of the research project and the analytical platform available. In the so-called "targeted" (also known as "closed") approach, methods for assaying a preselected list of known metabolites are defined and quantified across a number of biological samples. Targeted metabolite profiling has, in essence, existed for decades and can represent an assay as simple as a glucose test. Although clearly useful in certain circumstances, the primary disadvantage of a targeted method is that knowledge of the targets of interest is required *a priori* and is therefore not capable of providing novel metabolite discoveries. "Nontargeted" (otherwise known as "open" or "comprehensive") approaches are intended to identify as many metabolites as possible in a biological sample, without any prior selection of a metabolite panel. The advantages of such an approach outweigh targeted systems for many applications, particularly discovery of novel metabolites and metabolic pathways. Given our relatively limited understanding of metabolism and the general lack of success in translating transcriptomic and proteomic data into clinically valuable discoveries, the capacity for comprehensive metabolomics to contribute to our understanding and manipulation of diseases is promising. Furthermore, comprehensive metabolomics has the potential to characterize large numbers of novel metabolites that targeted platforms are incapable of identifying. By convention, the term metabolomics should be reserved for those analyses comprising, at least in part, a nontargeted component for identifying small molecules.[3]

Several of the same analytical tools can be used to perform both targeted and nontargeted metabolomic analysis. The two most commonly used technologies are MS and nuclear magnetic resistance (NMR). There are many in-depth reviews available on these topics[4–8]; therefore, the objective of the following section is to touch briefly on the principles of MS and NMR in relation to the field of comprehensive metabolomics.

MS is based on the principle that molecules can be ionized into charged particles, which can subsequently be detected and the data used to determine the mass of and structural data about the ion. There are several methods of generating ionized molecules; the most commonly used for metabolomics include electrospray ionization (ESI), atmospheric pressure chemical ionization (APCI), and electron impact (EI). Selected ionization methods are summarized in Table 11.1. Once ions have

Table 11.1 Summary of Some Common Ionization Methods

Ionization Method	Description
Electrospray ionization (ESI)	Ions are produced by the evaporation of charged droplets that are generated by forcing a liquid-based analyte through a fine needle, to which a voltage is applied that results in uniformly charged droplets. Evaporation of the solvent causes the droplets to shrink and break into smaller droplets until only individually charged ions are left.
Martix-assisted laser desorption ionization (MALDI)	Analytes (traditionally peptides) are combined with a matrix that absorbs ultraviolet light. Ions are produced following desorption of the sample with an ultraviolet laser under low pressure.
Atmospheric pressure chemical ionization (APCI)	Analyte is nebulized with a co-axial flow of nitrogen and heat resulting in gas-phase species. Ions are generated by a corona discharge, which creates reagent ions from the solvent vapor with minimal fragmentation.
Electron Impact (EI)	Analyte is delivered as gas through heated vaporization, and passed through an electron beam resulting in election ejection and fragmentation.

been generated, there are a number of methods available to analyze the ionized species; examples are time-of-flight (TOF), quadrupole, and cyclotron resonance. Table 11.2 summarizes the principles of some common mass detectors.

The capability of certain mass spectrometers can be greatly enhanced if the analyte components can be separated prior to ionization. Coupling gas or liquid chromatography with MS can achieve this goal by improving sensitivity and resolution. Fourier transform mass spectrometry (FTMS), however, has a resolving power high enough (>300,000) that individual components of complex biological mixtures can be precisely identified and quantified without any prior separation.[9] The ability to analyze samples using FTMS in the absence of chromatography reduces the sample processing time and eliminates issues associated with retention time, such as dynamic matrix suppression effects and precise alignment of spectra. FTMS also offers the advantage that because of the high resolution, spectra can be aligned with accuracy well below one part per million (ppm), in which case molecular formulas (and therefore putative identities) can be computationally assigned to every peak.[9] These attributes make FTMS the preferred analytical tool for comprehensive metabolomics. Examples of FTMS-based data are presented later in the chapter.

NMR spectroscopy is based on the principle that hydrogen and carbon-13 atoms contain nuclei that can absorb energy when subjected to electromagnetic radiation in a strong magnetic field (through a process called magnetic resonance). Sweeping the magnetic field strength or electromagnetic radiation will produce a series of frequencies that can then be amplified and displayed as a series of signals. Although NMR can provide valuable structural information on a wide spectrum of metabolites, the sensitivity is significantly lower than MS, restricting it primarily to the analysis of the most abundant metabolites within samples. Furthermore, data acquisition time can be significantly longer than MS, limiting throughput to fewer than 10 samples per day. Analysis of intact tissues or whole surrogate tissues with NMR is possible; however, interpretation of the resulting spectra can be very complicated, even with the best computational tools available. Ideally, NMR is best used as a complement to MS-based comprehensive metabolomics where structural verification is required.

Table 11.2 Summary of Some Common Mass Analyzers

Detection Method	Description
Time-of-flight	Ions are accelerated through a vacuum tube to a detection plate located at a fixed distance. The speed of ion movement down the flight tube is proportional to its mass; ions with large masses travel more slowly than smaller masses, making the distinction between ions possible.
Quadropole	Ions are detected by scanning a radiofrequency for each mass across four parallel rods and measuring the number of ions passing through at each frequency.
Cyclotron resonance	Ions are detected by measuring an image current produced as result of their cyclotron motion in the presence of a magnetic field. Fourier transformation of the image current results in the mass-to-charge ratios of component frequencies.

11.3 SEARCHING FOR BIOMARKERS IN A SEA OF HUMAN VARIABILITY

The ease of obtaining biofluids such as serum, urine, and even CSF makes them the preferred choice for diagnostic applications. Identification of surrogate biomarkers for diseases in these tissues has been the focus of researchers for decades. However, with the advent of omics-based technologies, there has been enormous attention given to surrogate tissues for biomarker discovery. It was anticipated that gene chip-based approaches would identify expression patterns or particular genes involved in specific disease processes, and that such findings would translate into clinically applicable diagnostic markers; however, this goal has yet to be fully realized.

Significantly higher expectations have been placed on proteomic methods for the identification of serum peptide biomarkers. Although MS-based profiling of serum proteins has been shown to discriminate between normal and disease states, there is ongoing debate whether tumor proteins and fragments thereof are capable of entering the circulation, and whether such fragments would exist at concentrations high enough that they are detectable in the milieu of albumin and other serum proteins.[10] There are also technical issues that can influence the validity and reproducibility of serum proteomic profiles, such as sample collection and storage conditions, freeze–thaw cycles, protein chip manufacturer deviations, and overall experimental design. Such parameters can result in increased external noise, which has been shown to contribute to poor reproducibility, for example, among separate SELDI data sets produced from the same serum of patients with ovarian cancer.[11] Significant increases in sensitivity, reproducibility, and improved detection of post-translational modifications are required for proteomics to yield highly reproducible biomarkers suitable for clinical diagnostic applications.

Metabolomics-based analysis of serum and other surrogate biofluids offers certain unique advantages over transcriptomics- and proteomics-based approaches. For example, there is a higher possibility that small molecules can either be transported or diffuse across cell membranes and into the bloodstream more readily than transcripts, whole proteins, or peptides. Since metabolite levels are a direct reflection of enzymatic activity, using metabolites as biomarkers should, in theory, have higher levels of specificity and more accurately reflect the true physiological state of a cell or organism than transcript or protein profiles. Proteomics is further complicated by the fact that proteins, even if overexpressed, are subject to a myriad of regulatory processes (post-translations modifications, drug interactions, environmental carcinogens, etc.), which may or may not be important in a biological context, or produce phenotypic changes. Metabolites are not subject to these processes, since any chemical modification of a metabolite results in a new molecule, which usually can be detected as efficiently as the parent molecule. Furthermore, metabolomics offers the advantage of having the capacity to identify drug-related metabolites, nutritional metabolic markers, environmental toxins, and other exogenous entities. From a technical perspective, the removal of protein from serum prior to metabolite analysis circumvents the proteomic issues surrounding albumin and low-abundance peptides.

Metabolites carried on albumin and other carrier proteins can be released by various protein denaturation methods, and isolated for analysis.

Independent of the -omics strategy employed or the analytical platform used, there is a fundamental issue that has surprisingly been given little attention, but which needs to be critically evaluated for the successful interpretation of any human -omics data: human biological variability. As biological entities, humans and other species comprise complex biochemical networks that can exhibit high degrees of variability as a result of homeostatic oscillations associated with interactions between the organism's genome and its environment. Age, health, gender, diet, genetics, fitness level, weight, geography, race, circadian status, drug exposure, and even psychosomatic state are some of the factors that can contribute to human biological variability. A significant effort has been exerted in trying to understand the genetic components of human variability, as is evident by the characterization and creation of comprehensive single-nucleotide polymorphism (SNP) databases.[12,13] There have been significantly fewer achievements in understanding and characterizing human phenotypic variability, due primarily to the lack of capable technologies. It is anticipated that comprehensive metabolomics technologies will provide the means necessary to begin human phenotyping at the biochemical level.

11.4 COMPREHENSIVE METABOLOMIC PROFILE ANALYSIS OF HUMAN SERUM

The analysis of human serum has traditionally involved very limited measurements of well-known metabolites and proteins, for example, glucose, cholesterol, albumin, and alkaline phosphatase. Recently, there has been an intense application of proteomics-based technologies toward the identification of disease-specific proteins in serum (for review, see References 14 through 17). Comprehensive metabolomic analysis of serum remains relatively uncharted territory, with little if any data available on the true small molecule "untargeted" composition of serum. Several private sector entities have begun to measure serum metabolites in the search for disease-related biomarkers, but in most cases such information cannot be put into the public domain until patents are filed detailing the utility of such discoveries. However, as the field of metabolomics continues to advance, reports of serum metabolite profiles from both industry and academia will begin to emerge in the public domain.

Analysis of both normal and disease human serum using FTMS at Phenomenome Discoveries, Inc., has provided several important insights into the use of serum as a surrogate tissue for biomarker identification, independent of health status or experimental objectives. Aside from using metabolomics for identifying metabolite biomarkers, a key finding is the significant level of variability that exists between individuals, and even between multiple samples from the same individual. Although blood has traditionally been thought of as a homeostatic tissue incapable of tolerating large changes in concentrations of metabolites (such as glucose), in fact this is not the case for many serum molecules. For example, significantly different metabolite profiles were observed in two separate serum samples collected 3 months apart from healthy male individuals between the ages of 25 and 35. Protocols were in place to ensure consistent sample collections, and that the individuals fasted systematically prior to sampling. Overall, nearly 70,000 data points representing metabolite intensities for all samples were detected in 40 samples from 20 individuals. This corresponds to approximately 1500 common metabolites measured across the 40 samples. When matched against a database of more than half a million compounds, including common metabolic pathway intermediates, approximately 10% of these metabolites matched to previously identified molecules. The remaining molecules represent novel metabolites, or novel chemical transformations of known metabolites. Thus, there is an enormous untapped resource of unidentified molecules present in the serum that remains to be further identified and characterized.

One method of visualizing metabolomics data is in a metabolite array format, which is similar to a gene chip cDNA microarray. For example, a partial metabolite array of the serum metabolomes for the 20 male individuals tested is shown in Figure 11.1. It is evident from this figure that not all columns show the same pattern, and that certain columns differ quite dramatically from others. Other regions of the array, in contrast, appear quite consistent across all tested individuals. A closer examination of selected regions of the array, as expanded in Figure 11.2A, shows metabolites that have minimal changes between visits of the same individual, as well as minimal changes between individuals. There are also several distinct regions of Figure 11.1 (for example, the region expanded in Figure 11.2B), which show dramatic differences between two temporally separated samples from the same person, as well as between different individuals. For example, individual 1 is positive for a subset of metabolites at both collection times, although a slight decrease in intensity in some of the metabolites is apparent by the second collection. In contrast, individuals 2 and 4 are negative for this cluster of metabolites at both collections. Interestingly, individuals 3 and 6 are negative for the same metabolites on the first collection, but then show a strong positive profile on the second collection. Individual 5 appears to display the opposite pattern, showing elevated metabolite levels on the first collection (although not to the same degree as the second collection of patient 6), and absent levels of the same metabolites on the second collection. Although samples taken from different individuals would be expected to show some variability, significant levels of variability between two samples taken as consistently as possible from the same individual within a couple of months, at first glance, might seem surprising. The question is whether or not this is really so surprising. It is not irrational to suggest that the existence of such variability could occur even within hours or minutes. As previously stated, there are numerous factors that can contribute to such variability, including nutritional status, acute drug usage, etc. Since the individuals in this example were healthy male subjects, these data represent an approximation of the natural variability that exists between individuals, as well as the variability that can materialize over 3 months within the same individual. Therefore, when using serum as a medium for pathology-associated biomarker discovery, it is important that such variability be considered in an appropriate manner. This may include very tight control of sample collection protocols, ensuring high participant compliance with the study design, or filtering of data to exclude metabolites that fluctuate independently of disease-related variables, such as dietary or chronic drug-related effects.

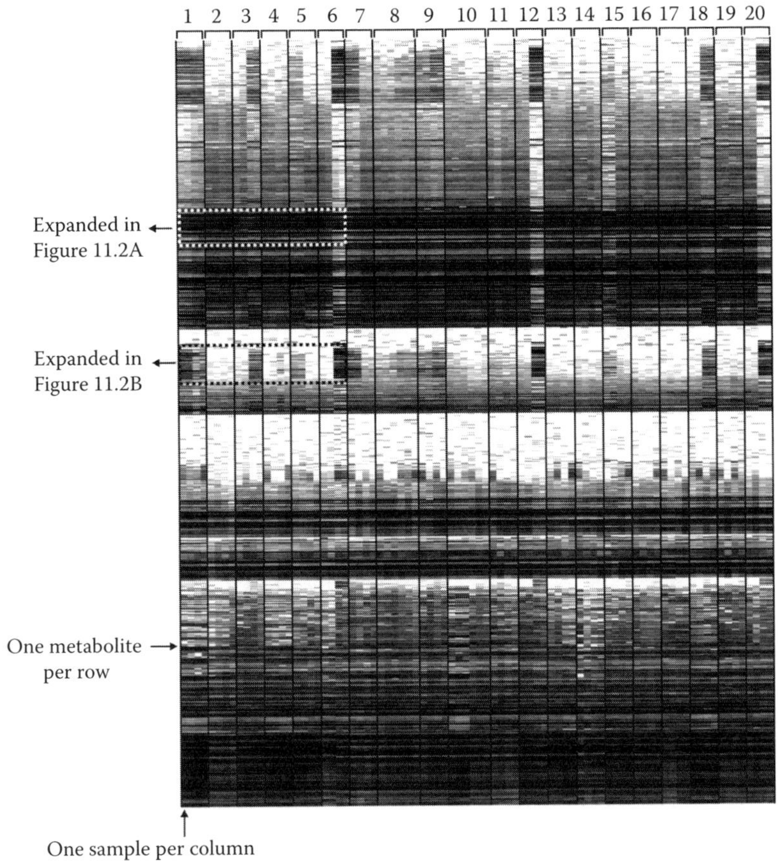

Figure 11.1 Metabolomic profiles of 40 serum samples from 20 individuals shown in an array format. The first two columns of each individual (and the first three columns for patients 8 and 10) represent duplicate (or triplicate) analysis of a baseline serum sample. The last two columns of each individual (last three columns for patients 8 and 10) represent duplicate (or triplicate) analysis of serum from the same individual 3 months later. Each row of the array represents a single metabolite. The regions highlighted by dotted boxes are expanded in Figure 11.2 (see text for explanations). Darker shades of gray represent metabolites with increased intensity.

Although biological variability among individual serum samples would traditionally be thought of as an impediment for biomarker discovery, knowledge of human biological variability in general can be highly valuable. For example, cluster analysis of the serum data suggests that the variability may not necessarily be random, and that such information could be used to stratify individuals into discrete subpopulations. This observation has been shown to hold true, even for very small studies. This is illustrated in Figure 11.3, which shows the metabolite array from Figure 11.1 clustered hierarchically by sample. As shown, there are three relatively distinct clusters of samples. Cluster 1 contains a group of samples which, for the most part, are absent for metabolites indicated by regions A and B, while cluster 2,

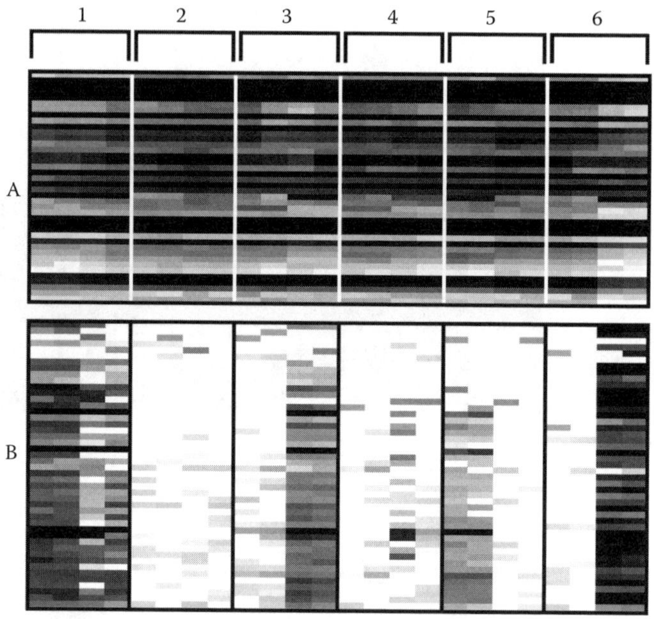

Figure 11.2 A subset of the array from Figure 11.1 for individuals 1 to 6. (A) Metabolites showing little deviation across the six individuals; (B) metabolites showing differential abundance between individuals and between collections from the same individual (see text for explanation). Darker shades of gray represent metabolites with increased intensity.

on the other hand, begins to show relatively low intensities for these metabolites. Cluster 3 is distantly related to clusters 1 and 2 and contains samples that have high, albeit still variable, levels of metabolites in clusters A and B. The fact that some of the temporally separated samples from the same individual fall within different clusters suggests that there has been an equilibrium shift of the individual to a state that resembles that of other individuals, and that this occurred within the time span between collections. These observations suggest common regulatory, compensatory, or other phenomena occurring within and between subpopulations of individuals. Such discoveries may be harnessed to stratify patients for the individualized improvement of personal health through the association of metabolic profiles with drug efficacy, disease susceptibility, or other phenomena.

11.5 COMPREHENSIVE METABOLOMIC PROFILE ANALYSIS OF HUMAN CEREBROSPINAL FLUID

CSF is found within the subarachnoid space that surrounds and protects the brain and spinal cord. In addition to physical support, the CSF functions to control excretory processes, transport metabolites within the intracerebral environment of the central nervous system (CNS), and regulate intracranial pressure. The composition of CSF is dependent on secretory processes as the fluid derives from the choroid

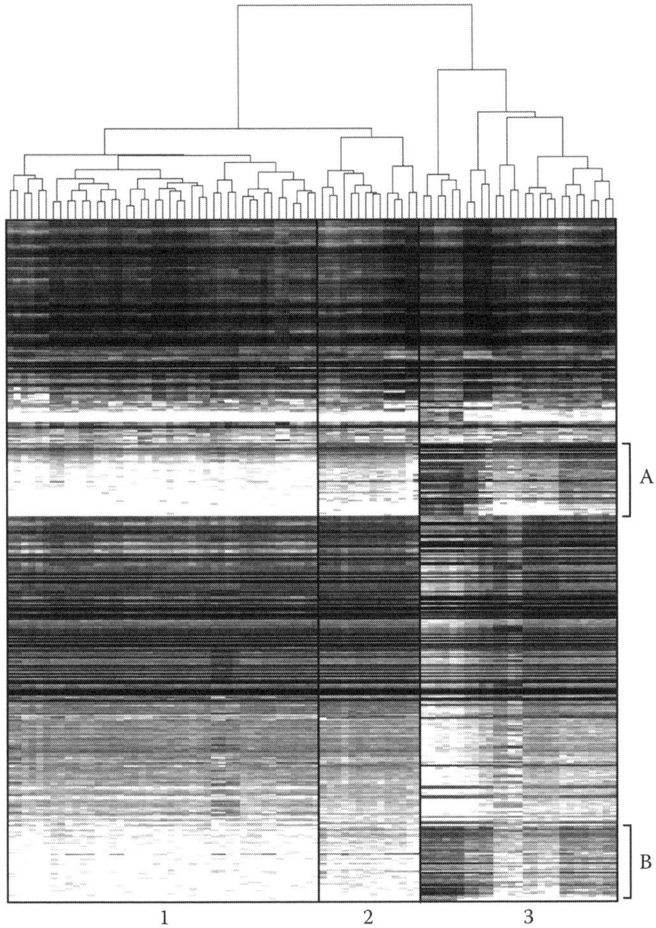

Figure 11.3 Metabolomic serum profiles of the 20 individuals (40 duplicate samples as shown in Figure 11.1) clustered hierarchically by sample using a Euclidean distance metric. Three resulting distinct clusters are labeled along the bottom. Regions A and B highlight examples of metabolites showing significant differences in intensity between the clusters (see text for detailed explanation). Darker shades of gray represent metabolites with increased intensity.

plexus, the ependymal lining of the ventricular system, and blood vessels in the pia-arachnoid. In addition, ultrafiltration of blood plasma and various transport mechanisms contributes to the composition of CSF. Analysis of CSF has been used as the gold standard for diagnosing many neurological and CNS disorders. CSF can only be collected following a lumbar puncture (spinal tap) and is therefore a significantly less accessible surrogate tissue. The complexity and discomfort of the procedure limit the diagnostic utility of CSF to individuals who have already begun to show disease-related clinical symptoms or have a strong genetic predisposition to a neurological disorder. In fact, a lumbar puncture should be carried out only after the analysis of serum and urine, and following careful clinical evaluation and assessment of neuroimaging results.[18] It is therefore unlikely to see widespread analysis of CSF for population-based epidemiological screening.

On the other hand, CSF may hold the key to understanding complex neurological disorders such as Alzheimer's, Creutzfeldt-Jacob, multiple sclerosis, and meningitis. Currently, there is a relatively standard set of CSF clinical parameters that is used for diagnostics. These parameters include supernatant color, cell counts, histological examination, total protein concentration, cell culture, latex agglutination, polymerase chain reaction (PCR), and measurements of certain metabolites.[18,19] For example, increased protein can be attributed to a decreased turnover of CSF (known as the CSF flow rate), which is often associated with disease onset. Decreased turnover can result in a nonspecific accumulation of molecules (both proteins and metabolites), and to date, total CSF protein is still claimed to be the most sensitive indicator of nonspecific CNS pathology.[19,20] However, the average adult CSF protein concentration can range anywhere between 18 and 58 mg/dl, and the average concentrations of protein in multiple sclerosis, epilepsy, and aseptic meningitis are 43, 31, and 77 mg/dl, respectively.[19] For certain conditions, such as bacterial infections, cerebral hemorrhaging, and select brain tumors, protein concentrations may surpass 100 mg/dl.[19] Since the average protein concentrations for many neurological conditions fall near or within the normal range, the diagnostic informativeness of total CSF protein is far from high.

In addition to protein concentration, several primary metabolites have been traditionally used to investigate CNS disorders, including specific amino acids, biogenic monoamines, GABA metabolites and other neurotransmitters, neuroendocrine substances, organic acids (such as glutaric acid), and methylation pathways. A detailed discussion about the utility of these for understanding CNS pathology is beyond the scope of this chapter, and can be found in Reference 18.

The application of nontargeted proteomics and metabolomics to the analyses of CSF affords an opportunity to identify and correlate, in an unbiased manner, biomolecules with specific neurological disorders. For example, SELDI-TOF has been used to identify statistically significant peptides in Alzheimer's disease[21] and MALDI-TOF for the identification of brain-tumor-related peptides.[22] However, because the concentration of protein in the CSF is very low and comprises mainly albumin derived from the blood, detection of pathology-specific peptides for many CNS disorders may not be feasible.

The advantages of comprehensive metabolomic analysis for serum also apply to the analysis of CSF. It can be speculated that since neurotransmitters and other small molecules can more readily cross the blood–brain barrier though passive diffusion compared to proteins, the potential for pathology-specific metabolite markers to exist in the CSF is theoretically higher than for proteins. Furthermore, lumbar CSF is only a distant representation of brain-related metabolic processes, which means that the detection of even minute perturbations in CSF metabolic composition could be highly informative with regard to CNS pathology.[18] Metabolomic analysis of CSF also offers the advantage of monitoring the efficiency with which patients respond to drug intervention by measuring drug uptake, metabolism, and toxicity.

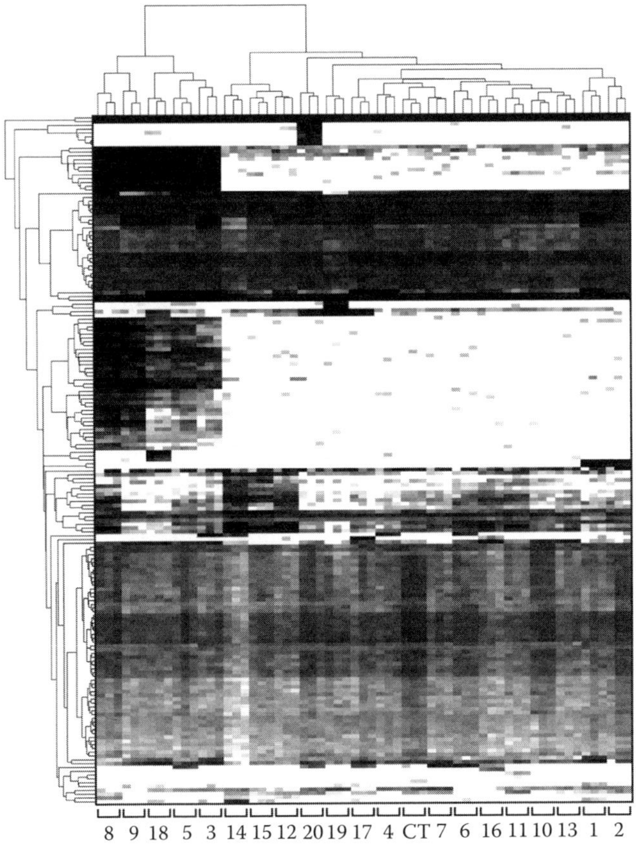

Figure 11.4 Metabolomic profiles of CSF from 21 patients afflicted with varying degrees of encephalopathy. The array is clustered hierarchically by sample (Euclidean distance metric) and by metabolite (Chebychev distance metric). Triplicate samples from patients 1 to 20 are indicated along the bottom. CT; control patient (no encephalopathy). Darker shades of gray represent metabolites with increased intensity.

FTMS-based comprehensive metabolomic analysis has been performed on a number of human CSF samples spanning a multitude of diseases. For example, Figure 11.4 shows a metabolite heat map of several hundred metabolites identified among 20 patients displaying a wide spectrum of influenza-associated encephalopathies. In this study, there was very low experimental deviation and relatively large biological variability, resulting in data that, when clustered in an unsupervised manner, were able to precisely resolve all 21 patients (63 triplicate samples). Distinct subfamilies of patients can be differentiated within the array, particularly one cluster of five patients (8, 9, 18, 5, and 3) who are positive for several metabolites that are absent from the other patients. All the patients enrolled in this study showed varying degrees of encephalopathy and were on a wide spectrum of drugs including diazepam, midazolam, tamuflu, phenobarbital, herbal supplements, and other drugs that were identified but not reported in the patient's clinical information. The advantage

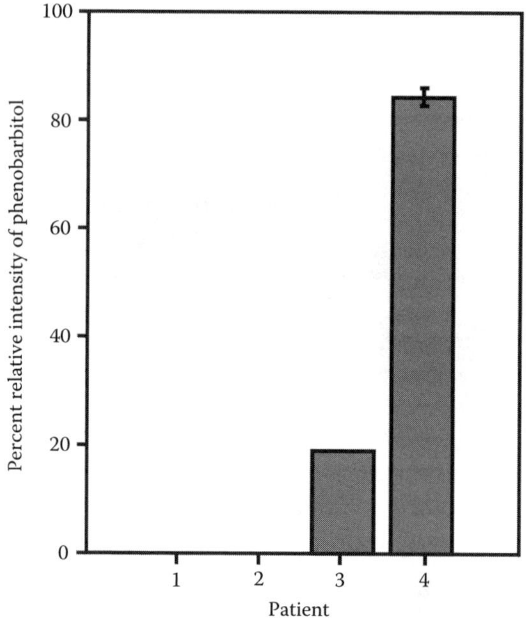

Figure 11.5 Relative levels of phenobarbitol across four individuals with encephalopathy. All four patients were reported to have been taking phenobarbitol during the collection time (see text for further explanation).

of a comprehensive method is evident in this example, which shows how endogenous metabolic changes can be associated with the presence of specific drugs or natural products.

The analysis of these samples also illustrates that there can be patient-specific differences in the CSF uptake of particular drugs. For example, the plot in Figure 11.5 shows the relative intensity of phenobarbital among four patients who were reported as taking the drug during the time of sampling. Detection of the drug in only two of the four patients has important implications in understanding and interpreting drug efficacy, and reinforces the importance of designing and administering individualized drug treatments for the future. First, detection of the parent drug in a nontargeted way within the CSF is proof in itself that the compound is stable and transportable across the blood–brain barrier. In this particular study, the metabolomic data were further searched for common chemical transformations of phenobarbital, which were not detected. Therefore, one could quickly hypothesize that the lack of phenobarbital in patients 1 and 2 could have resulted from inhibited uptake into the CSF, or accelerated excretion of the drug. The analysis of serum and urine in conjunction with the CSF would clearly resolve this issue. A second implication, albeit known for some time but inadequately addressed and understood, is that not all patients respond the same way to a given drug. Hence, comprehensive metabolomic analysis of CSF can be used not only to monitor drug delivery, but also to identify profiles specific for drug efficacy and toxicity. Such markers could be used to stratify patients as either appropriate or inappropriate for a given treatment.

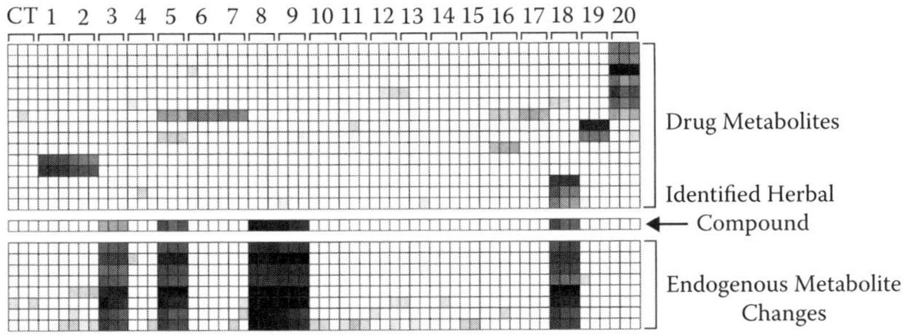

Figure 11.6 Subsets of metabolites from the CSF metabolomic profiles. The top metabolite array (labeled drug metabolites) shows relative intensity levels for several drugs and drug-related metabolites across all triplicate analyses of the 21 individuals. The middle row (labeled identified herbal compound) shows the intensity of a natural product detected in five of the patients. The bottom array (labeled endogenous metabolite changes) shows endogenous metabolite intensities in the CSF of the patients that correlate with the presence of the natural product. Darker shades of gray represent metabolites with increased intensity.

To further illustrate the power of uniting metabolomic analysis with surrogate tissues such as CSF, a heat map containing a subset of 24 CSF metabolites detected across the 21 individuals is shown in Figure 11.6. As mentioned previously, drugs and drug-related metabolites were clearly identified in the CSF of particular patients. However, a specific natural product originating from a herbal extract was also identified in five of the patients (3, 5, 8, 9, and 18), who all showed dramatically upregulated levels of many endogenous metabolites, as depicted in the bottom panel of Figure 11.6. In this example, the association of an exogenous metabolite with specific endogenous metabolic changes affords the opportunity to assign functions to specific natural products. These associations would have been nearly impossible to ascertain using a targeted platform, as it was unknown prior to analysis that certain patients were taking herbal supplements.

In summary, serum and CSF hold promise as surrogate tissues for detecting and diagnosing diseases. Our current lack of understanding regarding the composition of these tissues can largely be attributed to the lack of suitable technologies that can capture and identify metabolites comprehensively. The fact that hundreds of uncharacterized molecules can be easily detected in these tissues using comprehensive metabolomics reveals the potential for partnering such methodologies with surrogate tissue analysis for the discovery of disease and drug activity markers, as well as for expanding our understanding of human variability.

11.6 APPLICATION OF METABOLOMICS TO THE DISCOVERY OF TOXICOLOGIC MARKERS

In addition to the identification of disease- and health-specific biomarkers, comprehensive metabolomics is ideally suited for toxicological applications. First, toxic events such as those associated with adverse drug reactions result largely from unknown mechanisms; this makes them amenable to investigative approaches like comprehensive metabolomics and nontargeted gene expression analysis such as serial analysis of gene expression (SAGE). Second, toxic events themselves are phenotypic end points resulting from a cascade of signals induced by some exogenous agent. Although toxicogenomics studies have shown that gene expression patterns can correlate with the toxicity of certain drugs,[23–26] it is not the actual presence of increased mRNA, or even protein, that is the contributing factor to the toxicity. Rather, toxicity is a reflection of enzyme activity, which is ultimately responsible for altering the biochemical composition of a cell or organism. In essence, toxicogenomics aims to understand the genetic potential of an agent to induce a toxic response. Physiologically, a final manifested toxic state often transpires from a myriad of occurrences stemming ultimately from deregulated biochemical pathways and shifts in the equilibriums of metabolic pools. These biochemical perturbations need to be accurately identified and cataloged in both animals and humans to begin making intelligent predictions and drawing correlations regarding compounds and their effects. It is the new field of "toxicometabolomics" that will address these issues and provide meaningful insights into toxicology in the future. A summary of how toxicogenomics and toxicometabolomics can be utilized to investigate mechanisms of toxicity is presented in Figure 11.7.

The aim of toxicogenomics and toxicometabolomics is to correlate patterns of gene expression with specific toxic activity and metabolite abundances, respectively. There is a large literature base available that covers in great detail many of these issues as well as applications of toxicogenomics to drug discovery.[23–26] Toxicometabolomics can provide value for the pharmaceutical industry during many stages of drug development. Early in the discovery stage, comprehensive toxicometabolomic profiles can be used to screen candidate molecules. The advantage of using a metabolomics-based approach is that *in vitro* toxicity profiles can be simultaneously acquired with efficacy data on combinatorial libraries. Alternatively, a comprehensive metabolomic profile of a drug known to induce a toxic response can be generated to identify a panel of toxicity-associated metabolite biomarkers for the particular drug class. A subset of markers from the comprehensive profile can then be integrated into a high-throughput assay to screen related compounds for toxicity. Furthermore, having identities of the toxicity markers allows one to connect an adverse biological event directly to a specific mechanism of drug pathology. Such insight could dramatically improve the direction and throughput of preclinical lead selection by engineering or screening for new candidates that are efficacious but that do not provoke toxic responses.

In later-stage preclinical trials, toxicometabolomics can provide valuable insights into the relationships between animal and human toxicity. The identification of metabolomic toxicity markers common to both mouse and human, for example, could reduce the percentage of toxicity-related drug failures following enrollment into clinical trials. Given that fewer than 1 in 1000 promising lead compounds actually enter clinical trials, and that greater than 20% of these fail in clinical development for toxicity-related reasons,[27] better informed decisions early in preclinical development represent an opportunity for cost savings in the millions of dollars.

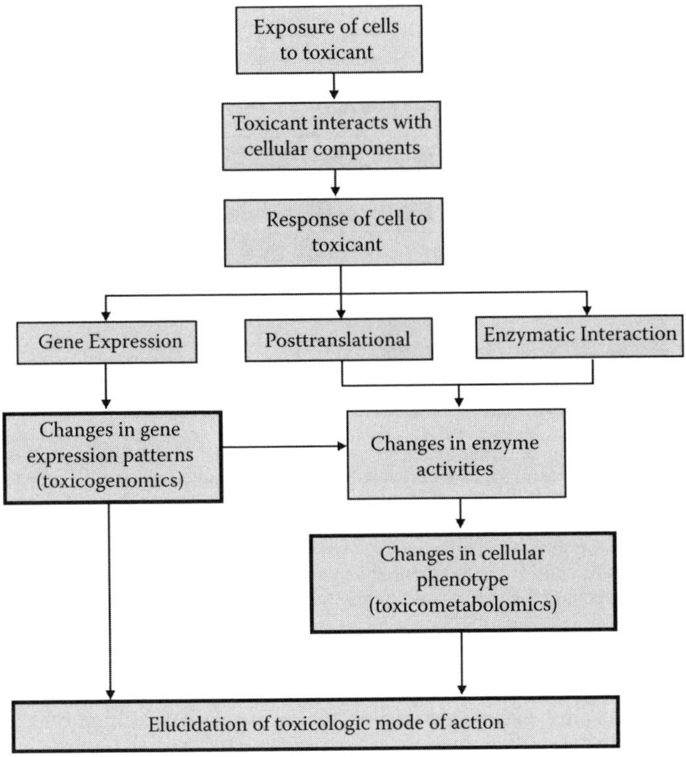

Figure 11.7 Diagrammatic representation showing the interaction between toxicogenomics and toxicometabolomics for elucidating toxicologic modes of action. See text for details.

During clinical trials, toxicometabolomics and the discipline of pharmacometabolomics, which better describes the application of metabolomics to understanding and improving drug efficacy, become more closely integrated. During clinical development, a large emphasis is often placed on the prediction of both efficacy and toxicity. The identification of biomarkers of toxicity and efficacy, which can stratify patients as appropriate or inappropriate candidates for a given therapy, would further reduce attrition in clinical trials, significantly shorten the clinical trial duration, and require fewer enrollments.

An example showcasing a pharmacometabolomic application to preclinical drug development is illustrated by the metabolite heat map in Figure 11.8. In this example, markers of histone deacetylase (HDAC)-specific efficacy, as well as drug-specific toxicity, were simultaneously identified. HT29 colon adenocarcinoma cells were treated with two HDAC-inhibitory drugs: sodium butyrate and Trichostatin A (TSA). Sodium butyrate is a four-carbon fatty acid produced naturally in the human gut and is associated with cell cycle arrest, differentiation, and apoptosis through an HDAC inhibitory activity. TSA was a subsequently identified compound found to induce a similar response to sodium butyrate, but at a much lower concentration and with higher specificity. Comparing the metabolite profiles of both drugs over 24 hours

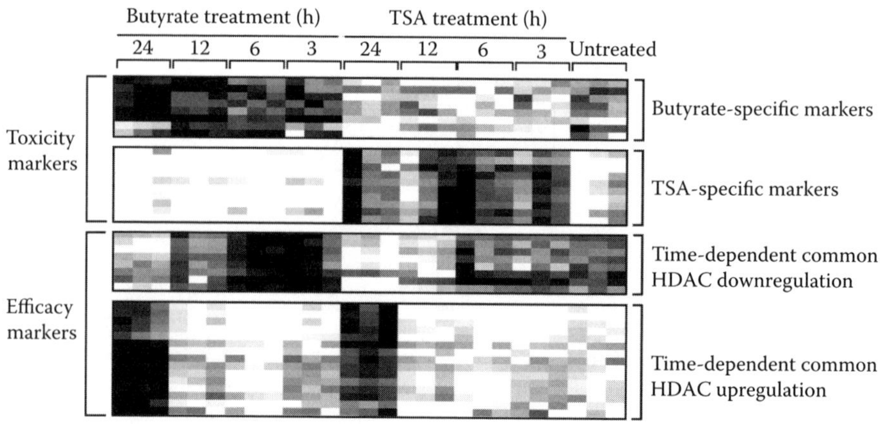

Figure 11.8 Application of comprehensive metabolomics for investigating drug efficacy and toxicity. The upper two metabolite arrays show butyrate-specific and TSA-specific metabolite changes, respectively. The bottom two arrays show metabolites that decrease consistently with both drugs, and increase in intensity following 24 hours of treatment with both drugs, respectively. The drug treatment and times of exposure are indicated along the top of the figure. Darker shades of gray represent metabolites with increased intensity.

allowed for the identification of four clusters of metabolite markers indicative of efficacy and toxicity for both drugs. The top two clusters show metabolites that change exclusively for either butyrate or TSA, respectively. Since both agents are classified as HDAC inhibitors, these markers are indicative of biological processes separate from the intended drug effect, and therefore represent toxicologic markers specific for each drug. The bottom two clusters, on the other hand, show metabolite markers that either decrease (third cluster) or increase (fourth cluster) consistently across both drugs. These represent markers of HDAC-specific efficacy, as this is a common biological activity shared by both drugs. The identification of such markers from preclinical *in vitro* screens can be deployed to screen compound libraries for HDAC activity or used to monitor efficacy and/or toxicity in animal models or clinical trials. In addition, the identification of the specific molecules implicated provides an indication of the underlying toxic or efficacious mechanism, which can be further exploited for the development of drugs with reduced toxicity and higher efficacy.

11.7 CONCLUDING REMARKS

Combining comprehensive profiling methods with surrogate tissues offers exciting opportunities for biomarker identification. These can include disease-specific diagnostic markers, drug efficacy or toxicity markers, and long-term chronic health- or lifestyle-associated markers. One of the key findings provided by FTMS-based comprehensive metabolomic analysis of surrogate tissue is the realization that many uncharacterized molecules exist in human biospecimens and that our current understanding of metabolism is far from complete. A second insight afforded by these analyses is that considerable biochemical variability exists not only within the human population, but also between multiple samples collected from the same individual over time. Understanding these individual fluctuations will build the foundation for real-time biochemical monitoring of individual health status, and provide important insights regarding disease management in the future.

REFERENCES

1. Lander, E.S. et al., Initial sequencing and analysis of the human genome. *Nature,* 409(6822), 860–921, 2001.
2. Venter, J.C. et al., The sequence of the human genome. *Science,* 291(5507), 1304–1351, 2001.
3. Goodacre, R. et al., Metabolomics by numbers: acquiring and understanding global metabolite data. *Trends Biotechnol.,* 22(5), 245–252, 2004.
4. Barker, J., *Mass Spectrometry.* 2nd ed. John Wiley & Sons, New York, 1999.
5. Dass, C., *Principles and Practice of Biological Mass Spectrometry.* John Wiley & Sons, New York, 2001.
6. Hoffmann, E.D., *Mass Spectrometry: Principles and Applications*, 2nd ed. Wiley, New York, 2001.
7. Reo, N.V., NMR-based metabolomics. *Drug Chem. Toxicol.*, 25(4), 375–382, 2002.
8. Shockcor, J.P. and Holmes, E., Metabonomic applications in toxicity screening and disease diagnosis. *Curr. Topics Med. Chem.,* 2(1), 35–51, 2002.
9. Aharoni, A. et al., Nontargeted metabolome analysis by use of Fourier Transform Ion Cyclotron Mass Spectrometry. *Omics*, 6(3), 217–234, 2002.
10. Diamandis, E.P., Mass spectrometry as a diagnostic and a cancer biomarker discovery tool: opportunities and potential limitations. *Mol. Cell Proteomics,* 3(4), 367–378, 2004.
11. Baggerly, K.A., Morris, J.S., and Coombes, K.R., Reproducibility of SELDI-TOF protein patterns in serum: comparing datasets from different experiments. *Bioinformatics,* 20(5), 777–785, 2004.
12. Chen, X. and Sullivan, P.F., Single nucleotide polymorphism genotyping: biochemistry, protocol, cost and throughput. *Pharmacogenomics J.*, 3(2), 77–96, 2003.
13. Jiang, R. et al., Genome-wide evaluation of the public SNP databases. *Pharmacogenomics*, 4(6), 779–789, 2003.
14. Conrads, T.P. and Veenstra, T.D., The utility of proteomic patterns for the diagnosis of cancer. *Curr. Drug Targets Immune Endocr. Metab. Disord.*, 4(1), 41–50, 2004.
15. Petricoin, E.F. et al., Use of proteomic patterns in serum to identify ovarian cancer. *Lancet*, 359(9306), 572–577, 2002.
16. Yu, L.R. et al., Diagnostic proteomics: serum proteomic patterns for the detection of early stage cancers. *Dis. Markers*, 19(4–5), 209–218, 2003.
17. Veenstra, T.D. and Conrads, T.P., Serum protein fingerprinting. *Curr. Opin. Mol. Ther.*, 5(6), 584–593, 2003.
18. Hoffmann, G.F., Surtees, R.A., and Wevers, R.A., Cerebrospinal fluid investigations for neurometabolic disorders. *Neuropediatrics*, 29(2), 59–71, 1998.
19. Seehusen, D.A., Reeves, M.M., and Fomin, D.A., Cerebrospinal fluid analysis. *Am. Fam. Physician*, 68(6), 1103–1108, 2003.
20. Reiber, H. and Peter, J.B., Cerebrospinal fluid analysis: disease-related data patterns and evaluation programs. *J. Neurol. Sci.*, 184(2), 101–122, 2001.

21. Carrette, O. et al., A panel of cerebrospinal fluid potential biomarkers for the diagnosis of Alzheimer's disease. *Proteomics*, 3(8), 1486–1494, 2003.
22. Zheng, P.P. et al., Identification of tumor-related proteins by proteomic analysis of cerebrospinal fluid from patients with primary brain tumors. *J. Neuropathol. Exp. Neurol.*, 62(8), 855–862, 2003.
23. Aardema, M.J. and MacGregor, J.T., Toxicology and genetic toxicology in the new era of "toxicogenomics": impact of "-omics" technologies. *Mutat. Res.*, 499(1), 13–25, 2002.
24. Guerreiro, N. et al., Toxicogenomics in drug development. *Toxicol. Pathol.*, 31(5), 471–479, 2003.
25. Lord, P.G., Progress in applying genomics in drug development. *Toxicol. Lett.*, 149(1–3), 371–375, 2004.
26. Suter, L., Babiss, L.E., and Wheeldon, E.B., Toxicogenomics in predictive toxicology in drug development. *Chem. Biol.*, 11(2), 161–171, 2004.
27. Frank, R. and Hargreaves, R., Clinical biomarkers in drug discovery and development. *Nat. Rev. Drug Discov.*, 2(7), 566–580, 2003.

CHAPTER 12

Lipidomic Analysis of Plasma and Tissues: Lipid-Derived Mediators of Inflammation and Markers of Disease

Clary B. Clish and Charles N. Serhan

CONTENTS

12.1 INTRODUCTION

Lipidomics, the systematic decoding of lipid-based information in biosystems, comprises identification and profiling of lipids and lipid-derived mediators. As practiced today, lipidomics can be subdivided into the study of lipids involved in (1) energy metabolism, (2) architecture/membranes, and (3) lipid-derived mediator-lipidomics. The mapping of structural components and their relation to cell activation as well as generation of potent lipid mediators involves a combined quantitative profiling and informatics approach[1] to appreciate inter-relationships and complex mediator networks important for cell homeostasis. Cell membranes are composed of a lipid bilayer that contains species such as phospholipids and sphingolipids (Figure 12.1) as well as integral membrane proteins and membrane-associated proteins. Membrane composition of many cell types is established. However, their organization and how they affect cell function remain areas of interest and hold the promise of designing novel therapeutic approaches that target specific subcellular components. Membranes serve barrier functions — separating the inside from outside or compartments within cells — regulating passage of nutrients, gasses, and specific ions. Membranes also generate signals to the intracellular milieu by their ability to interact with key proteins. Fatty acids are key components of membranes and membrane lipids; in addition to playing important roles in energy generation via their catabolic metabolism, they are also integral to signaling pathways and serve as substrates for lipid mediator generation. The promise of lipidomics is to deconvolute the complex web of structural, energy metabolism, signaling, and mediator functions in which lipids are involved and reveal the nature of the interplay among these functions. Understanding these complex structures and the networks of local chemical signals generated in lipid microdomains can unlock new vistas for cellular and molecular therapeutics.

The determination of structure–activity relationships (SARs) for bioactive lipids conceptually preceded the current appreciation of biological mass spectrometry as well as so-called chemical biology and genetics. A case in point are the prostaglandins, which were first discovered in the 1930s for their ability to stimulate uterine contractions and lower blood pressure, and were isolated and structurally characterized in the 1950s. Prostaglandins (Figure 12.1C) are potent, fatty acid-derived, local-acting mediators important in a wide range of processes, such as inflammation, parturition, labor, hemodynamics, and renal function.[2] Gas chromatography and mass spectrometry (MS) methods are well established and useful in probing bioactive lipid mediators.[3] Liquid chromatography (LC)-MS/MS technologies permit profiling of closely related compounds without the need for prior derivatization of samples, reducing the potential for workup-induced artifacts. Advancement and commercialization of MS and separation technologies over the last several decades have enabled increasing numbers of investigators to engage in both qualitative and quantitative profiling of lipids, currently termed lipidomics,[4] as well as other endogenous metabolites, activities falling within the broader scope of the term *metabolomics*.

12.2 MEMBRANE ARCHITECTURE/STRUCTURE–FUNCTION

Figure 12.1 outlines the diversity of lipid structures and the scope of lipidomics. Cell membranes comprise a phospholipid bilayer depicted split down its hydrophobic region as envisioned with results from freeze-etched electron micrograms and the widely appreciated Singer–Nicholson model.[5] The bilayer is illustrated as a sea of phospholipids. Their organization and the precise compositions of microdomains surrounding key integral membrane proteins, subcellular membranes, and other lipid-enriched domains within cells remain to be fully appreciated. We have little information on the organization of discrete lipid patches and microdomains in the struc-

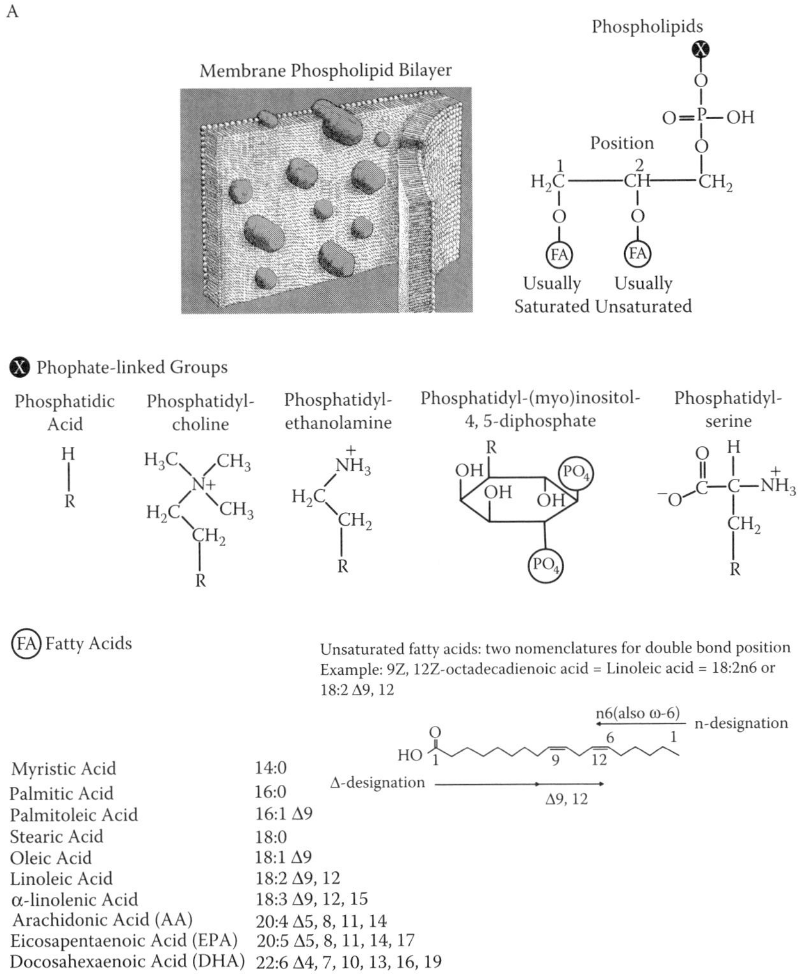

Figure 12.1 Scope of the problem for lipidomics: diverse structures of lipids and lipid mediators. (A) Depiction of the major phospholipid subunits building on the glycerol framework. Addition of specific polar groups and fatty acids in the 1 or 2 position (see text for details) yields specific phospholipids. It should be noted that, for example, phosphatidic acid can represent multiple molecular species given the possibility of many different fatty acids placed in its number 1 or 2 position. That is, phosphatidic acid containing two molecules of steric acid in its 1 and 2 position is distinctly different from the properties of phosphatidic acid that contains steric acid in the 1 position and arachidonic acid in its second position. Each of these individual molecular species (>1000 distinct molecules) gives rise to unique physical properties and therefore different molecular ions, retention times, and fragmentation profiles on LC-MS/MS analysis. (B) The basic structures of sphingolipids, diacylglercides, and triacylglycerides. Note, each of these is a basic unit and as depicted can represent specific molecular species that can comprise many different types of individual molecules. (C) Families of bioactive lipid autacoids. Arachidonic acid is the precursor for many of the known bioactive mediators, epoxyeicosatetraenoic acids (EETs), prostaglandins, leukotrienes, and lipoxins. Also, EPA (C22.5 and C22.6) are precursors to potent new families of mediators termed resolvins and neuroprotectins.

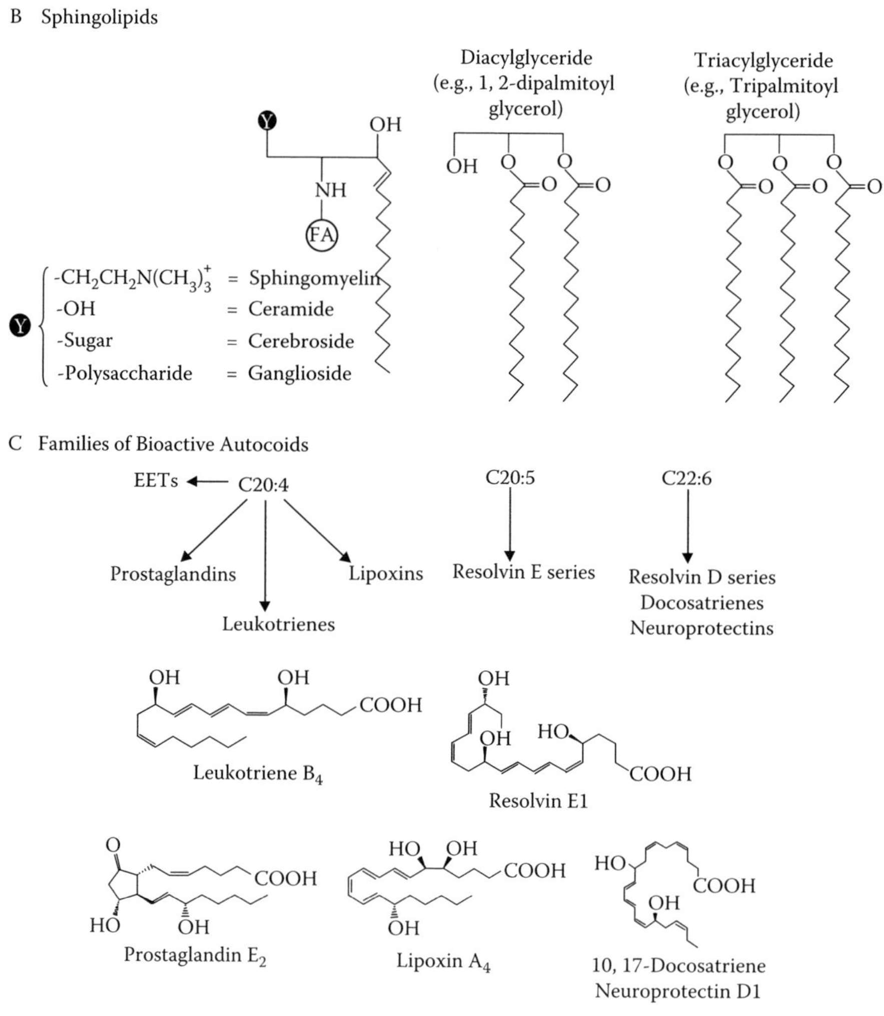

Figure 12.1 (continued)

ture of plasma membranes. Nonetheless, it is these microdomains and patches that are of considerable significance in regulating the "outside to inside." A further understanding may be gained from compositional analysis of microdomains and the ability to identify each of the major phospholipid structures; such as phosphatidic acid (PA), phosphatidylcholine (PC), phosphatidylethanolamine (PE), phosphatidylinositol (PI), and phosphatidylserine (PS) (Figure 12.1A). Contributing to the diversity of both lipid structure and function are the numerous combinations of esterified fatty acid that are possible. For example, phospholipids may contain up to two fatty acid moieties esterified to the 1 and/or 2 positions of the glycerol backbone. The fatty acid composition defines the physical properties of the molecule and different acyl chain compositions can have dramatic effects in cell function. Phospholipids comprising saturated fatty acids can make membrane regions more crystalline-like or rigid compared to those containing fatty acids with greater numbers of double bonds (i.e., polyunsaturated fatty acids) that decrease the acyl chain packing density in membranes and have comparatively fewer van der Waals interactions with other molecules.

Shown in Figure 12.1A is a partial list of fatty acids of biological significance: myristic acid, palmitic acid, palmitoleic acid, stearic acid, oleic acid, linoleic acid, α-linolenic acid, arachidonic acid (AA), eicosapentaenoic acid (EPA), and docosahexaenoic acid (DHA), each named by its carbon chain length and degree of unsaturation (i.e., double bonds and their position). Sphingolipids (Figure 12.1B) are an important class of complex lipids that are derived from amino alcohols. They are appreciated for their roles in insulating neurons as well as for acting directly as or giving rise to signaling molecules.[6] One of the major groups focusing on these compounds using lipidomics is at the Medical University of South Carolina (see http://hcc.musc.edu/research/shared_resources/lipidomics.cfm). Given the considerable diversity in triglyceride structure and the importance of phosphatidylinositol as well as other sugar-linked phospholipids in cell signaling, a systematic analysis is under way via the NIGMS Lipid MAPS consortium (see http://www.nigms.nih.gov/funding/gluegrants.html and http://www.lipidmaps.org). This consortium plans to assemble maps of analytical profiles as cells are activated in experimental settings.

12.3 LIPID SIGNALS AND AUTOCOIDS IN DISEASE

Diacylglycerol (DAG, Figure 12.1B) is an intracellular second messenger that helps to illustrate the potential for second messenger lipidomics. Recently, we linked a genetic abnormality in patients with localized aggressive periodontal disease to impaired DAG kinase activity in their peripheral blood neutrophils. This familial disorder is characterized by destruction of the supporting structures of the dentition. In these patients, neutrophils, the first line of defense to host infection, display reduced chemotaxis toward pathogenic microbes and reduced ability to generate reactive oxygen species. Using a comparative lipidomics approach to profiling unique DAG species extracted from the neutrophils of these patients, we found alterations in levels and specific molecular species. LC-MS/MS-based lipidomics analyses were performed to identify and quantitate individual species of DAGs involved in second messenger signaling (Figure 12.2). Profiles of specific DAG species were identified by their physical properties, including: molecular ion, specific MS/MS product ions, as well as by co-elution with authentic standards of the major species. We demonstrated both molecular and temporal differences in DAG signaling species between neutrophils sampled from healthy individuals and subjects with localized aggressive periodontal disease.[7] These results exemplify the importance of structure–function profiling of lipid intracellular messengers to improving our appreciation of signaling pathways and their alterations in disease.

Another powerful use of lipidomics specifically focuses on the area of mediators — coined mediator-lipidomics. In general, unsaturated double bonds present in polyunsaturated fatty acids, such as arachidonic acid, are nonconjugated, making

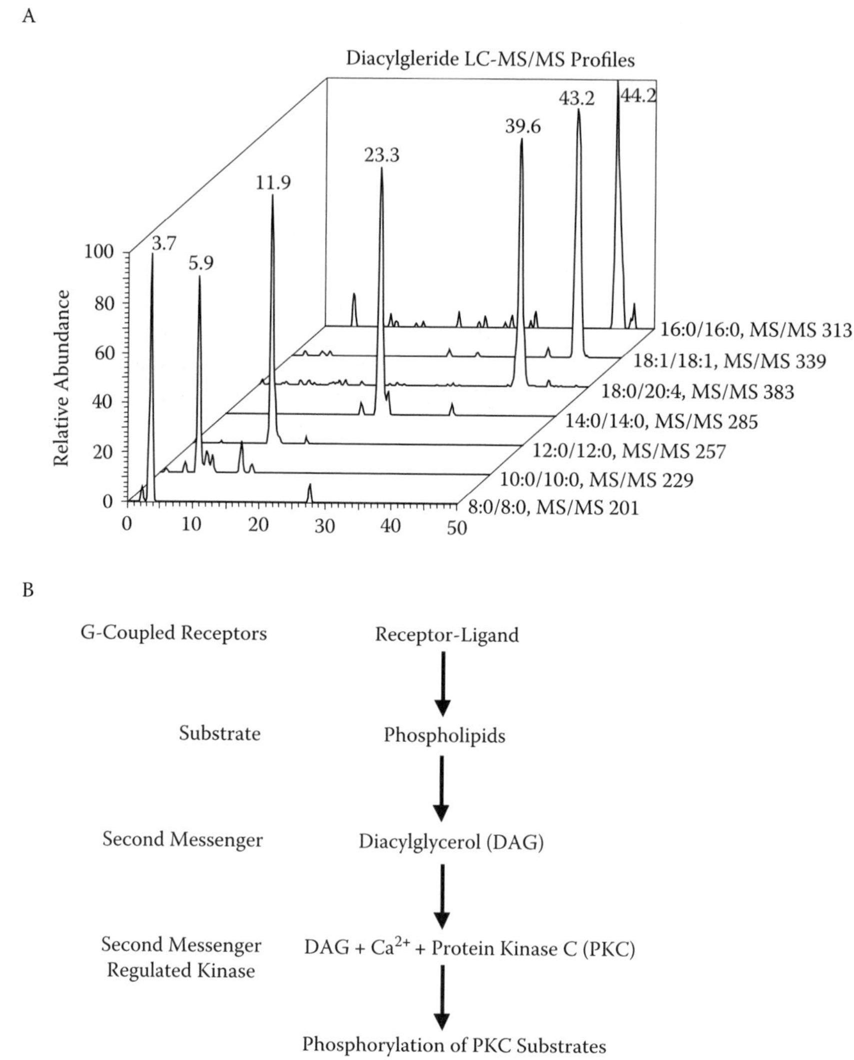

Figure 12.2 (A) Elevated DAG levels in PMN from patients. Left panel: Selected MS ion chromatograms of synthetic 1,2 diacyl-sn-3-glycerol molecular species. DAG species were resolved and identified by LC-MS/MS using specific retention time and unique MS/MS signature product ions for each molecular species as indicated. (B) Role of DAG in signal transduction and PKC activation.

the fatty acids essentially devoid of a characteristic ultraviolet (UV) spectra. During the release of arachidonic acid and its transformation to bioactive eicosanoids (a term derived from the Greek word *eicosa,* meaning 20, which corresponds to the number of carbon atoms in the molecule), stereoselective hydrogen abstraction leads to formation of conjugated diene-, triene-, or tetraene-containing chromophores, particularly with respect to the lipoxygenase pathway products leukotrienes and lipoxins (Figure 12.1C). These compounds can be characterized on the basis of both specific UV chromophores and characteristic MS/MS spectra and possess profound, stereospecific bioactivity in the nano- to picomolar concentration range. In general, lipid-derived mediators are rapidly formed within seconds to minutes, act on cells locally in either a paracrine or autocrine fashion, and then are rapidly inactivated. As shown in Figure 12.3, for example, the endogenous anti-inflammatory eicosanoid lipoxin A_4 (LXA_4) is inactivated via a sequence of reactions catalyzed by endogenous eicosanoid oxidoreductases that generate oxo- and dihdydro- products.[8] Eicosanoids usually act as extracellular mediators within their local milieu and therefore are classed with the broader group of autacoids (such as serotonin, histamine, etc.). Differences in physical properties of each of these related structures permit identification and profiling of the cellular milieu. Unlike phospholipids or other structural lipids that keep a barrier function, those derived from arachidonic acid, including prostaglandins, leukotrienes, and lipoxins (Figure 12.1C), have unique and potent actions on neighboring cells. This makes it very important for profiling efforts to clearly separate these compounds for their identification, as closely related structures may be biologically devoid of actions. Accurate profiling and determination of relationships between products within a snapshot of a biological process or disease state can give valuable information.[9] Also, when specific drugs are taken, such as aspirin, the relationship between individual pathway products can be altered and their relationship may be directly linked to the drug's action while *in vivo*.[4,10] Hence, mediator-lipidomics provides a valuable means to understanding the phenotype in many prevalent diseases, particularly ones in which inflammation has an important pathologic basis.

12.4 COMPARATIVE MEDIATOR-LIPIDOMIC PROFILING OF ENGINEERED EXPERIMENTAL ANIMALS

The powerful approach of transgenics (TG), namely, deletion and overexpression of a gene product coupled with lipidomics, can give valuable insights into the role of select pathways in disease processes. We recently used the mediator lipidomics approach to evaluate transgenic rabbits overexpressing human 15-lipoxygenase (LOX) type 1 in their leukocytes.[11] We can take a lipidomic snapshot of cell activation and examine the difference between the transgenic and the nontransgenic rabbits, where the key enzyme is not overproduced, but rather is in its normal state, to evaluate the impact of overexpression of a key enzyme in a pathway. In this case, 15-LOX overexpression leads to enhanced LXA_4, as well as enhanced 5,15-diHETE formation with reduced leukotriene B_4 (LTB_4) formation (Figure 12.4). Because LTB_4 is a potent chemoattractant and LXA_4 is a counter-regulatory anti-inflammatory within the eicosanoid family, the relationship between these mediators and the overproduction of LXA_4 is a key index to appreciating the overall role of the 15-LOX type 1 in inflammation. In short, overexpression of this enzyme yields upregulation of its pathway products, such as LXA_4, in these transgenic rabbits that also displayed a generally reduced inflammation and protection from tissue damage.

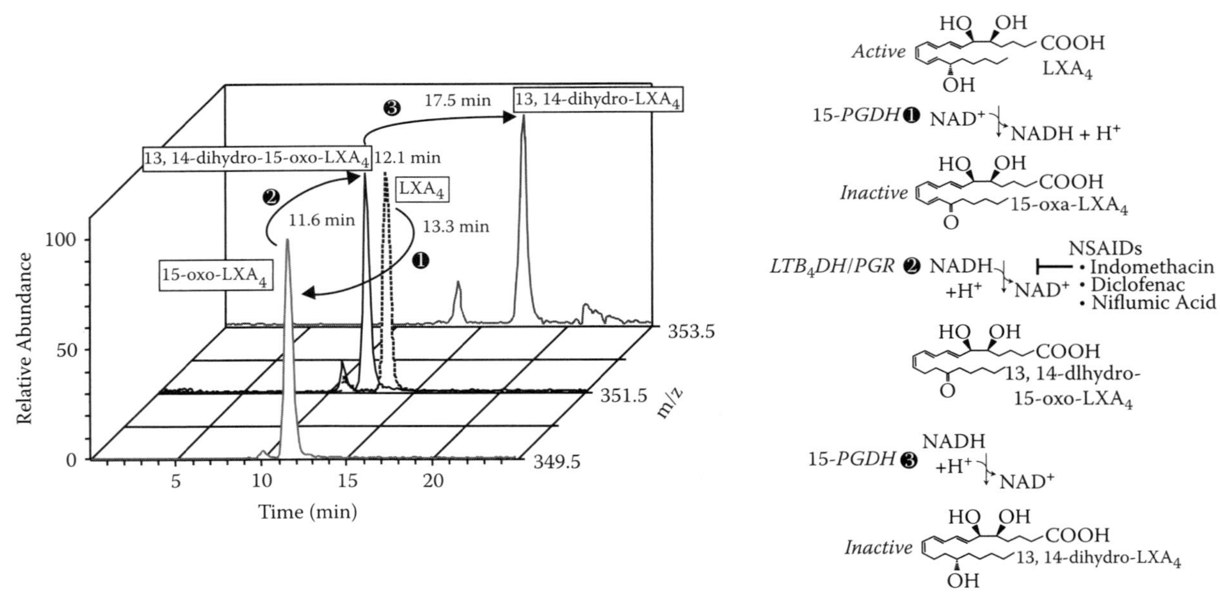

Figure 12.3 Lipoxin local inactivation route: LC-MS chromatograms of LXA_4 further metabolites. The initial step in LXA_4 (*m/z* 351.5, retention time 13.3 min) inactivation is dehydrogenation of the 15-hydroxyl group catalyzed by an enzyme that was first characterized as 15-hydroxyprostaglandin dehydrogenase (15-PGDH) to yield 15-oxo-LXA_4 (*m/z* 349.5, retention time 11.6 min). A multifunctional eicosanoid oxidoreductase (LTB_4DH/PGR), which has been named both leukotriene B_4 12-hydroxydehydrogenase and 15-oxoprostaglandin 13-reductase as result of independent findings that the enzyme can convert these substrates, catalyzes the reduction of the 13,14 double bond of 15-oxo-LXA_4 to give 13,14-dihydro-15-oxo-LXA_4 (*m/z* 351.5, retention time 12.1 min). This product then serves as a substrate for the 15-hydroxy/oxo-eicosanoid oxidoreductase, which catalyzes the reduction of the C15 oxo-group to give 13,14-dihydro-LXA_4 (*m/z* 353.5, retention time 17.5 min). Neither 15-oxo-LXA_4 nor 13,14-dihydro-LXA_4 binds to the LXA_4 receptor and, unlike the parent compound, they do not inhibit the generation of reactive oxygen species in human neutrophils.[8]

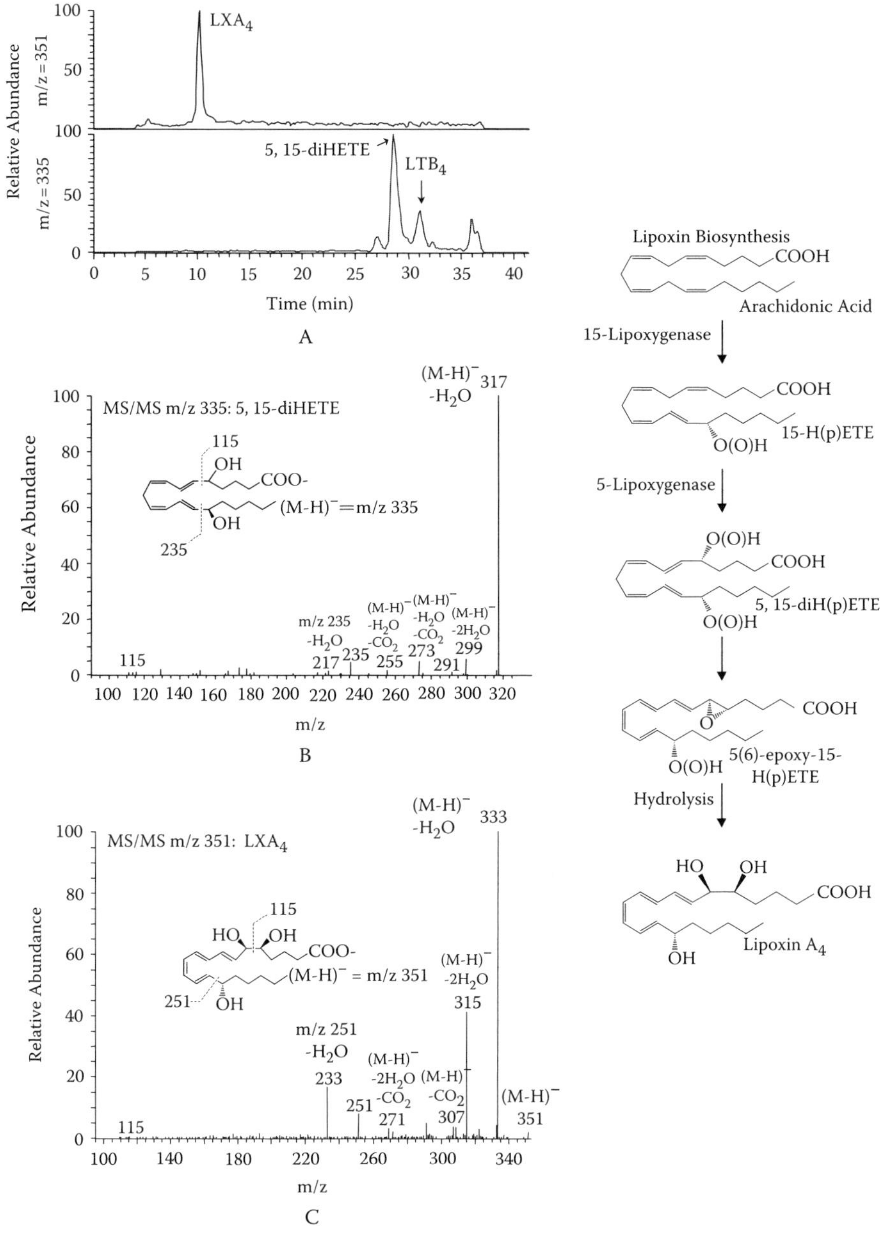

Figure 12.4 Mediator-lipidomic profiling of engineered experimental animals. Leukocytes were isolated from both 15-LOX type 1 transgenic rabbits and nontransgenic rabbits and incubated with ionophore A23187 (15 m*M*, 20 min, 37°C). Products were extracted and identified.[7] (A) LC profiles from 15-LOX type 1 transgenic rabbits. (B) MS/MS spectrum of 5,15-diHETE with diagnostic ions as indicated. (C) MS/MS spectrum of LXA_4 with diagnostic ions as indicated.

12.5 NOVEL EXTRACELLULAR BIOSIGNALS FROM LIPIDS: PATHWAYS OF INFLAMMATION-RESOLUTION

It is now appreciated that inflammation plays an important role in many prevalent diseases in the Western world. In addition to the chronic inflammatory diseases, such as arthritis, psoriasis, and periodontitis, as noted above, it is now increasingly apparent that diseases such as asthma, Alzheimer's disease, and even cancer have an inflammatory component associated with the disease process. Therefore, it is important for us to gain more detailed information on the molecules and mechanisms controlling inflammation and its resolution. Toward this end, we recently identified new families of lipid mediators generated from fatty acids during resolution of inflammation; termed *resolvins and docosatrienes.* Using systematic analysis of resolving inflammatory exudates, we sampled exudates during resolution as leukocytic infiltrates were declining to determine whether there were indeed new mediators generated. Figure 12.5 schematically represents our *functional mediator-lipidomics* approach using LC tandem MS (LC-MS/MS-based analyses) to evaluate and profile temporal production of compounds at defined points during experimental inflammation and its resolution. We constructed libraries of physical properties for known mediators, i.e., prostaglandins, epoxyeicosatetraenoic acids (EETs), leukotrienes, and lipoxins (Figure 12.1C), as well as theoretical compounds and potential diagnostic fragments as signatures for specific enzymatic pathways. When novel compounds were pinpointed within chromatographic profiles, we carried out complete structural elucidation as well as retrograde chemical analyses that involve both biogenic and total organic synthesis, which permitted scaling up of the compound of interest and its evaluation *in vitro* and *in vivo*. These *in vivo* models include the murine air pouch model of inflammation as well as peritonitis. *In vitro* cell assays focused on regulation of cytokines and leukocyte migration across transepithelial or

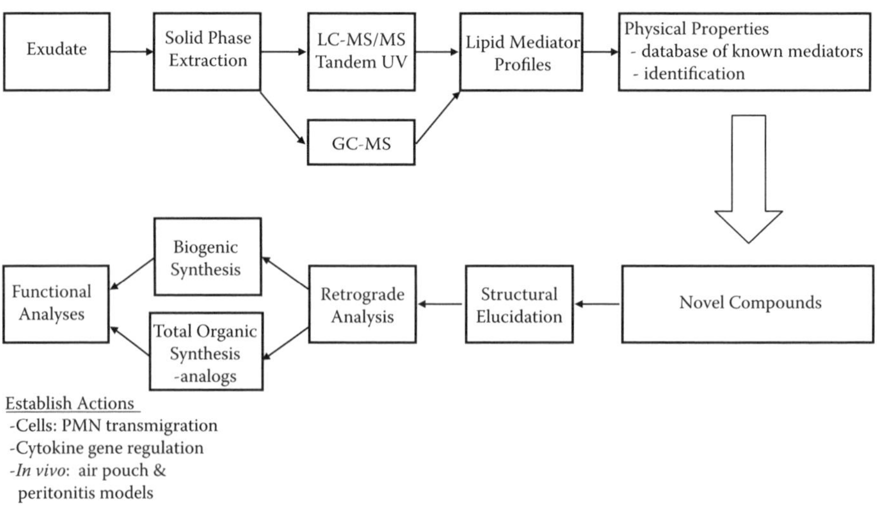

Figure 12.5 Elucidation of the cycle of mediator-lipidomics.

transendothelial monolayers. This full cycle of events defines *mediator-lipidomics* because it is important to establish both the structure and function of bioactive molecules. With this new lipidomics-based approach that combined LC-PDA-MS/MS, a novel array of endogenous lipid mediators were identified[4,10] during the multicellular events that occur during resolution of inflammation. The novel biosynthetic pathways uncovered use omega-3 fatty acids, eicosapentanoic acid and docosahexanoic acid, as precursors to new families of protective molecules, termed resolvins. These include resolvin E (18*R* series from EPA) and resolvin D (17-series from DHA).[12] In humans, the vasculature — particularly endothelial cells during cross-talk with leukocytes — generate these products via transcellular biosynthesis pathways.[4] In this novel cell–cell interaction, endothelial cells generate the first biochemical step and then pass this intermediate 18*R*-HEPE to leukocytes, which transform this to a potent molecule termed resolvin E1 (RvE1), as depicted in Figure 12.6. RvE1 is ~100 to 1000 times more potent than native EPA as a down-regulator of neutrophils and stops their migration into inflammatory loci.[4,10] DHA, which is enriched in neural systems, is also released and transformed to potent bioactive molecules denoted 10,17-docosatriene (neuroprotectin D1) and resolvins of the D series (Figure 12.1C). Human brain, synapses, and retina are rich with DHA, a major omega-3 fatty acid. Deficiencies in DHA are associated with altered neural functions, cancer, and inflammation in experimental animals.[13] Employing our mediator-lipidomics approach, we learned that on activation neural systems release DHA to produce neuroprotectin D1, which in addition to stopping leukocyte-mediated tissue

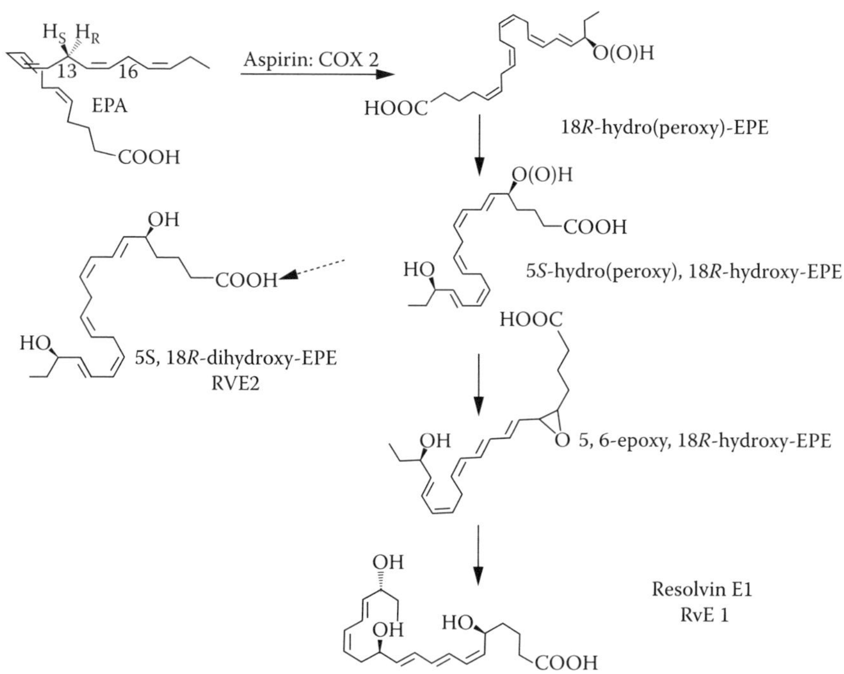

Figure 12.6 Biosynthesis of resolvin E1 derived from EPA.

damage in stroke also maintains retinal integrity.[14] Figure 12.7 gives an example of resolvin PDA profiles and MS/MS spectra.

12.6 BIOMARKER LIPIDOMICS

The discovery of biomarkers of disease is another important application for lipidomics. Comparative profiling of specific lipid species in peripheral fluids such as plasma or experimentally available tissues can reveal significant differences between diseased and normal control subjects. This is of particular interest in human disease states where lipid metabolism or utilization are altered, such as in certain cardiovascular and metabolic diseases. Biomarkers may be of utility in either early diagnosis of chronic disease at the molecular level (i.e., atherosclerosis, pulmonary disease, Alzheimer's disease) or in potentially monitoring the progress of patients receiving therapeutics and/or nutritional supplementation.

Effective discovery-oriented approaches to LC-MS-based comparative lipidomics have been developed. Using LC-MS, it is possible to profile multiple lipid classes within a single sample analysis to produce a two-dimensional array of peaks, each of which can be distinguished by the combination of its mass to charge ratio (*m/z*) and retention time. Discovery lipidomic methods are optimized to give quantitative data for the broadest range of lipid moieties that can be accommodated within the limitations of the instrumentation, which is typically limited by the dynamic range of the mass spectrometer or loading capacity of the chromatography column. Figure 12.8 outlines a typical discovery lipidomics workflow. An LC-MS chromatogram is acquired for each sample in the study. Next, peak information (for example, the *m/z*, retention time, and integrated area for each peak in the data set) is extracted from the raw data using customized software tools, followed by further data preprocessing to adjust minor shifts in chromatographic retention times so that the data sets may be compared across all samples. Statistical analyses are then applied to identify significantly changing lipid peaks.

An example is described here where lipidomic biomarker profiling was applied to samples taken from transgenic animals engineered to be susceptible to cardiovascular disease, the apolipoprotein E3-Leiden (APOE*3-Leiden) transgenic mouse. Apolipoprotein E is a component of very low density lipoproteins (VLDL) and VLDL remnants and is required for receptor-mediated reuptake of lipoproteins by the liver. Transgenic mice over-expressing human APOE*3-Leiden are highly susceptible to diet-induced hyperlipoproteinemia and atherosclerosis due to diminished hepatic LDL receptor recognition, but when fed a normal chow diet they display only mild type I (macrophage foam cells) and II (fatty streaks with intracellular lipid accumulation) lesions.[15] Plasma and liver samples were taken for comparative analysis from APOE*3-Leiden and wild-type mice that were 9 weeks of age and that were fed a normal chow diet to elucidate molecular markers of predisposition to disease well before any clinical symptoms were apparent.[16] Figure 12.9A shows the results of an unsupervised multivariate analysis, specifically principal components analysis, of LC-MS lipid peak profiles of the APOE*3-Leiden and wild-type mice from two distinct clusters. Results of deconvolution of the relative weighting of each

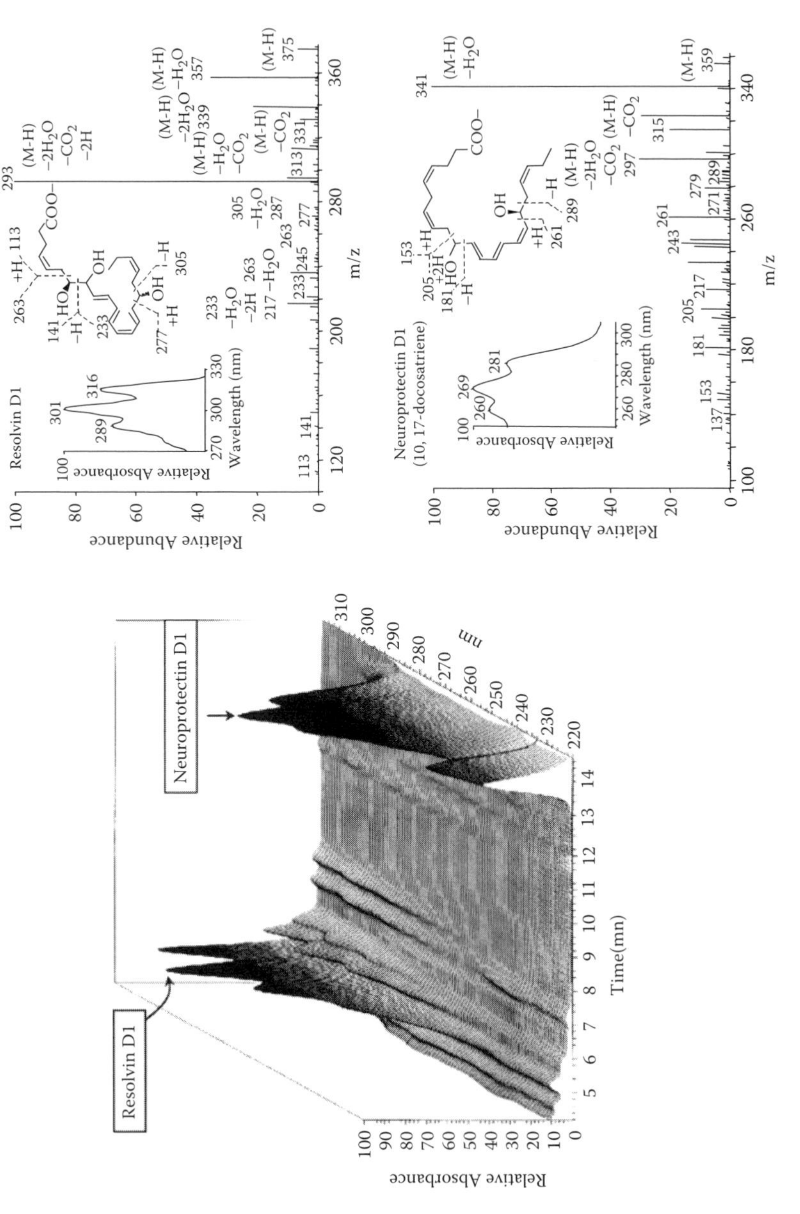

Figure 12.7 LC lipidomic profiles of resolvin D1 and neuroprotectin D1 and tandem MS/MS spectra. Right upper panel: MS/MS of RvD1; right lower panel: MS/MS of NPD1. Each corresponds to the materials beneath the UV chromatophore (left).

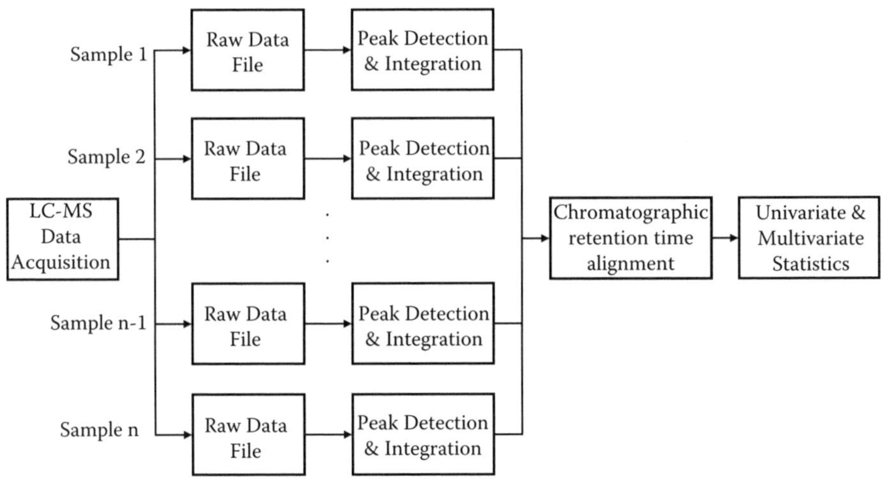

Figure 12.8 Lipidomics LC-MS discovery workflow.

lipid peak to the separation of the two clusters are plotted in Figure 12.9B. Each bar in the plot corresponds to a measured LC-MS peak, and each is coded by an index number that is a combination of the mass and retention time. There is a general trend of lower levels of lysophophatidylcholine (LysoPC) moieties and higher levels of triglycerides (TG) in APOE*3-Leiden compared to wild-type mice. In contrast, phosphatidylcholines (PC) show differences in both directions, demonstrating the importance of profiling unique molecular species within each lipid class.

Similar lipidomic analyses were applied to liver tissue as part of a multi-omic, or systems biology, study in an attempt to add further biological context. One approach to decoding relationships that might exist among profiled components is a systems biology analysis to generate correlation networks.[16] The association between two entities i and j (for example, a lipid and an mRNA transcript) can be determined by calculating their Pearson correlation coefficient, C_{ij}. The coefficients for all combinations of pairs of biomolecules within the data set are calculated to generate an array of values. By imposing a correlation value threshold, weaker associations can be filtered out, leaving behind a network containing only highly correlated biomolecules. By layering existing experimental knowledge onto the network, such as biochemical pathways or regulatory mechanisms, novel relationships among individual entities or groups may then be identified. This method is similar to relevance networks introduced by Butte et al.[17] Figure 12.10 is an example of a correlation network of a subset of biomolecules measured in a "multi-omic" analysis — i.e., lipidomic, proteomic, and transcriptomic — comparing liver tissues from APOE*3-Leiden to wild-type mice under conditions where the transgenic mice do not display any clinical symptoms of disease.[16] Of the many lipid moieties profiled, two lipid molecules, C16:0 LysoPC (1-palmitoyl-2-hydroxy-sn-glycero-3-phosphocholine) and C38:1 DAG (1-octadecanoyl-2-eicosenoyl-sn-glycerol), which were upregulated in the livers of APOE*3-Leiden mice, showed a high degree of correlation to significantly increasing mRNA transcript and protein expression levels

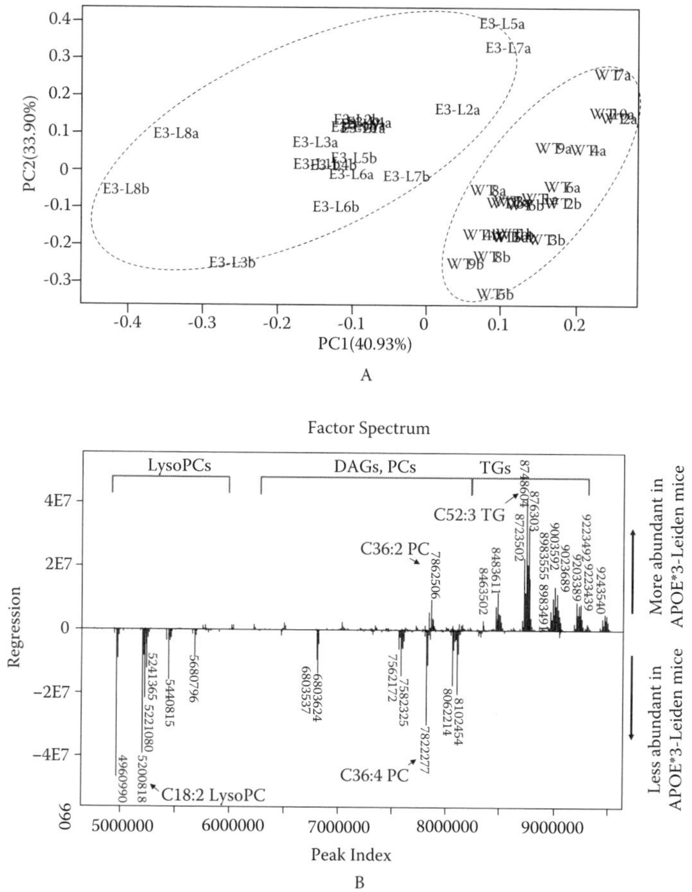

Figure 12.9 Principal components analysis (PCA) of a nonpolar and polar lipid profiles of APOE*-Leiden transgenic and wild-type mice plasma samples. Nonpolar and polar lipids were extracted from plasma samples taken from transgenic ($n = 9$) and wild-type ($n = 10$) mice. Lipid profiles were acquired in duplicate using LC-MS, and data were processed via the workflow illustrated in Figure 12.6 prior to comparative analyses. (A) PCA of lipid profiles. Each point, denoted by "E3-L" for an APOE*3-Leiden and "WT" for wild-type, on the PCA score plot represents an LC-MS data set. This scatter plot shows distinctly separted clusters of APOE*3-Leiden and wild-type data sets. Within each cluster there are several highly similar data sets that are overlapped in this two-dimensional representation; particularly among the APOE*3-Leiden data sets, 1a, 1b, 2b, 4a, 4b, 9a, and 9b are overlapped. Among the wild-type data sets, 1a, 1b, 2b, 3a, 3b, 4b, 5a, 6a, 6b, 7b, and 10b are overlapped. (B) The PCA factor spectrum shows relative weighting of each measured LC-MS peak in the separation between the APOE*3-Leiden and wild-type mice clusters in A as well as the direction of the difference in peak intensity between the two groups. There is a general trend of lower levels of lysophophatidylcholine (lysoPC) moieties and higher levels of triglycerides (TG) in APOE*3-Leiden compared to wild-type mice. In contrast, phosphatidylcholines (PC) show differences in both directions. Note that total carbon and double bond content is given for PC and TG species, from 2 and 3 acyl groups in each molecule, respectively.

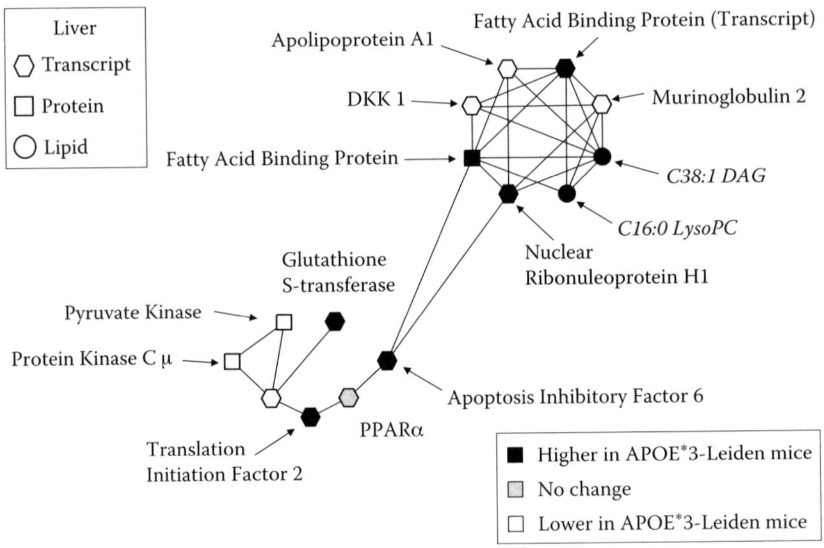

Figure 12.10 Diagram of correlated mRNA transcripts, proteins, and lipids changes in APOE*3-Leiden mouse livers. mRNA transcript, proteomic, and lipidomic profiles were acquired from liver samples of APOE*3-Leiden ($n = 4$) and wild-type ($n = 4$) mice. Among the molecules depicted here is a selection of those that showed significant-fold differences between the two groups. The shading inside the polygon indicates the direction of the difference between the transgenic and wild-type control animals (black fill = higher level, gray fill = no change, and white fill = lower level) and a line connecting two polygons indicates a high level of correlation (a Pearson correlation coefficient greater than or equal to 0.8). Significant increases in liver C16:0 LysoPC and C38:1 DAG were highly correlated to increases in liver fatty acid protein message and protein levels, while these molecules were also correlated with a lowering of apolipoprotein AI (ApoAI) message.

of fatty acid binding protein (FABP), as well as a high correlation with *decreasing* apolipoprotein A1 (ApoA1) mRNA expression. Furthermore, the decrease in apolipoprotein A1 expression was also highly correlated to the changes in FABP mRNA and protein expression levels. Prior to this analysis, there were no documented reports of a direct connection between these specific lipids and FABP and/or apolipoprotein A1, and thus such results provide a basis for the generation of new hypotheses and further experimentation.

12.7 SUMMARY

In conclusion, lipidomics applied in comparative analyses of diseased and non-diseased experimental or clinical subjects provides a powerful means of uncovering specific biomarkers of disease. By incorporating lipidomics into a wider multi-omic, or systems biology, analysis, we can begin to elucidate interconnected relationships among changes across a range of biomolecule classes and provide broader insight into the pathophysiology of disease. At this juncture we can also begin to appreciate the temporal differences as well as spatial components within sites of inflammation that are responsible for generating specific local-acting lipid-derived mediators. Mapping of the local biochemical mediators and the impact of drugs, diet, stress (e.g., hypoxia and ischemia reperfusion) in these bionetworks constitute exciting research terrain. Transient and quantitatively fleeting members of lipid mediator pathways and their temporal relationship change extensively during the course of a physiologic or pathophysiologic response. The application of mediator-lipidomic profiling technologies to quantify these changes over time enables us to decode the network of relationships among autocoids/local mediators. Moreover, its utility in finding novel approaches to understanding the basis of complex human diseases and search for new therapeutic interventions will accelerate the growth of lipidomics.

ACKNOWLEDGMENTS

We thank C. Gitlin and M. Halm Small for assistance with manuscript preparation, and E. Tjonahen, Center for Experimental Therapeutics and Reperfusion Injury, for assistance with graphics. The work in the C.N.S. lab was supported in part by National Institutes of Health Grant Nos. GM38765 and P50-DEO16191.

REFERENCES

1. Lu, Y., Hong, S., and Serhan, C.N., Mediator-lipidomics: databases and search algorithms for PUFA-derived mediators, *J. Lipid Res.*, 46, 790, 2005.
2. Samuelsson, B., From studies of biochemical mechanisms to novel biological mediators: prostaglandin endoperoxides, thromboxanes and leukotrienes, in *Les Prix Nobel: Nobel Prizes, Presentations, Biographies and Lectures*, Almqvist & Wiksell, Stockholm, 1982, 153.
3. Bergström, S., The prostaglandins: from the laboratory to the clinic, in *Les Prix Nobel: Nobel Prizes, Presentations, Biographies and Lectures*, Almqvist & Wiksell, Stockholm, 1982, 129.
4. Serhan, C.N. et al., Novel functional sets of lipid-derived mediators with antiinflammatory actions generated from omega-3 fatty acids via cyclooxygenase 2-nonsteroidal antiinflammatory drugs and transcellular processing, *J. Exp. Med.*, 192, 1197, 2000.
5. Singer, S.J. and Nicholson, G.L., The fluid mosaic model of the structure of cell membranes, *Science*, 175, 720, 1972.
6. Hannun, Y.A. and Obeid, L.M., The ceramide-centric universe of lipid-mediated cell regulation: stress encounters of the lipid kind, *J. Biol. Chem.*, 277, 25847, 2002.
7. Gronert, K. at al., A molecular defect in intracellular lipid signaling in human neutrophils in localized aggressive periodontal tissue damage, *J. Immunol.*, 172, 1856, 2004.
8. Clish, C.B. et al., Oxidoreductases in lipoxin A_4 metabolic inactivation: 15-oxoprostaglandin 13-reductase/ leukotriene B_4 12-hydroxydehydrogenase is a multifunctional eicosanoid oxidoreductase in inflammation. *J. Biol. Chem.*, 275, 25372, 2000.
9. Levy, B.D. et al., Lipid mediator class switching during acute inflammation: signals in resolution, *Nature Immunol.*, 2, 612, 2001.

10. Serhan, C.N. et al., Resolvins: a family of bioactive products of omega-3 fatty acid transformation circuits initiated by aspirin treatment that counter proinflammation signals, *J. Exp. Med.*, 196, 1025, 2002.
11. Serhan, C.N. et al., Reduced inflammation and tissue damage in transgenic rabbits overexpressing 15-lipoxygenase and endogenous antiinflammatory lipid mediators, *J. Immunol.*, 171, 6856, 2003.
12. Hong, S. et al., Novel docosatrienes and 17S-resolvins generated from docosahexaenoic acid in murine brain, human blood and glial cells: autacoids in antiinflammation, *J. Biol. Chem.*, 278, 14677, 2003.
13. Burr, G.O. and Burr, M.M., A new deficiency disease produced by the rigid exclusion of fat from the diet, *J. Biol. Chem.*, 82, 345, 1929.
14. Mukherjee, P.K. et al., Neuroprotectin D1: a docosahexaenoic acid-derived docosatriene protects human retinal pigment epithelial cells from oxidative stress, *Proc. Natl. Acad. Sci. U.S.A.*, 101, 8491, 2004.
15. Lutgens, E. et al., Atherosclerosis in APOE*3-Leiden transgenic mice: from proliferative to atheromatous stage, *Circulation*, 99, 276, 1999.
16. Clish C.B. et al., Intregrative biological analysis of the APOE*3-Leiden transgenic mouse, *OMICS*, 8, 3, 2004.
17. Butte, A.J. et al., Discovering functional relationships between RNA expression and chemotherapeutic susceptibility using relevance networks, *Proc. Natl. Acad. Sci. U.S.A.*, 97, 12182, 2000.

CHAPTER 13

Molecular Detection and Characterization of Circulating Tumor Cells and Micrometastases in Solid Tumors

Ronald A. Ghossein, Hikmat Al-Ahmadie, and Satyajit Bhattacharya

CONTENTS

13.1 INTRODUCTION

The detection of circulating tumor cells (CTC) has interested physicians since the 19th century, when Ashworth described a case of cancer in which cells similar to those in the tumor were found in the blood after death.[1] However, the detection of CTC first gained widespread attention in 1955 when Engell reported the detection of CTC in patients with various types of carcinomas using a cell block technique.[2] Subsequently, between 1955 and 1965, several thousand patients with cancer (most with solid malignancies) were tested for CTC by 40 investigative teams using 20 different cytologic methods.[3] These early studies reported very high positivity rates of CTC among patients with cancer (up to 100%).[3] However, these results were soon shown to be due to false positives since circulating hematopoietic elements, especially megakaryocytes, were often confused with tumor cells. When cell preservation techniques were improved, allowing a better morphological analysis, the detection of true CTC by light microscopy was shown to have a very low sensitivity (approximately1%) in patients with cancer.[3] Routine cytologic examination of blood specimens for CTC was therefore abandoned in the mid-1960s. The issue of CTC and micrometastases reappeared 20 years later with the advent of immunocytochemistry. Sensitive immunocytologic assays were developed to detect tumor cells in the bone marrow (BM) and peripheral blood (PB) of patients with neuroblastoma, breast, and lung carcinomas.[4–6] Immunostains were shown to identify BM micrometastases with much greater sensitivity than conventional techniques.[5,6] Indeed, these immunocytological assays were said to detect a single tumor cell seeded among 10,000 to 100,000 mononuclear cells. Despite evidence of the prognostic value of this determination in some studies,[6–9] the detection of micrometastases by immunocytochemistry was not routinely used in cancer staging protocols.[10] This was due to a combination of factors, such as the absence of clinical significance in some studies.[11–14] loss of antigen expression in poorly differentiated tumors, and reports of false positives with epithelial markers such as cytokeratin and epithelial membrane antigen.[15,16] Meanwhile, there was hope for the development of an ever better assay for the detection of occult tumor cells using nucleic acid analysis. This hope was fulfilled by the development of the highly sensitive polymerase chain reaction (PCR) technique in the mid-1980s.[17] Since 1987, a variety of PCR-based techniques have been devised for the identification of CTC and micrometastases in leukemias, lymphomas, and various types of solid malignancies.[18–23] The focus of this chapter is the detection and characterization of CTC in five major types of solid tumors, namely, malignant melanoma and carcinomas of the prostate, breast, lung, and gastrointestinal tract.

13.2 PCR TECHNOLOGY

PCR is an *in vitro* method that enzymatically amplifies specific DNA sequences using oligonucleotide primers (short DNA sequences composed of 18 to 25 nucleotides) that flank and therefore define the region of interest in the target DNA.[24] PCR amplification can be accomplished using RNA as starting material. This procedure is known as reverse transcriptase PCR (RT-PCR). It is similar to standard PCR with the modification that PCR amplification is preceded by reverse transcription of RNA into cDNA.

One major strategy for the detection of occult tumor cells is PCR amplification of tumor-specific abnormalities present in the DNA or mRNA of these cells. This approach was mostly used in hematological malignancies. It was first applied to the detection of the t(14;18) translocation associated with follicular lymphomas.[22] The primers used hybridize to the region flanking the translocation and will therefore amplify the DNA only when the translocation is present. If the translocation is not present, the primers

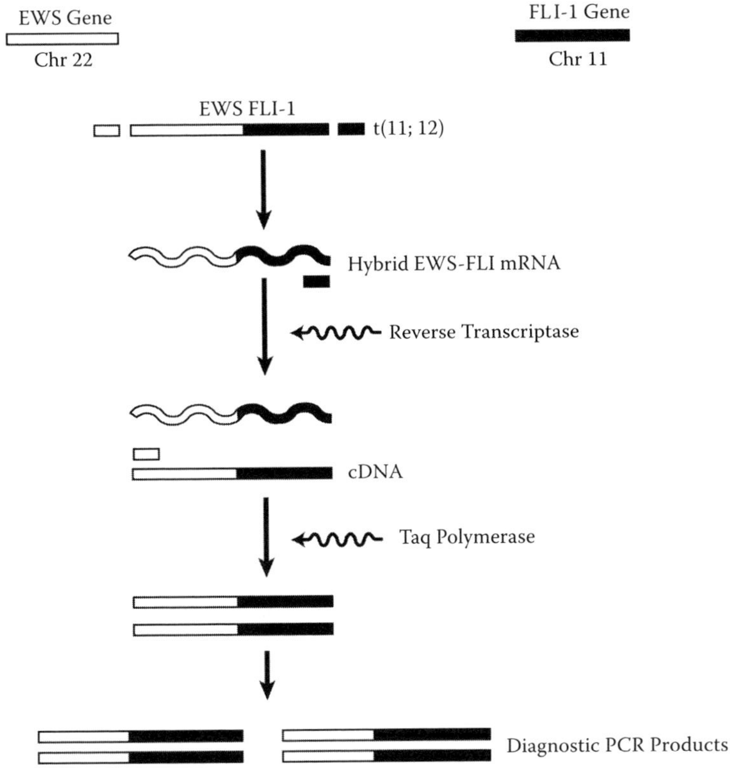

Figure 13.1 Detection of occult tumor cells by RT-PCR amplification of tumor-specific abnormalities in the mRNA. In this example, the primers are chosen to flank the t(11;22) translocation present in Ewing's sarcoma (EWS). This translocation juxtapose the FLI-1 gene on chromosome 11 to the EWS gene on chromosome 22. The primers will therefore anneal to and amplify the hybrid EWS/FLI-1 transcript when the translocation is present. (From Ghossein, R.A. et al. *Clin. Cancer Res.* 5, 1950–1960, 1999. With permission.)

anneal to different chromosomes and PCR is impossible. The detection of occult tumor cells by RT-PCR of chimeric tumor-specific mRNA has been performed in a few solid tumors such as Ewing's sarcoma[25] (Figure 13.1).

The other main PCR strategy for the detection of occult tumor cells involves amplification of tissue-specific mRNA by RT-PCR. This has been mainly used for the detection of CTC and micrometastases in solid tumors since tumor specific abnormalities are rare in nonhematopoietic malignancies (Table 13.1). This approach is based on the fact that malignant cells often continue to express markers that are characteristic or specific of the normal tissue from which the tumor originates. It is the appearance of these tissue-specific mRNAs at a body site where these transcripts are not normally present that implies tumor spread (e.g., the melanocytic tissue-specific marker tyrosinase mRNA in BM). From a technical standpoint, RT-PCR detection of any tissue-specific marker requires knowledge of its gene sequence and specifically of intron–exon junctions, which facilitates the selection of oligonucleotide primers for RT-PCR (Figure 13.2).

Table 13.1 PCR and RT-PCR Methods for the Detection of Occult Tumor Cells in Solid Tumors

Tumor Type	Molecular Target
Melanoma	Tyrosinase mRNA
	MART 1 mRNA
	GAGE mRNA ☒
Prostate	PSA mRNA
	PSMA mRNA
Breast carcinoma	Muc 1 mRNA
	CEA mRNA
	Cytokeratin 19 mRNA
	Mammaglobin mRNA
Hepatocellular carcinoma	AFP mRNA ☒
	Albumin mRNA
Gastrointestinal carcinomas	CEA mRNA
	Cytokeratin 20 mRNA
Lung carcinoma	CEA mRNA
	Muc 1 mRNA
	Cytokeratin 19 mRNA
	Surfactant protein mRNA
	EGFR ☒
Neuroblastoma	Tyrosine hydroxylase mRNA
	PGP 9.5 mRNA
	GAGE mRNA ☒
Ewing's sarcoma	EWS/FLI1 fusion transcript ☒
	EWS-ERG fusion transcript ☒
Uterine cervix carcinoma	SCC antigen mRNA
	HPV E6 mRNA ☒
Thyroid carcinomas of follicular origin	TGB mRNA
	TPO mRNA

Abbreviations: RT-PCR: reverse transcriptase-polymerase chain reaction; PSA: prostate-specific antigen; PSMA: prostate-specific membrane antigen; CEA: carcinoembryonic antigen; AFP: alpha fetoprotein; PGP 9.5: neuroendocrine protein gene product; EWS: Ewing sarcoma; SCC: squamous cell carcinoma; HPV: human papilloma virus; TGB: thyroglobulin; TPO: thyroid peroxidase. Except for those molecules labeled with ☒; all other markers are tissue specific.

13.2.1 Limitations of PCR Technology

13.2.1.1 False Positive PCR Results

The power of PCR resides in the extreme sensitivity of the technique. Many publications report the detection of one tumor cell per milliliter of whole blood (Figure 13.3).[26] It is this extreme sensitivity that confers an inherent tendency to produce false positive results if sufficient precautions are not taken to prevent contamination of samples. One study reported a wide variability of results from one laboratory to the next using coded samples.[27] Meticulous laboratory techniques have been developed to prevent contamination of samples.[24] False positives could be due to the general process of illegitimate transcription (i.e., transcription of any gene in any cell type). Although the number of these trancripts in inappropriate cells is very low (estimated at 1 mRNA molecule per 100 to 1000 cells),[28] it can result in the occurrence of false positives

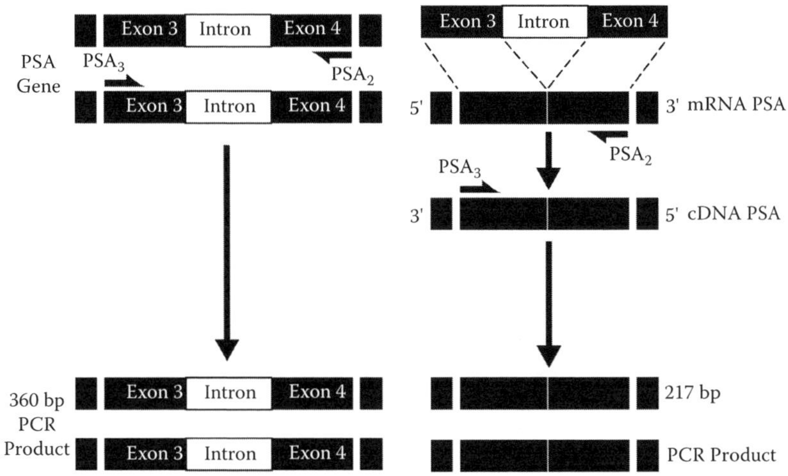

Figure 13.2 Detection of occult tumor cells by RT-PCR of tissue-specific mRNA. In this example, primer sets for PSA mRNA were selected to span the intronic sequence. This will allow discrimination, based on size, between RT-PCR products from mRNA targets (right) and PCR products from contaminating genomic DNA (left). (From Ghossein, R.A. et al. *Cancer* 78(1), 10–16, 1996. Copyright © 1996 American Cancer Society. With permission of Wiley-Liss, Inc., a subsidiary of John Wiley & Sons, Inc.)

because of the high sensitivity of RT-PCR. For example, a neuronal specific marker, neuroendocrine protein gene product (PGP 9.5), was shown to be present in scant amount in normal BM cells.[29] In view of this problem, there has been much effort to find genes that display the least amount of illegitimate transcription in blood, BM, and lymph nodes.[30] Some authors have attempted to solve this issue by optimizing the PCR thermocycling conditions, as has been shown for tyrosinase mRNA, a marker of melanocytic lineage.[31] For example, the number of PCR cycles should be carefully selected to be high enough to detect occult tumor cells but low enough to avoid amplification of illegitimate transcripts.[30] Processed pseudogenes can also give rise to false positive results. Since they lack an intronic sequence, RT-PCR amplification of processed pseudogenes will lead to PCR products indistinguishable from those generated from the mRNA. Because most markers of CTC and micrometastases of solid tumors are tissue specific (i.e., expressed in tumor and their normal tissue of origin), the mechanical introduction of normal or benign cells in the circulation after invasive procedures may lead to false positive PCR results. For example, many studies showed that a significant number of patients hemoconverted from RT-PCR negative to RT-PCR positive after radical prostatectomy (RP).[32] However, the percentage of RT-PCR-negative patients hemoconverting after less invasive procedures (e.g., transrectal ultrasound, prostatic core biopsy) was much lower. These false positive PCR results can be averted by timing the RT-PCR assays weeks after any invasive procedure. In principle, venipuncture by itself may generate false positives because of the introduction of normal keratinocytes or melanocytes in the circulation. We did not encounter false positive results while PCR testing PB and BM for melanocytic tissue-specific markers in our control population.[33] Our control group included dark-skinned individuals making it unlikely that venipunc-

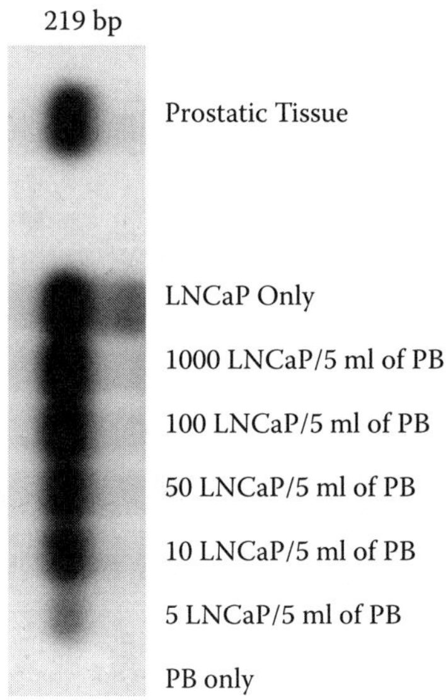

Figure 13.3 Immunobead nested RT-PCR for PSMA mRNA after Southern blot hybridization of the nested RT-PCR products. Results of sensitivity experiment. Samples containing only prostatic tissue and LNCap prostatic carcinoma cells are used as positive controls. A sample containing only PB from a healthy subject is used as negative control. The remaining samples are serial dilutions of LNCaP cells with whole blood from healthy volunteers. After nested immunobead RT-PCR (two rounds of amplification), PCR can produce a band corresponding to 5 LNCaP cells diluted in 5 ml of PB. The diagnostic fragment is indicated at left in base pairs. (From Ghossein, R.A. et al. *Diagn. Mol. Pathol.* 8, 59–65, 1999. With permission.)

ture is a cause of false positives by RT-PCR since these individuals harbor a high number of normal melanocytes in their skin**.** This is most probably due to the fact that PCR sensitivity *in vivo* is not as high as the one reported *in vitro* (see next paragraph). PCR is therefore not able to detect the rare skin melanocytes that are introduced in the sample after blood drawing or BM aspiration. To avoid this problem, some authors recommend discarding the first few milliliters of blood that are collected, as they may be contaminated with normal cells from the epidermis.[34]

13.2.1.2 False Negative PCR Results

The sensitivity of PCR is variable, and this can lead to false negative results, especially in the detection of occult tumor cells where low-level signals are expected. Inhibitors present in some tissues and fluids can diminish PCR sensitivity. Therefore, careful controls are necessary to ensure that there is amplifiable RNA or DNA in the sample. This is accomplished by demonstrating amplification of a constitutively present transcript such as beta actin. The reader should therefore be aware that the *in vitro* sensitivity reported in all articles on CTC and micrometastases (often expressed in number of cell line-derived tumor cells detected per million of white cells) does not reflect the *in vivo* sensitivity of PCR. The latter is most probably lower than the *in vitro* sensitivity because of inhibitors of the PCR reaction present in tissues and body fluids and because the tumor cell line chosen for these sensitivity experiments strongly express the marker of interest. In contrast, tumor cells *in vivo* may not necessarily express the marker of interest because of tumor cell heterogeneity. False negatives could also be due to a sampling problem or to intermittent shedding of tumor cells in the circulation since only a few milliliters of PB are analyzed at a certain time. The latter two problems could be minimized by sequential sampling, defined as the analysis of multiple blood samples at different time points.[35] False negative results could also be due to downregulation of the target gene by therapy (e.g., hormonal treatment) or to the presence of poorly differentiated subclones that do not express the tissue-specific marker being tested. For example, PSA mRNA expression was shown to be decreased by antiandrogen therapy[36] and in poorly differentiated prostatic carcinoma.[37] In this setting, a multiple marker PCR assay may help increase PCR positivity by overcoming the problem of tumor cell heterogeneity.

13.2.2 Quantitative PCR

It is now possible to quantify the amount of target nucleic acids present in a given sample with a user friendly automated real-time quantitative RT-PCR assay.[38] These quantitative PCR methods are, however, unable to estimate the number of tumor cells present in a sample, since the transcription rate (i.e., the amount of target mRNA) varies between individual tumor cells.[39] This fact limits the value of quantitative PCR in detecting occult tumor cells.

13.3 APPLICATIONS TO SPECIFIC TUMOR TYPES

13.3.1 Prostatic Carcinoma

RT-PCR detection of CTC and micrometastases has the potential to improve case selection in patients with localized prostatic carcinoma (PC) and to monitor disease activity more accurately in patients with metastatic disease. We and others have detected occult tumor cells in the PB and BM of patients with localized and metastatic PC using RT-PCR for PSA mRNA[40–51] (Table 13.2). We detected CTC in 16% of patients with clinically organ-confined (T1-2) disease and in 35% of patients with distant metastases.[40] In accordance with most other reports on the subject.[32] none of our controls was positive, indicating the specificity of the technique when applied to PB. The frequency of RT-PCR positivity increases with tumor stage and high serum PSA levels.[40] Unfortunately, a significant proportion of patients with metastatic disease was negative. Prostatic cells may be shed intermittently in the circulation, and this phenomenon leading to sampling errors. Other possibilities include (1) the presence in the circulation of tumor cells that express very low levels of PSA mRNA because of tumor cell heterogeneity; and (2) a difference in sensitivity between different sets of PCR primers for a given marker.[52,53] To avoid false positives due to mechanical introduction of benign prostatic epithelial cells in the circulation, our patient population was tested 8 weeks after any prostatic invasive procedure. One article reported the detection CTC in 20%

Table 13.2 RT-PCR Detection of CTC and BM Micrometastases in PC Using PSA and PSMA mRNA

Ref.	Marker	Sample	Localized PC*	Metastatic PC**
			RT- PCR Pos/Total (%)	
Katz et al.[42]	PSA mRNA	Blood	25/65 (38%)	14/18 (78%)
Israeli et al.[41]	PSA mRNA	Blood	0/18 (0%)	6/24 (25%)
	PSMA mRNA	Blood	13/18 (72%)	16/24 (67%)
Seiden et al.[43]	PSA mRNA	Blood	3/41 (7%)	11/35 (31%)
Ghossein et al.[40]	PSA mRNA	Blood	4/25 (16%)	26/76 (34%)
Sokoloff et al.[44]	PSA mRNA	Blood	43/69 (62%)	29/33 (88%)
	PSMA mRNA	Blood	12/69 (17%)	13/33 (39%)
Corey et al.[45]	PSA mRNA	Blood	12/63 (19%)	6/13 (46%)
		BM	45/63 (71%)	10/13 (77%)
Wood et al.[46]	PSA mRNA	BM	39/86 (45%)	—
Gao et al.[47]	PSA mRNA	Blood	25/84 (30%)	3/8 (37.5%)
Ennis et al.[48]	PSA mRNA	Blood	55/201 (27%)	—
Loric et al.[68]	PSMA mRNA	Blood	6/17 (35%)	28/33 (85%)
Zhang et al.[72]	PSA mRNA	Blood	6/48 (12.5%)	7/11 (64%)
	PSMA mRNA	Blood	11/48 (23%)	10/11 (98%)
	PSA/PSMA	Blood	14/48 (29%)	11/11 (100%)
Shariat et al.[51]	PSA mRNA	Blood	39/145 (27%)	—
Kantoff et al.[50]	PSA mRNA	Blood	—	75/156 (48%)

Abbreviations: RT-PCR: reverse transcriptase polymerase chain reaction; CTC: circulating tumor cells; BM: bone marrow; Pos: positive; PC: prostatic carcinoma; PSA: prostatic specific antigen; PSMA: prostatic specific membrane antigen; PSA/PSMA: combined assay with its positivity defined as positivity for either or both markers.

* Localized PC includes stage A,B, (clinically organ confined disease only).

** Metastatic PC includes stage D1–D3 patients (D1: pelvic lymph node metastases, D2: distant metastases without prior hormonal therapy, D3: D2 disease refractory to hormonal therapy) in all the listed studies except in that of Israeli et al.[41] In that article, three patients with "D0" disease (elevated serum tumor markers only) were also included as metastatic PC.

of previously RT-PCR-negative patients after needle biopsy.[54] The conversion rates were similar in patients regardless of biopsy results. Testing of serial post-biopsy samples revealed that most patients hemoconverting after biopsy reverted to an RT-PCR-negative PCR assay within 4 weeks.

Two groups of researchers showed that the presence of CTC by RT-PCR correlated with both capsular penetration and positive surgical margins.[42,55] They found RT-PCR to be superior to other staging modalities in predicting pathologic stage and proposed the use of this test as a staging modality for radical prostatectomy candidates. Since these exciting reports, all studies on the subject have not found a statistically significant and useful correlation between blood RT-PCR positivity and pathologic stage in patients with clinically organ-confined disease undergoing radical prostatectomy.[44,47,56–58] At the present time, RT-PCR detection of circulating prostatic tumor cells is not a useful staging tool for patients with radical prostatectomy. With regard to molecular prognosis, some groups have found a statistically significant correlation between preoperative RT-PCR positivity for PSA mRNA in PB and postoperative biochemical failure,[59,60] while other authors did not[47,51,56] (Table 13.3). Shariat et al.[51] found that early postoperative blood RT-PCR for PSA is an independent prognostic factor for poorer progression-free sur-

Table 13.3 Molecular Prognosis in PC Using RT-PCR for PSA and PSMA

Reference	Patient Population	Sample	Marker	End Point	P Value
de la Taille et al.[60]	Localized PC	Blood pre-RP	PSA	Failure-free survival	0.0002
Wood and Banerjee[46]	Localized PC	Bone marrow pre-RP	PSA	Failure-free survival	0.004
Gao et al.[47]	Localized PC	Blood pre-RP	PSA	Failure-free survival	0.598
Okegawa et al.[69]	Localized PC	Blood pre-RP	PSMA	Failure-free survival	<0.01
Shariat et al.[51]	Localized PC Localized PC	Blood pre-RP Blood post-RP	PSA PSA	Failure-free survival Failure-free survival	0.7221 0.022
Ghossein et al.[61]	Metastatic AIPC	Blood	PSA	Overall survival	0.028

Note: Failure was defined as serum PSA > 0.2 ng/ml on one occasion after RP in de la Taille's article and on two occasions in Gao's and Shariat's articles. Recurrence was defined as serum PSA ≥ 0.4 ng/ml in Okegawa's article and serum PSA > 0.4 ng/ml or local recurrence on digital rectal exam after RP in Wood and Banerjee's article. Except for Gao's article and Shariat's pre-RP samples, RT-PCR positivity did correlate with poorer survival. Only those articles using Kaplan–Meier survival analysis are included in this table. PC: Prostatic carcinoma; AI: Androgen independent; RT-PCR: Reverse transcriptase polymerase chain reaction; PSA: prostatic specific antigen; PSMA: Prostatic specific membrane antigen.

vival. We assessed the prognostic value of RT-PCR for PSA in metastatic disease by analyzing the PB of 122 men with metastatic androgen independent (AI) PC. Of these patients, 64 were tested in our institution, while the remainder were assayed at the Dana Farber Cancer Institute. We found that RT-PCR positivity correlates with decreased overall survival in both institutions. We also showed that RT-PCR is superior to a single serum PSA measurement in predicting survival in both groups of patients.[61]

RT-PCR for PSA mRNA has also been used to detect occult tumor cells in lymph nodes and, as stated earlier, in BM of patients with PC.[45,46,62–64] This technique was shown to be more sensitive than immunohistochemistry and standard histopathology in detecting lymph node micrometastases in localized disease.[62] All control lymph nodes and BM tested negative for PSA RT-PCR.[32,45,64] Wood and Banerjee followed 86 patients with clinically localized disease in whom preoperative bone marrow PSA RT-PCR was performed.[46] These authors defined recurrence as a postoperative serum PSA > 0.4 ng/ml, or clinical evidence of locally recurrent disease by digital rectal examination. Of the RT-PCR negative patients, 4% suffered a recurrence after prostatectomy, while 26% of the RT-PCR positive patients failed postoperatively.[46] Edelstein and colleagues[64] found a similar correlation when they studied pelvic lymph nodes using RT-PCR for PSA mRNA. Of the PCR-negative patients, 30% failed compared to 87.5% of the PCR-positive patients within a 5-year follow-up period.

RT-PCR assays for two additional prostatic markers, prostatic-specific membrane antigen (PSMA) and prostatic stem cell antigen (PSCA), have been reported.[41,47,65,66] PSMA is a cell-surface protein with sequence homology to transferrin. PSMA is expressed in benign and malignant prostatic epithelium and is upregulated in hormone refractory states, in metastatic situations, or in other situations where there is tumor recurrence or extension.[67] PSMA transcripts were detected in the peripheral blood of patients with localized and metastatic PC using RT-PCR.[41,44,65,68,69] Some investigators reported a high PCR positivity rate for PSMA mRNA in the blood of healthy individuals.[70,71] We and others did not encounter any false positives with PSMA RT-PCR.[26,72] When a combined PSA and PSMA RT-PCR test was used for CTC, this resulted in an increase in sensitivity and prognostic significance compared to a one-marker assay in metastatic androgen-independent PC (R. Ghossein, Memorial Sloan-Kettering Cancer Center). PSCA is a glycoprotein predominantly expressed in the basal cells of the normal prostatic glands, in placenta, and in > 80% of prostatic carcinomas.[66] In patients with extra-prostatic disease, PSCA RT-PCR in blood was shown to predict progression-free survival.[66] However, no multivariate analysis was performed in that study.

At the present time, most of the data suggest that RT-PCR assays for CTC and micrometastases in PC are predictors of outcome. However these assays are still unable to address the most important problem in the management of PC, which is to help better stage patients for radical prostatectomy (RP). One of the many reasons accounting for the limitation of RT-PCR as clinical assays in PC is the fact that it has to "compete" against a clinically very powerful marker, blood serum PSA. Indeed, only rare studies show a prognostic value for RT-PCR that is superior to serum PSA in multivariate analysis.

13.3.2 Breast Carcinoma

The majority of patients with mammary carcinoma (approximately 90%) present with tumors that are clinically confined to the breast and neighboring axillary lymph nodes. Essentially, all these patients are rendered free of measurable disease after primary surgery.[73] Despite this highly efficient locoregional therapy, 30 to 40% of these patients will develop clinically detectable metastases within 10 years if no further treatment is instituted.[73] The chief reason for these relapses is that breast carcinoma cells disseminate throughout the body early in tumor development.[74] To prevent the clinical progression of these micrometastases, about two thirds of the patients diagnosed with stage I to III breast cancer are candidates for adjuvant or neoadjuvant chemotherapy.[75] It has been reported that approximately 36% of these women would remain free of disease using locoregional therapy alone. Routine adjuvant chemotherapy would subject these patients to unnecessary and toxic treatment. To better identify those patients who will benefit from adjuvant chemotherapy, several groups have attempted the detection of BM micrometastases by immunohistochemistry.[7,76,77] Some authors have indicated the prognostic significance of these sensitive immunocytochemical assays,[7,77] but others failed to demonstrate such relevance.[11–14] Indeed, a significant minority of patients whose BM was positive by immunohistochemistry have remained free of clinically evident metastatic disease after relatively long intervals.[73] These findings could be due to several factors. Some micrometastases may be incapable of developing into clinically significant lesions.[73] Alternatively, the antibodies may have cross-reacted with normal marrow cells, leading to false positive results.

Table 13.4 RT-PCR and PCR Positivity Rate in Control Subjects Using Putative Markers for Breast Carcinoma

Ref.	Marker	Sample	Positive (%)	Total
Eltahir et al.[88]	Muc1 mRNA	Bl	21 (91%)	23
	CD44 variant	Bl	4 (40%)	10
Krisman et al.[90]	CK 19	Bl	13 (20%)	65
Mori et al.[84]	CEA	Bl	0 (0%)	22
Lopez-Guerrero et al.[91]	CK 19	Bl	0 (0%)	10
	CEA	Bl	0 (0%)	4
	Maspin	Bl	1 (20%)	5
De Graaf et al.[92]	*EGP-2*	Bl	10 (100%)	10
Ko et al.[89]	CEA	Bl	8 (33%)	24
Zach et al.[102]	Mammaglobin	Bl	0 (0%)	27
Silva et al.[103]	Mammaglobin	Plasma	3 (12%)	25
Bostick et al.[87]	Beta 1 4GalNAc-T	LN	0 (0%)	10
	C-Met	LN	1/10 (10%)	10
	P97	LN	1/10 (10%)	10

Note: Control subjects were defined as healthy volunteers only in all studies except Lopez-Guerrero's article. In this article,[91] control subjects were defined as "healthy volunteers and patients without any type of solid tumors." In all articles, non-immunobead RT-PCR techniques were used. RT-PCR: reverse transcriptase–polymerase chain reaction; CK: cytokeratin; CEA: carcinoembryonic antigen; Bl: blood; LN: lymph nodes; Beta 1-4GalNAc-T: beta 1 4-*N*-acetylgalactosaminyltransferase.

Several authors were able to detect tissue-specific transcripts in the PB, BM, and lymph nodes of patients with breast carcinomas using highly sensitive RT-PCR assays.[20,78–87] Unfortunately, almost all of the markers used were shown to have false positives (Table 13.4).[86–92] These false positives could be due to illegitimate transcription, the presence of pseudogene, or sample contamination. In the hope of improving RT-PCR specificity, several authors have lately attempted variations on previously published PCR protocols including the use of real-time quantitative RT-PCR or novel markers.[93–99]

One of these "popular" novel markers is mammaglobin, a tissue-specific marker that has homology with a family of secreted proteins that includes rabbit uteroglobin. This marker was found to be present only in adult mammary tissue and in 80 to 95% of primary breast carcinomas where it is frequently overexpressed.[100] According to one study, this marker was detectable by RT-PCR in breast carcinoma cell lines and absent in 20 normal lymph nodes.[85] In a small group of patients with breast carcinoma, Aihara et al.[101] found mammaglobin transcripts by RT-PCR in all histologically proven metastatic lymph nodes and in 31% of histologically negative lymph nodes. All their control lymph nodes were negative by mammaglobin RT-PCR. Zach et al.[102] were able to detect mammaglobin mRNA in the PB of 28% of patients with breast carcinoma of various stages, 5% of patients with nonbreast carcinoma malignancies, and in none of 27 healthy volunteers. However, one study showed the presence of mammaglobin mRNA in the plasma of healthy individuals.[103] Even the use of real-time quantitative RT-PCR did not eliminate the false positives encountered with the breast carcinoma-related markers. In one study, there was an overlap in the relative copies of cytokeratin 18 and 19 transcripts in BM between patients with benign tumors and those with breast carcinoma.[97] Despite this persistent specificity problem, many recent articles have shown a statistically significant correlation between RT-PCR detection of breast carcinoma related transcripts and survival.[94,98,104–107] In node-positive patients, bone marrow mammaglobin RT-PCR positivity significantly increased the risk of recurrence at a distant site.[94] The presence of cytokeratin 19 mRNA after surgery in the blood of patients with localized disease was a significant indicator of poor overall and disease-free survival.[104] Using four markers including cytokeratin 19, Weigelt et al.[96] were able to demonstrate that RT-PCR positivity in blood correlate with disease-free and overall survival in patients with distant metastases. These encouraging results demonstrate the potential clinical value of the detection of occult tumor cells in breast carcinoma. However, the use of these assays in the clinic awaits further improvement in specificity.

13.3.3 Malignant Melanoma

The main current criteria to assess prognosis in malignant melanoma are the histopathologic features of the primary tumor and the clinical presentation. However, these factors are of limited value in the advanced stages of the disease.[108] There is therefore a need for a better prognostic marker in those patients. The molecular detection of CTC and BM micrometastases has the potential for predicting outcome in patients with malignant melanoma. Smith et al.[31] were the first to propose that melanoma cells could be detected in PB using RT-PCR for tyrosinase mRNA. Tyrosinase is a key enzyme in melanin biosynthesis that catalyzes the conversion of tyrosine to dopa, and of dopa to dopaquinone. This test is presumed to detect circulating melanoma cells since tyrosinase is one of the most specific markers of melanocytic differentiation,[109] and melanocytes are not known to circulate. Furthermore, most studies show that tyrosinase mRNA is not present in the PB of healthy individuals.[33,108,110–112] Since the original study of Smith et al.,[31] many groups have attempted the detection of CTC in malignant melanoma using tyrosinase mRNA.[33,35,108,110–122] As shown in Table 13.5, the PCR positivity rates are extremely variable ranging from 0 to 100%. There is a correlation between blood tyrosinase RT-PCR results and stage in some but not all the studies. These disparate findings could in part be explained by differences in RNA extraction and PCR methodology.[109] They could also be due to unrecognized contamination leading to false positive results. Indeed, Foss et al.[113] acknowledged the presence of significant technical problems due to carry-over contamination that took 1 year to overcome. Despite these discrepancies, several authors have shown that RT-PCR for tyrosinase mRNA in PB is able to predict overall survival and disease-free survival in a statistically significant manner.[35,108,111,115–118,123] (Table 13.6). One study has demonstrated that blood RT-PCR is an independent prognostic marker for relapse-free survival in multivariate analysis in patients with advanced melanoma undergoing interferon therapy.[35] Mellado et al.[118] found an adverse prognostic effect for the presence of tyrosinase blood RT-PCR in patients with American Joint Committee on Cancer (AJCC) Stage II to IV melanoma undergoing similar therapy (Stage II: primary tumor > 1.5 mm in thickness with no metastases; Stage III: regional lymph node metastases; Stage IV: distant metastases). These results are, however, tempered by negative studies showing no prognostic value for blood tyrosinase RT-PCR in melanoma.[119] In an effort to improve the clinical value of RT-PCR for tyrosinase mRNA, Brossart et al.[124] developed a semiquantitative RT-

Table 13.5 Detection of CTC in the Peripheral Blood of Patients with Cutaneous Malignant Melanoma Using RT-PCR

Ref.	No. of RT-PCR-Positive Patients/Total No. of Patients Tested (%) According to AJCC Stage		
	I–II	III	IV
Brossart et al.[112]	1/10 (10%)	6/17 (35%)	29/29(100%)
Hoon et al.[110*]	13/17 (76%)	31/36 (86%)	63/66 (95%)
Battayani et al.[108]	2/10 (20%)	22/51 (43%)	16/32 (50%)
Foss et al.[113]	—	—	0/6 (0%)
Pittman et al.[114]	—	—	3/24 (12.5%)
Kunter et al.[111]	0/16 (0%)	0/16(0%)	9/34 (26%)
Mellado et al.[115]	8/44 (18%)	2/13 (15%)	—
Curry et al.[116***]	48/160 (30%)	60/116 (52%)	—
Farthman et al.[117]	6/46 (13%)	7/41 (17%)	16/36 (44%)
Cheung et al.[127****]	5/17 (29%)	4/54 (7%)	4/27 (15%)
Schitteck et al.[128***]	28/119 (24%)	14/48 (29%)	30/58 (52%)
Palmieri et al.[119**]	113/144 (78)	22/24 (92%)	23/23 (100%)

Abbreviations: CTC: circulating tumor cells; RT-PCR: reverse transcriptase polymerase chain reaction; AJCC: American Joint Committee on Cancer; AJCC Stage I: primary tumor < 1.5 mm in thickness with no metastases; AJCC Stage II: primary tumor > 1.5 mm in thickness with no metastases; AJCC Stage III: regional lymph node metastases; AJCC Stage IV: distant metastases.

* In this study, the peripheral blood was analyzed for four markers (tyrosinase, p97, Muc 18, MAGE-3).

** In this study, blood was analyzed for tyrosinase, p97, and MART-1.

*** In both reports, the samples were tested for tyrosinase and MART-1. In all four studies, [109,115,118,127] the presence of at least one marker defined positivity.

**** *GAGE* mRNA alone was used as a marker in this study. In the remaining studies listed, tyrosinase alone was used as a marker for melanoma cells.

PCR assay. According to these authors, the amount of tyrosinase transcripts increases with tumor burden in patients with metastatic disease and decrease in patients responding to immunotherapy. We are awaiting other studies using quantitative RT-PCR in blood to assess its potential clinical value in melanoma.

To increase our PCR positivity rate (only 19% tyrosinase positivity in blood and/or BM in advanced melanoma), we needed to detect those occult melanoma cells that do not express tyrosinase mRNA. For that purpose, we used an additional marker termed *GAGE*. *GAGE* was identified in a human melanoma cell line[125] and belongs to a family of genes coding for an antigen recognized by autologous cytotoxic T lymphocytes. *GAGE* gene expression was identified by RT-PCR in a variety of tumor types including melanoma, sarcoma, neuroblastoma, and non-small cell lung carcinoma.[125,126] It is silent in normal tissues except for the adult testis. Our group was able to detect *GAGE* mRNA in the PB and BM of patients with melanoma,[127] including some patients who were negative for tyrosinase mRNA, enhancing the sensitivity of our RT-PCR detection system to 45% in PB and/or BM (R. Ghossein, Memorial Sloan-Kettering Cancer Center). MART1/Melan A is a melanocytic tissue-specific marker recognized by cytolytic

Table 13.6 Molecular Prognosis in Melanoma Using RT-PCR

Ref.	AJCC Stage	Marker	Sample	End Point	*P* Value
Mellado et al.[115]	I–III	Tyrosinase	Blood	DFS	0.003
		Tyrosinase	Blood	OS	0.001
Kunter et al.[111]	IV	Tyrosinase	Blood	OS	≤0.0006
Gogas et al.[35]	IIB–III	Tyrosinase	Blood	DFS	0.03
	IIB-III	Tyrosinase	Blood	OS	0.61
Mellado et al.[118]	II–III	Tyrosinase	Blood	DFS	0.02
	II-III	Tyrosinase	Blood	OS	0.03
Wascher et al.[123]	III	Tyrosinase/	Blood	DFS	0.01
		uMAGE-A	Blood	OS	0.04
Curry et al.[116]	I–III	Tyrosinase/Mart 1	Blood	DFS	0.0022
Cheung et al.[127]	II–IV	GAGE	Blood/BM	OS	0.01
	III	GAGE	Blood/BM	OS	0.01
Shivers et al.[133]	I–II	Tyrosinase	SLN	DFS	0.006
			SLN	OS	0.02

Note: The relative risk was not available in all the above references. However, in each article RT-PCR positivity correlated with poorer survival time except for Gogas's study. Only those articles using Kaplan–Meier survival analysis are included in this table. In Curry's article the samples were tested for tyrosinase and Mart 1, and in Wascher's for tyrosinase and uMAGE-A. In both articles,[115,123] the presence of at least one marker defined positivity. In Cheung's article, RT-PCR positivity was defined as positivity for blood and/or BM. RT-PCR: reverse transcriptase polymerase chain reaction; AJCC: American Joint Committee on Cancer; AJCC Stage I: primary tumor < 1.5 mm in thickness with no metastases; AJCC Stage II: primary tumor > 1.5 mm in thickness with no metastases; AJCC Stage III: regional lymph node metastases; AJCC Stage IV: distant metastases; OS: overall survival; DFS: disease-free survival; BM: bone marrow; SLN: sentinel lymph node; uMAGE-A: universal melanoma antigen gene-A.

T lymphocytes and detected by RT-PCR in the PB of patients with melanoma.[116,128] Curry et al.[129] reported a difference in metastatic potential between CTC that were RT-PCR positive for MART-1/ Melan-A only, and those positive for tyrosinase alone. Patients with disseminated melanoma had a significantly lower incidence of MART-1 RT-PCR-positive CTC (16%) than of tyrosinase-positive CTC (63%). It was suggested that the lack of expression of MART-1/Melan A in CTC of patients with recurrent disseminated tumor is due to the higher immunogenicity of MART-1/Melan A compared to tyrosinase.[129] A multimarker RT-PCR assay seems therefore able to provide higher sensitivity and clinical value. Although this improvement was shown in some studies,[123] others have not found any prognostic utility for multimarker RT-PCR in multivariate analysis.[119] This could be explained by the fact that some of the targets used in multimarker RT-PCR assay such as p97 have very low specificity. The latter transcript is present in the blood of 90% of patients with Kaposi sarcoma.[119]

The above observations may have important clinical implications. RT-PCR may help define subsets of patients with poor prognosis for whom toxic forms of adjuvant therapies are justified. This test may help improve the stratification of patients for clinical trials into more homogeneous groups. This assay could also be used to measure treatment response in patients on current or novel therapeutic regimens like vaccine therapy.

The presence or absence of regional lymph node metastases is a powerful predictor of survival in patients with malignant melanoma. Standard histopathologic interpretation routinely underestimates the number of patients with lymph node metastases.[130] Indeed, routine histologic examination samples at most 1 to 5% of the submitted tissue.[131] Immunohistochemical staining with antibodies against S-100 protein or HMB-45 melanoma antigen increases the yield of occult lymph node metastases.[131] However, sampling error is a real possibility since it is completely impractical to examine the entire lymph node by immunohistochemistry. To circumvent these problems, Wang et al.[130] attempted the detection of lymph node micrometastases using RT-PCR for tyrosinase mRNA and showed this technique to be more sensitive than immunohistochemistry or morphology. Sentinel lymph node biopsy is an alternative to elective dissection or observation for managing lymph node basins in patients with cutaneous melanomas. Several groups including ours are testing sentinel lymph nodes for the presence of tyrosinase by RT-PCR, with the hope that this technique will help better stratify patients for elective lymphadenectomy.[132–134] Shivers et al.[133] reported that the probability of recurrence and overall survival is influenced by the RT-PCR detection of tyrosinase mRNA in sentinel lymph nodes. These authors found a statistically significant difference in overall and disease-free survival between patients with RT-PCR-negative histologically negative lymph nodes and those with RT-PCR-positive histologically negative specimens. However, they did not report on their RT-PCR results in control subjects without melanoma. In our laboratory, we were able to detect tyrosinase mRNA by RT-PCR in 73% of sentinel lymph nodes from patients at risk for regional nodal metastases, including all those with histologically positive sentinel lymph nodes and 65% of the histologically negative specimens.[132] Unfortunately, 2 of 18 control nodes without melanoma were tyrosinase PCR positive. We are currently following these patients to assess the prognostic value of this assay. Recently, a multimarker RT-PCR assay was developed for the detection of melanoma micrometastases in paraffin-embedded archival tissues.[134] In this study, 25% of histologically negative lymph nodes were upstaged by the presence of two or more markers by RT-PCR. In patients with histologically negative lymph nodes, the presence of two or more positive molecular markers correlated with worse overall and disease-free survival. If confirmed, the successful molecular detection of melanoma micrometastases in archival tissue will open tremendous research opportunities allowing the studies of large, clinically well characterized groups of patients.

RT-PCR assays for the detection of CTC and micrometastases in melanoma seem very promising in view of (1) the correlation between the RT-PCR assays results (especially blood tyrosinase) and outcome; and (2) the absence of accurate conventional prognostic marker in advanced melanoma. To clearly define the clinical usefulness of RT-PCR for occult melanoma cells including its reproducibility, methodological issues must be addressed using interlaboratory studies.[135] Longer follow-up is also needed since evidence is emerging that the prognostic value of tyrosinase RT-PCR decreases with longer follow-up of melanoma patients.[136]

13.3.4 Lung Carcinomas

The 5-year survival rate of Stage I to II non-small cell lung carcinoma is 30 to 50% after surgical resection. There is therefore a need to detect those patients with occult tumor cells who will recur and die. To better stratify patients with lung carcinomas, several groups have developed RT-PCR assays for cytokeratin

19, CEA, Muc-1, EGFR, and surfactant protein gene transcripts.[90,137–143] These assays were used to detect CTC and lymph node micrometastases. In one study, the CTC were semiquantified by taking the ratio of cytokeratin 19 band intensity from the second round of nested RT-PCR to the band intensity of a housekeeping gene (i.e., a widely expressed gene) after one round of PCR amplification.[140] In that article, serial measurement of the relative number of circulating cancer cells correlated with tumor burden and treatment response in non-small as well as small cell lung carcinomas. The relationship between CTC and therapy response was, however, analyzed in only a few patients. Unfortunately, all the above-mentioned markers including the surfactant protein gene products were shown to be expressed in control samples without carcinoma using RT-PCR[141,142] (Table 13.4). Many neuroendocrine markers (such as synaptophysin, gastrin, NCAM, and HuD) were used for the sole detection of CTC in small cell carcinoma. Only pre-progastrin-releasing peptide was shown to be specific.[144,145] Clearly, the molecular detection of CTC in lung carcinomas is in need of more specific markers and large follow-up studies.

13.3.5 Gastrointestinal Carcinoma

As with other solid tumors, the detection of early metastatic spread in gastrointestinal malignancies may help stratify patients for radical surgery and guide adjuvant therapies. Several authors reported the detection of CEA mRNA in the PB, BM, and lymph nodes of patients with gastric, colorectal, and pancreatic carcinomas but in none of the control subjects.[21,84,146–148] CEA mRNA was detected by RT-PCR in lymph nodes and BM specimens that were negative by immunohistochemistry for CEA and cytokeratin.[21,148] In patients with tumor node metastasis (TNM) Stage II colorectal carcinomas (who have no lymph node metastases by histology), the detection of CEA mRNA in regional lymph nodes was shown to correlate with a poorer 5-year survival rate.[148] However, in some studies, CEA mRNA was detected by RT-PCR in lymph nodes, blood, and BM samples from individuals without epithelial malignancies.[89,149–151] Cytokeratin 20 mRNA was used as a marker for colorectal carcinoma cells in lymph nodes, BM, and blood.[151,152] This marker was unfortunately detected by RT-PCR in 72% of blood samples and all BM specimens from healthy individuals.[151,153] Cytokeratin-19 RT-PCR was used to detect micrometastases in sentinel nodes from patients with gastric, esophageal, and rectal carcinomas and found to be positive in a significant number of histological negative nodes.[154] However, as previously stated RT-PCR for cytokeratin 19 has been shown in many studies to be nonspecific. Even the use of quantitative RT-PCR did not help improve the specificity of such markers as cytokeratin 8, 18, 19, 7, and 20. When maximal background values are used as a threshold to define positivity, the sensitivity of these markers dropped significantly. For example, cytokeratin 20 was positive in only 1 of 30 BM from gastrointestinal carcinomas.[155] Clearly, at the present time the detection of occult tumor cells in gastrointestinal carcinomas is hampered by significant specificity issues.

13.4 FUTURE TRENDS

Because of the limitations of PCR (e.g., contamination of samples, inability to quantify tumor cells or assess the cells for markers of disease progression), it is now clear that other approaches are needed for the detection and molecular characterization of occult tumor cells. In the past few years, we and others have used immunomagnetic separation technology as a means to improve the detection of CTC.[26,156–158] In this technique, the specimen is incubated with magnetic beads or ferrofluids coated with antibodies directed against a specific tissue type (e.g., Ber-EP4 antibody directed against carcinomas). The tumor cells are then isolated using a powerful magnet. The magnetic fraction can be used for downstream RT-PCR, *in situ* hybridization, or immunocytochemical analysis (Figure 13.4). The sample used for RT-PCR will therefore be considerably enriched in tumor cells with a minimal background of non-neoplastic cells, with the latter responsible for the false positives due to illegitimate transcriptions. Epithelial cell enrichment using magnetic beads will therefore render RT-PCR much more sensitive and specific. Immunocytochemical analysis of the specimen will allow

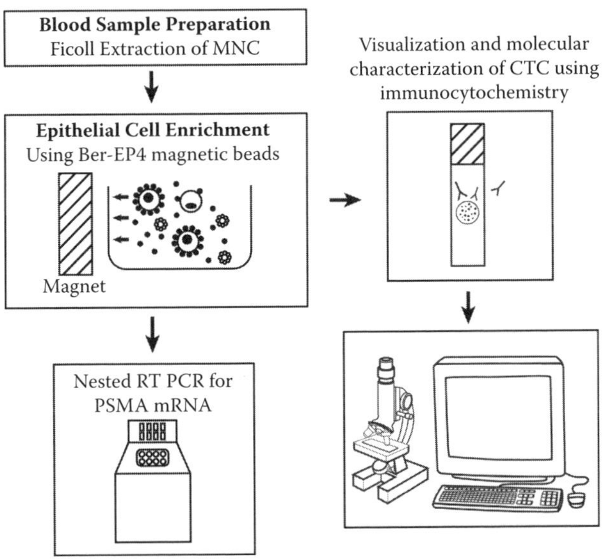

Figure 13.4 Immunobead-based assay for the detection and molecular characterization of CTCs. In this example, blood from a patient with prostatic carcinoma is subjected to a Ficoll separation of nucleated cells. The mononuclear cell (MNC) layer is incubated with magnetic beads coated with the Ber-EP4 anti-epithelial cell antibody directed against carcinomas. The magnetic fraction is then isolated using a powerful magnet and is rich in tumor cells. The cells present in the magnetic fraction are then lysed and their mRNA isolated. This preparation is then ready for RT-PCR for PSMA mRNA, a prostatic-specific marker (left). The isolated magnetic cell fraction can also be cytospun on glass slides and subjected to immunofluorescence, immunoperoxidase, and *in situ* hybridization to characterize and quantify the CTC using an image analyzer (a computerized semi-automated microscope) (right).

better quantification of the tumor cells,[159] and their assessment for various markers of tumor proliferation and progression using an image analyzer (a semi-automated computerized microscope). This will help monitor the effect of targeted therapy (e.g., the monoclonal antibody against HER-2, commercially known as Herceptin), better stratify patients with solid tumor, and shed more light on the dynamic process of metastases. The management of patients with solid malignancies will therefore become more rational, economical, and conservative.

REFERENCES

1. Ashworth, T.R. A case of cancer in which cells similar to those in the tumours were seen in the blood after death. *Aust. Med. J.* 14, 146, 1869.
2. Engell, H.C. Cancer cells in the circulating blood. *Acta Chir. Scand. Suppl.* 201, 1955.
3. Christopherson, W. Cancer cells in the peripheral blood: A second look. *Acta Cytol.* 9, 169, 1965.
4. Moss, T.J. and Sanders, D.G. Detection of neuroblastoma cells in blood. *J. Clin. Oncol.* 8, 736, 1990.
5. Redding, W.H., Coombes, R.C. and Monaghan, P. Detection of micrometastases in patients with primary breast cancer. *Lancet* 2, 1271, 1983.
6. Stahel, R.A. et al. Detection of bone marrow metastasis in small-cell lung cancer by monoclonal antibody. *J. Clin. Oncol.* 3, 455, 1985.
7. Diel, I.J. et al. Detection of tumor cells in bone marrow of patients with primary breast cancer: a prognostic factor for distant metastasis. *J. Clin. Oncol.* 10, 1534, 1992.
8. Pantel, K., Izbicki, J., and Passlick, B. Frequency and prognostic significance of isolated tumor cells in bone marrow of patients with non-small cell lung cancer without overt metastases. *Lancet* 347, 649, 1996.
9. Lindemann, F. et al. Prognostic significance of micrometastatic tumor cells in bone marrow of colorectal patients. *Lancet* 340, 685, 1992.
10. Pelkey, T.J., Frierson, H.F., and Bruns, D.E. Molecular and immunological detection of circulating tumor cells and micrometastases from solid tumors. *Clin. Chem.* 42, 1369, 1996.
11. Kirk, S.J. et al. The prognostic significance of marrow micrometastases in women with early breast cancer. *Eur. J. Surg. Oncol.* 16, 481, 1990.
12. Salvadori, B., Squicciarni, P., and Rovini, D. Use of monoclonal antibody Mbr1 to detect micrometastases in bone marrow specimens of breast cancer patients. *Eur. J. Cancer* 26, 865, 1990.
13. Singletary, S.E. et al. Detection of micrometastatic tumor cells in bone marrow of breast carcinoma patients. *J. Surg. Oncol.* 47, 32, 1991.
14. Courtemanche, D.J. et al. Detection of micrometastases from primary breast cancer. *Can. J. Surg.* 34, 15, 1991.
15. Miettinen, M. Keratin subsets in spindle cell sarcomas. Keratins are widespread but synovial sarcoma contains a distinctive keratin polypeptide pattern and desmoplakins. *Am. J. Pathol.* 138, 505, 1991.
16. Thomas, P. and Battifora, H. Keratins versus epithelial membrane antigen in tumor diagnosis-an immunohistochemical comparison of five monoclonal antibodies. *Hum. Pathol.* 18, 728, 1987.

17. Saiki, R.K., Bugawan, T.L., Horn, G.T., Mullis, K.B., and Erlich, H.A. Analysis of amplified beta-globin and HLA-DQ alpha DNA with allele-specific oligonucleotide probes. *Nature* 324, 163–166, 1986.
18. Campana, D. and Pui, C.H. Detection of minimal residual disease in acute leukemias: methodological advances and clinical significance. *Blood* 85, 1416, 1995.
19. Cave, H. et al. Prospective monitoring and quantitation of residual blasts in childhood acute leukemias by polymerase chain reaction of delta and gamma T-cell receptor genes. *Blood* 83, 1892, 1994.
20. Miyajama, Y. et al. Detection of neuroblastoma cells in bone marrow and peripheral blood at diagnosis by the reverse transcriptase-polymerase chain reaction for tyrosine hydroxylase mRNA. *Cancer* 75, 2757, 1995.
21. Gerhard, M. et al. Specific detection of carcinoembryonic antigen-expressing tumor cells in bone marrow aspirates by polymerase chain reaction. *J. Clin. Oncol.* 12, 724, 1994.
22. Lee, M.S. et al. Detection of minimal residual cells carrying the t(14;18) by DNA sequence amplification. *Science* 237, 175, 1987.
23. Komeda, T., Fukuda, Y., and Sando, T. Sensitive detection of circulating hepatocellular carcinoma cells in peripheral venous blood. *Cancer* 75, 2214, 1995.
24. Eeles, R.A. and Stamps, A.C. *Polymerase Chain Reaction (PCR): The Technique and Its Applications*. R.G. Landes, Austin, TX, 1993.
25. West, D.C. et al. Detection of circulating tumor cells in patients with Ewing's sarcoma and peripheral primitive neuroectodermal tumor. *J. Clin. Oncol.* 15, 583, 1997.
26. Ghossein, R.A., Osman, I., and Bhattacharya, S. Detection of prostatic specific membrane antigen mRNA using immunobead reverse transcriptase polymerase chain reaction. *Diagn. Mol. Pathol.* 8, 59, 1999.
27. Hughes, T., Janssen, J.G.W., and Morgan, G. False-positive results with PCR to detect leukemia-specific transcripts. *Lancet* 335, 1037, 1990 [letter].
28. Kaplan, J.C., Kahn, A., and Chelly, J. Illegitimate transcription: Its use in the study of inherited disease. *Hum. Mutat.* 1, 357, 1992.
29. Mattano, L.A., Moss, T.J., and Emerson, S.G. Sensitive detection of rare circulating neuroblastoma cells by the reverse transcriptase-polymerase chain reaction. *Cancer Res.* 52, 4701, 1992.
30. Sklar, J. and Costa, J.C. Principles of cancer management: molecular pathology, in *Cancer Principles and Practice of Oncology,* 5th ed., De Vita, V., Hellman, S., and Rosemberg, S.A., Eds. Lippincott-Raven, Philadelphia, 1997, 259.
31. Smith, B. et al. Detection of melanoma cells in peripheral blood by means of reverse transcriptase and polymerase chain reaction. *Lancet* 338, 1227, 1991.
32. Olsson, C.A. et al. Reverse transcriptase-polymerase chain reaction for prostate cancer. *Urol. Clin. North Am.* 24, 367, 1997.
33. Ghossein, R.A. et al. Prognostic significance of peripheral blood and bone marrow tyrosinase mRNA in malignant melanoma. *Clin. Cancer Res.* 4, 419, 1998.
34. Kuo, C.T. et al. Assessment of messenger RNA of beta 1→4-*N*-acetylgalactosaminyltransferase as a molecular marker for metastatic melanoma. *Clin. Cancer Res.* 4, 411, 1998.
35. Gogas, H. et al. Prognostic significance of the sequential detection of circulating melanoma cells by RT-PCR in high-risk melanoma patients receiving adjuvant interferon. *Br. J. Cancer* 87, 181, 2002.
36. Young, C.Y.F. et al. Hormonal regulation of prostate-specific antigen messenger RNA in human prostatic adenocarcinoma cell line LNCaP. *Cancer Res.* 51, 3748, 1991.

37. Qiu, S.D. et al. *In situ* hybridization of prostatic-specific antigen mRNA in human prostate. *J. Urol.* 144, 1550, 1990.
38. Heid, C.A. et al. Real time quantitative PCR. *Genome Res.* 6, 986, 1996.
39. Johnson, P.W.M, Burchill, S.A., and Selby, P.J. The molecular detection of circulating tumor cells. *Br. J. Cancer* 72, 268, 1995.
40. Ghossein, R.A. et al. Detection of circulating tumor cells in patients with localized and metastatic prostatic carcinoma prostatic carcinoma: clinical implications. *J. Clin. Oncol.* 13, 1195, 1995.
41. Israeli, R. et al. Sensitive nested reverse transcription polymerase chain reaction detection of circulating prostatic tumor cells: comparison of prostatic-specific membrane antigen and prostate-specific antigen-based assays. *Cancer Res.* 54, 6306, 1994.
42. Katz, A.E. et al. Molecular staging of prostate cancer with the use of an enhanced reverse transcriptase-PCR assay. *Urology* 43, 765, 1994.
43. Seiden, M.V. et al. Detection of circulating tumor cells in men with localized prostate cancer. *J. Clin. Oncol.* 12, 2634, 1995.
44. Sokoloff, M.H. et al. Quantitative polymerase chain reaction does not improve preoperative prostate cancer staging: a clinicopathological molecular analysis of 121 patients. *J. Urol.* 156, 1560, 1996.
45. Corey, E. et al. Detection of circulating prostatic cells by reverse transcriptase-polymerase chain reaction of human glandular kallikrein (Hk2) and prostatic-specific antigen (PSA) messages. *Urology* 50, 184, 1997.
46. Wood, D.P. and Banerjee, M. Presence of circulating prostate cells in the bone marrow of patients undergoing radical prostatectomy is predictive of disease free survival. *J. Clin. Oncol.* 15, 3451, 1997.
47. Gao, C.-L. et al. Blinded evaluation of reverse transcriptase-polymerase chain reaction prostate-specific antigen peripheral blood assay for molecular staging of prostate cancer. *Urology* 53, 714, 1999.
48. Ennis, R.D. et al. Detection of circulating prostate carcinoma cells via an enhanced reverse transcriptase-polymerase chain reaction assay in patients with early stage prostate carcinoma. *Cancer* 79, 2402, 1997.
49. Ellis, W.J. et al. The value of a reverse transcriptase polymerase chain reaction assay in preoperative staging and follow up of patients with prostate cancer. *J. Urol.* 159, 1134, 1998.
50. Kantoff, P.W. et al. Prognostic significance of reverse transcriptase polymerase chain reaction for prostate-specific antigen in men with hormone-refractory prostate cancer. *J. Clin. Oncol.* 19, 3025, 2001.
51. Shariat, S.F. et al. Early postoperative peripheral blood reverse transcription PCR assay for prostate-specific antigen is associated with prostate cancer progression in patients undergoing radical prostatectomy. *Cancer Res.* 63, 5874, 2003.
52. Mao, H. et al. Detection of PSA mRNA from the peripheral blood and pelvic lymph nodes in patients with prostatic cancer by means of reverse transcription-polymerase chain reaction (RT-PCR). *Nippon. Hinyokika. Gakkai. Zasshi* 89, 596, 1998.
53. Saimoto, A., Saito, S., and Murai, M. Prostate-specific membrane antigen-derived primers in a nested reverse transcription polymerase chain reaction for detecting prostatic cancer cells. *Jpn. J. Cancer Res.* 90, 233, 1999.
54. Price, D.K. et al. Detection and clearance of prostate cells subsequent to ultrasound-guided needle biopsy as determined by multiplex nested reverse transcription polymerase chain reaction assay. *Urology* 52, 261, 1998.

55. Grassa, Y.Z. et al. Combined nested RT-PCR assay for prostate-specific antigen and prostate-specific membrane antigen in prostate cancer patients: correlation with pathological stage. *Cancer Res.* 58, 1456, 1998.
56. Thomas, J. et al. Preoperative combined nested reverse transcriptase polymerase chain reaction for prostate-specific antigen and prostate-specific membrane antigen does not correlate with pathologic stage or biochemical failure in patients with localized prostate cancer undergoing radical prostatectomy. *J. Clin. Oncol.* 20, 3213, 2002.
57. Kurek, R. et al. Quantitative PSA RT-PCR for preoperative staging of prostate cancer. *Prostate* 56, 263, 2003.
58. Mejean, A. et al. Detection of circulating prostate derived cells in patients with prostate adenocarcinoma is an independent risk factor for tumor recurrence. *J. Urol.* 163, 2022, 2000.
59. Olsson, C.A. et al. Preoperative reverse transcriptase polymerase chain reaction for prostatic specific antigen predicts treatment failure following radical prostatectomy. *J. Urol.* 155, 1557, 1996.
60. de la Taille, A. et al. Blood based reverse transcriptase polymerase chain reaction assays for prostatic specific antigen: long term follow-up confirms the potential utility of this assay in identifying patients more likely to have biochemical recurrence (rising PSA) following radical prostatectomy. *Int. J. Cancer* 84, 360, 1999.
61. Ghossein, R. et al. Prognostic significance of detection of prostate specific antigen transcripts in the peripheral blood of patients with metastatic androgen independent prostatic carcinoma. *Urology* 50, 100, 1997.
62. Deguchi, T. et al. Detection of micrometastatic prostate cancer cells in lymph nodes by reverse transcriptase-polymerase chain reaction. *Cancer Res.* 53, 5350, 1993.
63. Meyers, F. et al. Bone marrow molecular staging of prostatic carcinoma: does it help? *Proc. Am. Soc. Clin. Oncol.* 14, 230, 1995.
64. Edelstein, R.A. et al. Implication of prostatic micrometastases to the pelvic lymph nodes: an archival tissue study. *Urology* 47, 370, 1996.
65. Price, D.K., Woodard, W.L., and Teigland, C.M. Simultaneous detection of prostate specific antigen-expressing and prostate-specific membrane antigen expressing cells by a multiplex reverse transcriptase polymerase chain reaction assay. *Urol. Oncol.* 1, 226, 1995.
66. Hara, N. et al. Reverse transcription-polymerase chain reaction detection of prostate-specific antigen, prostate-specific membrane antigen, and prostate stem cell antigen in one milliliter of peripheral blood: value for the staging of prostate cancer. *Clin. Cancer. Res.* 8, 1794, 2002.
67. Elgamal, A.A. et al. Prostate-specific membrane antigen (PSMA): current benefits and future value. *Semin. Surg. Oncol.* 18, 10, 2000.
68. Loric, S. et al. Enhanced detection of hematogenous circulating prostatic cells in patients with prostate adenocarcinoma by using nested reverse transcription polymerase chain reaction assay based on prostate-specific membrane antigen. *Clin. Chem.* 41, 1698, 1995.
69. Okegawa, T., Nutahara, K., and Higashihara, E. Preoperative nested reverse transcription-polymerase chain reaction for prostatic specific membrane antigen predicts biochemical recurrence after radical prostatectomy. *B.J.U. Int.* 84, 112, 1999.
70. Lintula, S. and Stenman, U.H. The expression of prostate-specific membrane antigen in peripheral blood leukocytes. *J. Urol.* 157, 1969, 1997.
71. Llanes, L. et al. The clinical utility of the prostate specific membrane antigen reverse-transcription/polymerase chain reaction to detect circulating prostate cells: an analysis in healthy men and women. *B.J.U. Int.* 89, 882, 2002.

72. Zhang, Y. et al. Combined nested reverse transcription-PCR assay for prostate-specific antigen and prostate-specific membrane antigen in detecting circulating prostatic cells. *Clin. Cancer Res.* 3, 1215, 1997.
73. Seiden, M. and Sklar, J.L. PCR and RT-PCR-based methods of tumor detection: potential applications and clinical implications, in *Important Advances in Oncology*, DeVita, V.T., Hellman, S., and Rosenberg, S.A., Eds., Lippincott-Raven, Philadelphia, 1996, 191–204.
74. Schlimock, G. and Riethmuller, G. Detection, characterization and tumorigenicity of disseminated tumor cells in human bone marrow. *Semin. Cancer Biol.* 1, 207, 1990.
75. Ghalie, R. The Osborne/Rosen article reviewed. *Oncology* 8, 36, 1994.
76. Osborne, M.P. et al. Immunofluorescent monoclonal antibody detection of breast cancer in bone marrow: sensitivity in a model system. *Cancer Res.* 49, 2510, 1989.
77. Cote, R.J. et al. Prediction of early relapse in patients with operable breast cancer by detection of occult bone marrow micrometastases. *J. Clin. Oncol.* 9, 1749, 1991.
78. Fields, K.K. et al. Clinical significance of bone marrow metastases as detected using the polymerase chain reaction in patients with breast cancer undergoing high-dose chemotherapy and autologous bone marrow transplantation. *J. Clin. Oncol.* 14, 1868, 1996.
79. Datta, Y.H. et al. Sensitive detection of occult breast cancer by the reverse-transcriptase polymerase chain reaction. *J. Clin. Oncol.* 12, 475, 1994.
80. Mori, M. et al. Detection of cancer micrometastases in lymph nodes by reverse transcriptase-polymerase chain reaction. *Cancer Res.* 55, 3417, 1995.
81. Brown, D.C. et al. Detection of intraoperative tumor cell dissemination in patients with breast cancer by use of the reverse transcription and polymerase chain reaction. *Surgery.* 117, 96, 1995.
82. Noguchi, S. et al. The detection of breast carcinoma micrometastases in axillary lymph nodes by means of reverse transcriptase-polymerase chain reaction. *Cancer* 74, 1595, 1994.
83. Noguchi, S. et al. Detection of breast cancer micrometastases in axillary lymph nodes by means of reverse transcriptase polymerase chain reaction. Comparison between MUC1 mRNA and keratin 19 mRNA amplification. *Am. J. Pathol.* 148, 649, 1996.
84. Mori, M. et al. Clinical significance of molecular detection of carcinoma cells in lymph nodes and peripheral blood by reverse transcription-polymerase chain reaction in patients with astrointestinal or breast carcinomas. *J. Clin. Oncol.* 16, 128, 1996.
85. Min, C.J., Tafra, L., and Verbanac, K.M. Identification of superior markers for polymerase chain reaction detection of breast cancer metastases in sentinel lymph nodes. *Cancer Res.* 58, 4581, 1998.
86. Merrie, A.E. et al. Analysis of potential markers for detection of submicroscopic node metastases in breast cancer. *Br. J. Cancer* 80, 2019, 1999.
87. Bostick, P.J. et al. Detection of metastases in sentinel lymph nodes of breast cancer patients by multiple-marker RT-PCR. *Int. J. Cancer.* 79, 645, 1998.
88. Eltahir, E.M. et al. Putative markers for the detection of breast carcinoma cells in blood. *Br. J. Cancer* 77, 1203, 1998.
89. Ko, Y. et al. Limitations of reverse transcription-polymerase chain reaction method for the detection of carcinoembryonic antigen-positive tumor cells in peripheral blood. *Clin. Cancer Res.* 4, 2141, 1998.
90. Krisman, M. et al. Low specificity of cytokeratin 19 reverse transcriptase polymerase chain reaction analyses for detection of hematogenous lung cancer dissemination. *J. Clin. Oncol.* 13, 2769, 1995.

91. Lopez-Guerrero, J.A. et al. Use of reverse-transcriptase polymerase chain reaction (RT-PCR) for carcinoembryonic antigen, cytokeratin 19, and maspin in the detection of tumor cells in leukapheresis products from patients with breast cancer: comparison with immunocytochemistry. *J. Hematother.* 8, 53, 1999.
92. De Graaf, H. et al. Ectopic expression of target genes may represent an inherent limitation of RT PCR assays used for micrometastasis detection: Studies on the epithelial glycoprotein gene EGP-2. *Int. J. Cancer* 72, 191, 1997.
93. Slade, M.J. et al. Quantitative Polymerase chain reaction for the detection of micrometastases in patients with breast cancer. *J. Clin. Oncol.* 17, 870, 1999.
94. Ooka, M. et al. Bone marrow micrometastases detected by RT-PCR for mammaglobin can be an alternative prognostic factor of breast cancer. *Breast Cancer Res. Treat.* 67, 169, 2001.
95. Berois, N. et al. Molecular detection of cancer cells in bone marrow and peripheral blood of patients with operable breast cancer. Comparison of CK19, MUC1 and CEA using RT-PCR. *Eur. J. Cancer* 36, 717, 2000.
96. Weigelt, B. et al. Marker genes for circulating tumour cells predict survival in metastasized breast cancer patients. *Br. J. Cancer* 88, 1091, 2003.
97. Tokunaga, E. et al. Correlation with bone metastasis and high expression of CK 19 mRNA measured by quantitative RT-PCR in the bone marrow of breast cancer patients. *Breast J.* 9, 440, 2003.
98. Ikeda, N. et al. Prognostic significance of occult bone marrow micrometastases of breast cancer detected by quantitative polymerase chain reaction for cytokeratin 19 mRNA. *Jpn. J. Cancer Res.* 91, 918, 2000.
99. Berois, N. et al. Detection of bone marrow-disseminated breast cancer cells using an RT-PCR assay of MUC5B mRNA. *Int. J. Cancer* 103, 550, 2003.
100. Watson, M.A. and Fleming, T.P. Mammaglobin, a mammary-specific member of the uteroglobin gene family, is overexpressed in human breast cancer. *Cancer Res.* 56, 860, 1996.
101. Aihara, T. et al. Mammaglobin B as a novel marker for detection of breast cancer micrometastases in axillary lymph nodes by reverse transcription-polymerase chain reaction. *Breast Cancer Res. Treat.* 58, 137, 1999.
102. Zach, O. et al. Detection of circulating mammary carcinoma cells in the peripheral blood of breast cancer patients via a nested reverse transcriptase polymerase chain reaction assay for mammaglobin mRNA. *J. Clin. Oncol.* 17, 2015, 1999.
103. Silva, J.M. et al. Detection of epithelial messenger RNA in the plasma of breast cancer patients is associated with poor prognosis tumor characteristics. *Clin. Cancer Res.* 7, 2821, 2001.
104. Stathopoulou, A. et al. Molecular detection of cytokeratin-19-positive cells in the peripheral blood of patients with operable breast cancer: evaluation of their prognostic significance. *J. Clin. Oncol.* 20, 3404, 2002.
105. Kahn, H.J. et al. RT-PCR amplification of CK19 mRNA in the blood of breast cancer patients: correlation with established prognostic parameters. *Breast Cancer Res. Treat.* 60, 143, 2000.
106. Xenidis, N. et al. Peripheral blood circulating cytokeratin-19 mRNA-positive cells after the completion of adjuvant chemotherapy in patients with operable breast cancer. *Ann. Oncol.* 14, 849, 2003.
107. Jung, Y.S. et al. Clinical significance of bone marrow micrometastasis detected by nested rt-PCR for keratin-19 in breast cancer patients. *Jpn. J. Clin. Oncol.* 33(4), 167–172, 2003.

108. Battayani, Z. et al. Polymerase chain reaction detection of circulating melanocytes as a prognostic marker in patients with melanoma. *Arch. Dermatol.* 131, 443, 1995.
109. Buzaid, A.C. and Balch, C.M. Polymerase chain reaction fro detection of melanoma in peripheral blood: too early to assess clinical value. *J. Natl. Cancer Inst.* 88, 569, 1996.
110. Hoon, D.S.B. et al. Detection of occult melanoma cells in blood with a multiple-marker polymerase chain reaction assay. *J. Clin. Oncol.* 13, 2109, 1995.
111. Kunter, U. et al. Peripheral blood tyrosinase messenger RNA detection and survival in malignant melanoma. *J. Natl. Cancer Inst.* 88, 590, 1996.
112. Brossart, P. et al. Hematogenous spread of malignant melanoma cells in different stages of disease. *J. Invest. Dermatol.* 101, 887, 1993.
113. Foss, A.J. et al. The detection of melanoma cells in peripheral blood by reverse transcriptase polymerase chain reaction. *Br. J. Cancer* 72, 155, 1995.
114. Pittman K. et al. Reverse transcriptase-polymerase chain reaction for expression of tyrosinase to identify malignant melanoma cells in peripheral blood. *Ann. Oncol.* 7, 297, 1996.
115. Mellado, B. et al. Prognostic significance of the detection of circulating malignant cells by reverse transcriptase-polymerase chain reaction in long-term clinically disease-free melanoma patients. *Clin. Cancer Res.* 5, 1843, 1999.
116. Curry, B.J., Myers, K., and Hersey, P. Polymerase chain reaction detection of melanoma cells in the circulation: relation to clinical stage, surgical treatment, and recurrence from melanoma. *J. Clin. Oncol.* 16, 1760, 1998.
117. Farthman, B. et al. RT PCR for tyrosinase mRNA positive cells in peripheral blood: evaluation strategy and correlation with known prognostic markers in 123 melanoma patients. *J. Invest. Dermatol.* 110, 263, 1998.
118. Mellado, B. et al. Tyrosinase mRNA in blood of patients with melanoma treated with adjuvant interferon. *J. Clin. Oncol.* 20, 4032, 2002.
119. Palmieri, G. et al. Prognostic value of circulating melanoma cells detected by reverse transcriptase-polymerase chain reaction. *J. Clin. Oncol.* 21, 767, 2003.
120. Jin, H.Y. et al. Detection of tyrosinase and tyrosinase-related protein 1 sequences from peripheral blood of melanoma patients using reverse transcription-polymerase chain reaction. *J. Dermatol. Sci.* 33, 169, 2003.
121. Szenajch, J. et al. Prognostic value of multiple reverse transcription-PCR tyrosinase testing for circulating neoplastic cells in malignant melanoma. *Clin. Chem.* 49, 1450, 2003.
122. Keilholz, U. Quantitative detection of circulating tumor cells in cutaneous and ocular melanoma and quality assessment by real-time reverse transcriptase-polymerase chain reaction. *Clin. Cancer Res.* 10, 1605, 2004.
123. Wascher, R.A. et al. Molecular tumor markers in the blood: early prediction of disease outcome in melanoma patients treated with a melanoma vaccine. *J. Clin. Oncol.* 21, 2558, 2003.
124. Brossart, P. et al. A polymerase chain reaction based semi-quantitative assessment of malignant melanoma cells in peripheral blood. *Cancer Res.* 55, 4065, 1995.
125. Van den Eynde, B. et al. A new family of genes coding for an antigen recognized by autologous cytolytic T lymphocytes on a human melanoma. *J. Exp. Med.* 182, 689, 1995.
126. Cheung, I.Y. and Cheung, N.K.V. Molecular detection of GAGE expression in peripheral blood and bone marrow: utility as a tumor marker for neuroblastoma. *Clin. Cancer Res.* 3, 821, 1997.

127. Cheung, I.Y. et al. Association between molecular detection of GAGE and survival in patients with malignant melanoma: a retrospective cohort study. *Clin. Cancer Res.* 5, 2042, 1999.
128. Schitteck, B. et al. Amplification of Melan A messenger RNA in addition to tyrosinase increases sensitivity of melanoma cell detection in peripheral blood and is associated with the clinical stage and prognosis of malignant melanoma. *Br. J. Dermatol.* 141, 30, 1999.
129. Curry, B.J., Meyers, K., and Hersey, P. MART-1 is expressed less frequently on circulating melanoma patients who develop distant compared with locoregional metastases. *J. Clin. Oncol.* 17, 2562, 1999.
130. Wang, X. et al. Detection of submicroscopic lymph node metastases with polymerase chain reaction in patients with malignant melanoma. *Ann. Surg.* 220, 768, 1994.
131. Busam, K.J. Advances in molecular staging of melanoma patients: multimarker analysis of archival lymph node tissue. *J. Clin. Oncol.* 21, 3559, 2003.
132. Biegliek, S.C. et al. Detection of tyrosinase mRNA by reverse transcriptase polymerase chain reaction (RT-PCR) in melanoma sentinel nodes. *Ann. Surg. Oncol.* 6, 232, 1999.
133. Shivers, S.C. et al. Molecular staging of malignant melanoma. *J. Am. Med. Assoc.* 280, 1410, 1998.
134. Kuo, C.T. et al. Prediction of disease outcome in melanoma patients by molecular analysis of paraffin-embedded sentinel lymph nodes. *J. Clin. Oncol.* 21, 3566, 2003.
135. Keiljolz, U. New prognostic factors in melanoma: mRNA tumour markers. *Eur. J. Cancer* 34, S37, 1998.
136. Kammula, U.S. et al. Serial follow up and the prognostic significance of RT PCR staged sentinel lymph nodes from melanoma patients. *J. Clin. Oncol.* 22, 3929, 2004.
137. Castaldo, G. et al. Lung cancer metastatic cells detected in blood by reverse transcriptase-polymerase chainreaction and dot-blot analysis. *J. Clin. Oncol.* 15, 3388, 1997.
138. Dingemans, A.M.C. et al. Detection of cytokeratin 19 transcripts by reverse transcriptase-polymerase chain reaction in lung cancer cell lines and blood of lung cancer patients. *Lab. Invest.* 77, 213, 1997.
139. Salerno, C.T. et al. Detection of occult micrometastases in non-small cell carcinoma by reverse transcriptase-polymerase chain reaction. *Chest* 113, 1526, 1998.
140. Peck, K. et al. Detection and quantification of circulating cancer cells in the peripheral blood of lung cancer patients. *Cancer Res.* 58, 2761, 1998.
141. Betz, C. et al. Surfactant protein gene expression in micrometastatic pulmonary adenocarcinoma and other non-small cell carcinomas: detection by reverse transcriptase-polymerase chain reaction. *Cancer Res.* 55, 4283, 1995.
142. De Luca, A. et al. Detection of circulating tumor cells in carcinoma patients by a novel epidermal growth factor receptor reverse transcription-PCR assay. *Clin. Cancer Res.* 6, 1439, 2000.
143. Clarke, L.E. et al. Epidermal growth factor receptor mRNA in peripheral blood of patients with pancreatic, lung, and colon carcinomas detected by RT-PCR. *Int. J. Oncol.* 22, 425, 2003.
144. Saito, T. et al. Sensitive detection of small cell lung carcinoma cells by reverse transcriptase-polymerase chain reaction for prepro-gastrin-releasing peptide mRNA. *Cancer* 15, 2504, 2003.
145. Lacroix, J. et al. Sensitive detection of rare cancer cells in sputum and peripheral blood samples of patients with lung cancer by preproGRP-specific RT-PCR. *Int. J. Cancer* 92, 1, 2001.

146. Funaki, N.O. et al. Identification of carcinoembryonic antigen mRNA in circulating peripheral blood of pancreatic carcinoma and gastric carcinoma patients. *Life Sci.* 59, 2187, 1996.
147. Mori, M. Molecular detection of circulating solid carcinoma cells in the peripheral blood: the concept of early systemic disease. *Int. J. Cancer* 68, 739, 1996.
148. Liefers, G.J. et al. Micrometastases and survival in stage II colorectal cancer. *N. Engl. J. Med.* 339, 223, 1998.
149. Bostick, P.J., Hoon, D.S.B., and Cote, R.C. Detection of carcinoembryonic antigen messenger RNA in lymph nodes from patients with colorectal cancer. *N. Engl. J. Med.* 339, 1643, 1998 [letter].
150. Zippelius, A. et al. Limitations of reverse transcriptase polymerase chain reaction analyses for detection of micrometastatic epithelial cancer cells in bone marrow. *J. Clin. Oncol.* 15, 2701, 1997.
151. Wharton, R.Q. et al. Increased detection of circulating tumor cells in the blood of colorectal carcinoma patients using two reverse transcriptase-PCR assays and multiple blood samples. *Clin. Cancer Res.* 5, 4158, 1999.
152. Weitz, J. et al. Detection of disseminated colorectal cancer cells in lymph nodes, blood and bone marrow. *Clin. Cancer Res.* 5, 1830, 1999.
153. Champelovier, P., Mongelard, F., and Seigneurin, D. CK 20 gene expression: technical limits for the detection of circulating tumor cells. *Anticancer Res.* 19, 2073, 1999.
154. Matsuda, J. et al. Significance of metastasis detected by molecular techniques in sentinel nodes of patients with gastrointestinal cancer. *Ann. Surg. Oncol.* 11, 250S, 2004.
155. Dimmler, A. et al. Transcription of cytokeratins 8, 18, and 19 in bone marrow and limited expression of cytokeratins 7 and 20 by carcinoma cells: inherent limitations for RT-PCR in the detection of isolated tumor cells. *Lab. Invest.* 81, 1351, 2001.
156. Martin, V.M. et al. Immunomagnetic enrichment of disseminated epithelial tumor cells from peripheral blood by MACS. *Exp. Hematol.* 26, 252, 1998.
157. Racila, E. et al. Detection and characterization of carcinoma cells in blood. *Proc. Natl. Acad. Sci. U.S.A.* 95, 4589, 1998.
158. Benez, A. et al. Detection of circulating melanoma cells by immunomagnetic cell sorting. *J. Clin. Lab. Anal.* 13, 229, 1999.
159. Witzig, T.E. et al. Detection of circulating cytokeratin-positive cells in the blood of breast cancer patients using immunomagnetic enrichment and digital microscopy. *Clin. Cancer Res.* 8, 1085, 2002.

CHAPTER 14

Methylation Profiling of Tumor Cells and Tumor DNA in Blood, Urine, and Body Fluids for Cancer Detection and Monitoring

Ivy H.N. Wong

CONTENTS

14.1 INTRODUCTION

DNA methylation takes place after DNA synthesis by the enzymatic transfer of a methyl group from the methyl donor *S*-adenosylmethionine to the carbon-5 position of cytosine. Cytosines (Cs) usually located 5′ to guanosines (Gs) are differentially methylated in the human genome.[1] Non-CpG-rich sequences are interspersed by CpG islands, which are approximately 500 bp long with G to C contents 55% and observed frequencies over expected frequencies of CpG dinucleotides 0.65. The CpG islands of an increasing number of human genes are differentially methylated in human tissues.

DNA methyltransferases such as DNMT1, DNMT3a, and DNMT3b catalyze genomic DNA methylation. DNMT1 is mainly responsible for the maintenance of DNA methylation, whereas DNMT3a and DNMT3b have been shown to catalyze methylation of hemimethylated and unmethylated DNA.[1,2] Epigenetics is the inheritance of information at the level of gene expression without any changes in the DNA sequence. Epigenetic DNA modification takes place after DNA synthesis, resulting in heritable chromatin structure. Histone deacetylases and histone methyltransferases may also alter the chromatin structure to become transcriptionally inactive.

Overexpression of both DNMT1 and DNMT3 mRNAs has been found in human cancers.[2] Owing to DNA methylation imbalance in cancer cells, specific methylation profiles have been identified in different cancer types.[3,4] In Knudson's two-hit hypothesis, loss of heterozygosity (LOH), homozygous deletion, or promoter methylation could lead to loss of gene function.[5] Early detection of cancer can help reduce the mortality. As discussed in this chapter, methylation changes detected in tumor cells and tumor DNA isolated from surrogate tissues such as blood, urine, and other body fluids may have important clinical implications for the differential cancer diagnosis and monitoring and the selection of therapy.

14.2 DNA HYPERMETHYLATION AND CANCER PROGRESSION

In human cancers, epigenetic alterations include global genomic hypomethylation and hypermethylation of tumor suppressor genes, DNA repair genes, and metastasis inhibitor genes. *p16* and *p15*, which encode cyclin-dependent kinase inhibitors as upstream regulators of pRb phosphorylation, have been recognized as tumor suppressor genes in many solid tumors and hematologic malignancies.[6–10] Frequent *p16* and *p15* hypermethylation has been detected in many human cancers.[9–11] During leukemic transformation and progression, *p15* methylation arises universally and *de novo* in different lineages and differentiation stages, in hematopoietic progenitors developing in the myeloid/lymphoid pathway or primitive stem cells with multilineage potential.[12] Since *p15* expression is induced by extracellular growth inhibitors, interferon-α, and transforming growth factor β (TGF-β),[13–15] *p15* inactivation via hypermethylation could possibly abrogate the cell cycle control and confer resistance to the growth-inhibitory effect of TGF-β that is usually overexpressed in different tissues.[15] *APC, BRCA-1, E-cadherin, LKB1, RB, VHL, hMLH1,* and *MGMT* are selectively hypermethylated in human cancers.[3] The latter two genes are particularly crucial for DNA repair. Lack of mismatch repair function owing to *hMLH1* hypermethylation can lead to microsatellite instability in colon, endometrial, and gastric cancers,[3] while *MGMT* hypermethylation can also lead to loss of repair of alkylating damage in human cancer.[16] Recently, ubiquitous *RASSF1A* promoter methylation has been found in various cancer types in association with cell cycle deregulation.[17]

It is noteworthy that no significant differences in methylation patterns were detected between high- and low-grade tumors,[4] suggesting that epigenetic abnormalities could occur in early stages of cancer development or tumor initiation. Deregulation of critical pathways, such as *Rb/p16* and *p53/p14/MDM2* pathways, can very often lead to neoplastic growth. Early loss of cell cycle control via *p16/p15* hypermethylation, deregulation of transcription factors via *RUNX3* hypermethylation, disruption of cell adherence/cell–cell interaction via *E-cadherin* hypermethylation, and bypassing of cellular mortality check points via *p53* hypermethylation can all contribute to uncontrolled proliferation and cellular immortalization.[3] In addition, *DAPK* hypermethylation can be attributed to metastasis development.[18] Taken together, these phenomena highlight the significance of DNA hypermethylation during cancer progression.

14.3 CONCURRENT HYPERMETHYLATION, TRANSCRIPTIONAL SILENCING, AND LOSS OF FUNCTION

DNA hypermethylation plays an important role in epigenetic regulation by enhancing the binding of methylcytosine-binding proteins and the recruitment of histone deacetylases and co-repressors.[5] The methylated DNA–protein complex formation is critical for constructing transcriptionally repressive chromatin structure. Among human cancers, some genes are epigenetically altered as a group in a tumor-type-specific manner.[19] On the other hand, some methylation patterns are shared among different tumor types.[4] It is possible that individual CpG islands are differentially susceptible to hypermethylation under growth selection pressures, which may drive characteristic pathways leading to the development of different tumor types.

Concurrent hypermethylation of a panel of genes has previously been reported in human cancers.[1,2,9,12] To augment selective growth advantage, *p16* methylation may possibly act in concert with *p15* methylation during carcinogenesis. Differential methylation of CpG islands and the methylation density may vary with the developmental stage of a specific cancer.[20] During tumor progression, the frequency of *MGMT, RASSF1A,* and *DAPK* hypermethylation in metastatic melanomas was higher than that in primary melanomas.[21] Differential methylation could be caused by heterogeneity between different clonally derived tumor cells. Furthermore, the level of transcriptional repression is directly associated with the methylation density.[22] During tumor progression, methylation changes may possibly be accumulated until critical CpG sites are methylated, leading to transcriptional silencing and the complete loss of gene functions and hence contributing to selective growth advantage.

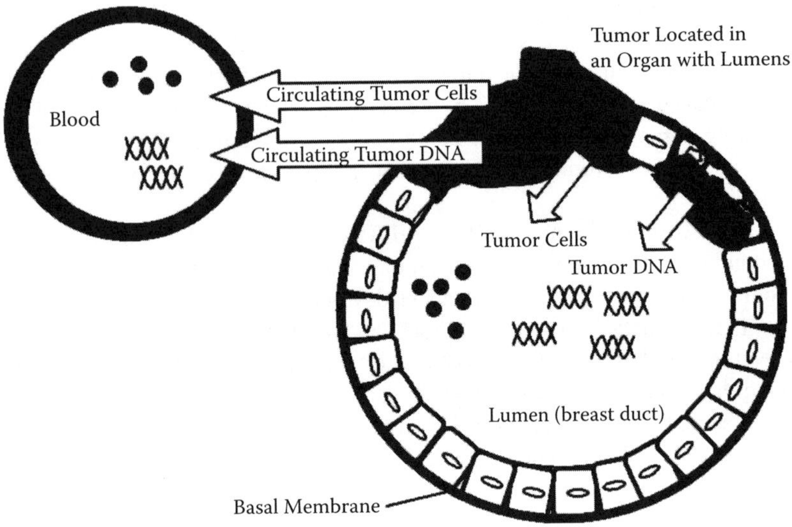

Figure 14.1 Methylation profiling of tumor cells and tumor DNA in blood and body fluids of patients with cancer. In some patients, tumor cells and tumor DNA appear to be disseminated from primary tumors located in organs bearing lumens such as the breast ducts, where ductal lavage fluids can be collected for methylation analyses. In addition to this scenario, CTCs and circulating tumor DNA can be detected in the bloodstream of patients with cancer, where blood cells, plasma, and serum samples can be collected for methylation profiling.

Hypermethylation in the 5′ upstream region (potentially the promoter domain) was observed to be critical in causing transcriptional loss. Dense methylation and methylation of critical sites within the 5′ upstream region can completely block gene transcription. Increasing hypermethylation of tumor suppressor genes, DNA repair genes, and metastasis suppressor genes would therefore be expected to promote the process of tumor progression, stepwise transformation, and metastasis development.

14.4 METHYLATION PROFILES IN CIRCULATING TUMOR CELLS ISOLATED FROM BLOOD OF PATIENTS WITH CANCER AND BIOLOGICAL IMPLICATIONS

Hematogenous dissemination is presumably a major route of metastasis, and circulating tumor cells (CTCs) may remain dormant for long times before recurrence or the development of metastasis.[23–27] After necrosis of some of these CTCs, tumor DNA may also possibly be released (Figure 14.1). To improve the current diagnostic procedures in identifying high-risk patients, we need to develop new approaches for early diagnosis of human cancers or premalignancies.

DNA methylation profiling is a new molecular diagnostic approach in which cancer-specific DNA methylation patterns can be detected in surrogate tissues such as blood or other body fluids. Tumor-specific and age-related DNA hypermethylation of different genes has been described for different cancer types.[4,28] The detection

TABLE 14.1 Methylation Profiles of Multiple Genes in Tumors, CTCs, Plasma, and Serum Samples from Patients with Different Cancer Types

Gene	Cancer	Detection Rate (%)				Relative Sensitivity of MSP	Ref.
		Tumor	CTCs	Plasma/ serum	Serum		
p16	Lung cancer	41	ND	ND	33	10^{-3}	31
	HCC	73	13	81	ND	10^{-5}	29, 32, 51
	Breast cancer	23	ND	14	ND	Non-MSP	55
	Head and neck cancer	27	ND	ND	31	10^{-3}	35
p15	Acute leukemia	58	ND	92	ND	10^{-4}	12
	HCC	64	100	25	ND	10^{-4}	33
MGMT	Lung cancer	27	ND	ND	66	10^{-3}	31
	Head and neck cancer	33	ND	ND	48	10^{-3}	35
DAPK	Lung cancer	23	ND	ND	80	10^{-3}	31
	Head and neck cancer	18	ND	ND	18	10^{-3}	35
GSTP1	Lung cancer	9	ND	ND	50	10^{-3}	31
	Prostate cancer	94	30	72	ND	10^{-5}	30
hMLH1	Colon cancer	47	ND	ND	33	10^{-2}	34

ND = not done.

rates for tumor cells and tumor DNA in blood are exceptionally high in patients with liver cancer or prostate cancer.[29,30] With high specificity and sensitivity, molecular analyses of tumor-derived epigenetic alterations in blood of cancer patients may possibly create a profound impact on noninvasive diagnosis of cancers among high-risk populations, cancer monitoring, and prognostication.[29,31,32] Also, the circulating tumor burden is much greater among patients with cancer with metastases as compared with those without metastases.

Biologically, the epigenetic change is heritable and associated with transcriptional silencing and hence the loss of function. As mentioned previously, a panel of genes encoding tumor suppressors, DNA repair proteins, and metastasis inhibitors has been found hypermethylated in multiple human cancers. Epigenetic changes of *p16*, *p15*, and *GSTP1* have been detected in blood cells from patients with cancer with tumoral methylation (Table 14.1).[12,29,30,33] Methylated *p15* and *p16* sequences were tumor specific and readily detected in blood from patients with hepatocellular carcinoma (HCC) or acute leukemia.[12,29,33] Also, *GSTP1* methylation was detectable in CTCs from 30% of patients with prostate cancer.[30] A panel of epigenetic markers including *p15*, *p16*, and *RASSF1A* methylation may enable specific detection of CTCs from patients with different tumor types for assessing cancer progression.[12,17,29]

It is important to note that the mechanism and timing of tumor cell dissemination from the primary tumor into circulation remain largely unknown. The biological characteristics of a tumor, such as the growth rate, histological grade, and metastatic potential, may affect the quantity of tumor cells released into bloodstream of patients with cancer during the clinical course. As an approach for cancer diagnosis and prognostication, the methylation analysis of blood cells may also be further applied for studying the pathophysiological basis of tumor cell dissemination into patients' circulation.

14.5 METHYLATION PROFILING OF CIRCULATING TUMOR DNA IN PLASMA AND SERUM FROM PATIENTS WITH CANCER

Detection of methylation abnormalities may form a novel basis for noninvasive diagnosis of very small to large tumors among high-risk populations and for disease monitoring. Methylated DNA sequences can be successfully detected in plasma and serum of patients with cancer (Table 14.1).[12,29–31,34,35] Aberrant *p16* methylation appears to be a common biomarker for the noninvasive diagnosis of multiple cancers among high-risk populations at early stages. Tumors that have not developed to metastasize may not shed many cells into blood, but would possibly release tumor DNA into the circulation (Figure 14.1). Of note, nearly all patients showing concurrent *p15* and *p16* methylation in primary HCCs also had detectable methylation abnormalities in surrogate tissues including plasma and serum.[33] Regardless of the tumor size, the author's team found *p15/p16* methylation in peripheral blood of 87% of HCC patients with tumoral *p15/p16* methylation. Detection of tumor-derived epigenetic changes in plasma and serum, even from patients with very small HCCs, may open the possibility of noninvasive cancer screening. Also, methylated *p15* sequences were detected in plasma of 92% of patients with acute leukemia who possessed identical alterations in blood cells and bone marrow.[12]

Elevated levels of circulating nucleic acids have been shown in patients with cancer.[36] DNA concentrations in plasma and serum are especially low in healthy individuals (2 to 30 ng/ml) as compared to those in patients with cancer (20 to 1200 ng/ml), attributable to the presence of circulating tumor DNA.[36–38] In plasma of patients with breast cancer, angiosarcoma, melanoma, or head and neck cancer, the fractional concentration of tumor DNA was determined to range from 3 to 93% (mean = 53%) of the total amount of circulating DNA.[38] Tumor DNA may possibly be enriched in plasma by selective DNA release from tumor cells or inhibition of tumor DNA degradation as protected from DNase digestion. *MGMT, RAR*β*2,* and/or *RASSF1A* hypermethylation was demonstrated in circulating DNA isolated from preoperative plasma of patients with cutaneous melanoma.[21] In serum, the DNA concentration is also much higher among patients with cancer as compared to that among healthy individuals.[39] Elevated DNA levels were detected in serum from patients with metastases as compared to those in patients with nonmetastatic cancer.[39]

In patients with lung cancer, *APC* hypermethylation was commonly detected in plasma or serum by quantitative real-time methylation-specific PCR (MSP).[40] DNA methylation was also found in plasma or serum from patients with colorectal cancer, liver cancer, or esophageal cancer.[29,34,41,42] Circulating tumor DNA was detectable in plasma or serum of patients with bladder cancer, renal cell carcinoma, or prostate cancer,[43,44] but the presence of circulating tumor DNA in plasma/serum was not associated with the tumor stage.[43] Interestingly, circulating nucleosomes were also detectable in serum of patients with cancer.[45] However, tumor DNA in plasma, serum, and body fluids needs to be further characterized in terms of the stability, sizes, and biological implications.

The DNA content in serum may possibly be higher than that in plasma, owing to cell lysis during coagulation. However, the fractional concentration of tumor DNA may still be low, due to the relatively higher amount of normal DNA in serum. Although the origin of circulating DNA remains unclear at present, the detection of tumor-specific DNA methylation markers enables cancer diagnosis with high sensitivity and specificity.[46] The mechanism of DNA release from the tumor into plasma or serum may possibly be related to cellular turnover, necrosis, or apoptosis as proven *in vitro* and *in vivo*.[38] In particular, a spectrum of multiples of 180-bp fragments in plasma is reminiscent of the oligonucleosomal DNA of chromatin degraded by caspase-activated DNase, indicative of cellular apoptosis. Conversely, DNA fragments of > 10 kb could possibly originate from cells dying via necrosis.[38] It is clear that the biological characteristics of a tumor such as the growth rate, histological grade, rates of apoptosis and necrosis may all affect the ultimate amount of tumor DNA released into the bloodstream.[47]

14.6 COMBINATORIAL ANALYSES OF DNA HYPERMETHYLATION IN PLASMA/SERUM AND CONVENTIONAL PROTEIN TUMOR MARKERS IN SERUM

Serum alpha-fetoprotein (AFP), a conventional marker for HCC, is not completely reliable for cancer screening or tumor staging. High-risk individuals with normal AFP levels may have already developed HCC at an advanced stage.[49,50] Of diagnostic interest, the *p16* methylation status in circulating DNA was significantly associated with the preoperative serum AFP level.[51] All HCC cases with serum AFP levels > 45 ng/ml showed *p16* methylation in circulating DNA.[51] Clinically, the serum AFP level alone cannot reliably differentiate HCC from benign liver diseases. The MSP assay for *p16* alone could identify 53% of patients with HCC, including those who were not detectable by the AFP screening, and the combination of the MSP assay with the AFP test further improved the rate of HCC detection.[51] These findings demonstrate the usefulness of methylation abnormalities for noninvasive cancer diagnosis. The additional analysis of methylation markers in surrogate tissues such as serum would likely permit earlier and more reliable detection of cancers among high-risk populations.

For cancer monitoring, an elevated serum AFP level (>10 ng/ml) is generally applied as one of the useful criteria. However, clinical metastasis and recurrence may occur in patients with HCC with normal serum AFP levels. In this regard, the peripheral blood MSP analysis may greatly enhance the sensitivity and specificity to enable the monitoring of minimal residual tumors or micrometastases during clinical follow-up. The methlyation analysis may open a new dimension for further investigations on the correlations of other molecular alterations in circulating DNA and levels of serum markers among patients with different cancers.

14.7 MOLECULAR MONITORING OF HUMAN CANCERS IN BLOOD AND PROGNOSTIC IMPLICATIONS

The presence of minimal residual disease (MRD) and tumor recurrence may be monitored by quantifying epigenetic changes that can progressively lead to gene silencing and eventually clinical metastasis. The identified epigenetic markers serve as surrogate end-point biomarkers to evaluate patients' response to therapies. Aberrant *p15* methylation, which is frequently found in adult and childhood acute leukemias, appears to be a very useful molecular prognostic marker for risk assessment and early detection of MRD or relapse.[12] *p15* hypermethylation was in good agreement with morphological relapse or active and residual leukemia, whereas *p15* hypomethylation was concordant with morphological remission or lack of residual leukemia.[12] This suggests that *p15* methylation may be a specific molecular abnormality largely associated with tumor recurrence. Sequential monitoring of *p15* methylation status may be useful for anticipating the stage of clinical remission.

Molecular analysis of aberrant *p15* methylation in blood may possibly enable disease monitoring and risk assessment.[12] *p15* methylation may play a role in cancer progression in addition to leukemogenesis[52] and aberrant *p15* methylation appears to have important prognostic implications. The median survival time of patients with acute leukemia with *p15* methylation at diagnosis was notably reduced as compared with those carrying unmethylated *p15* alleles.[12] This molecular approach may be applied to many other tumor suppressor genes, DNA repair genes, or metastasis suppressor genes, which are methylated in different tumor types. For example, *DAPK* hypermethylation has been associated with shortened survival in patients with lung cancer.[18] *RASSF1A* and *APC* hypermethylation in blood of patients with breast cancer has also been associated with reduced survival.[53]

The functional significance of *p15* and *p16* methylation may thus be implicated in tumor progression, in that methylation could be an initiating event leading to progressive inactivation of the cell cycle regulatory genes.[9,22,54] Impaired *p15* and *p16* expression, which confers selective growth advantage to tumor cells capable of clonal expansion, would possibly promote stepwise transformation and neoplastic progression. Aberrant methylation profiles in CTCs may be associated with tumor recurrence or metastasis development. For further investigation, it will be interesting to study the methylation profiles of CTCs in association with patient's response to treatment. This methylation-based approach may also be widely employed for monitoring the biological behavior and the clinical course of many malignancies.

Moreover, the MSP analysis of plasma samples may potentially be applied for risk assessment and early detection of human cancers. *p15* and *p16* methylation abnormalities in plasma and serum can be generally employed as diagnostic and prognostic markers for a wide variety of cancers. *p16* methylation was detectable in plasma and serum from patients with lung cancer, colorectal cancer, breast cancer, or head and neck cancer.[31,35,38,55] *hMLH1* hypermethylation was detected in serum from patients with colon cancer, and *MGMT, DAPK,* or *GSTP1* hypermethylation was found in plasma and serum from patients with lung cancer, prostate cancer, or head and neck cancer (Table 14.1).[30,31,34,35] The peripheral blood MSP analysis can be easily conducted for widespread cancer screening and the monitoring of patients' response to therapies, such as surgical resection, chemotherapy, or radiotherapy.[56]

The combination of epigenetic markers may prove valuable for noninvasive cancer diagnosis and prognostic assessment. The presence of aberrant *p16* methylation in plasma and serum may possibly be associated with tumor recurrence or metastasis development. Nearly half of the patients with HCC with *p16* methylation in plasma or serum developed local recurrence and distant metastasis.[32] The methylation profiles in CTCs and circulating tumor DNA may be associated with patients' response to treatment. This methylation-based approach may also be widely employed for monitoring the biological behavior and the clinical course of many malignancies.

14.8 METHYLATION PROFILING OF TUMOR CELLS AND TUMOR DNA IN URINE FROM PATIENTS WITH CANCER

As compared to urinary cytological analysis, the methylation analysis of DNA in urine from patients with bladder cancer may predict tumor recurrence with higher sensitivity.[57,58] Moreover, tumor DNA may be enriched in urine of patients with bladder cancer, enabling the detection of tumor cells at very early stages. Methylated *DAPK, RAR*β*, E-cadherin,* and *p16INK4a* sequences were frequently found in the urine of patients with bladder cancer (Table 14.2).[59] Tumor DNA was also found in urine from patients with renal cell carcinoma or prostate cancer,[44] and hypermethylation of six genes was detected in urine of patients with kidney cancer with very high specificity and sensitivity.[60]

Unexpectedly, tumor DNA was detected in the urine of patients with nonurological malignancies, providing evidence that small amounts of short tumor DNA fragments might possibly bypass the kidney barrier to enter the urine with implications for diagnosis of cancers.[61] Noninvasive detection of epigenetic alterations in tumor cells or released tumor DNA in body fluids, including serum, plasma, sputum, saliva, urine, nipple aspirate, synovial fluid, ascite, pleural effusion, ejaculate, and stool (Table 14.1 and Table 14.2), may ultimately allow the discovery of powerful and easily monitored molecular markers for human cancers.[62]

14.9 METHYLATION PROFILING OF TUMOR CELLS AND TUMOR DNA IN OTHER BODY FLUIDS FROM PATIENTS WITH CANCER

With diagnostic, prognostic, and therapeutic implications, tumor DNA has been detected in plasma, serum, and other body fluids from patients with tumors initiating in virtually all organs, including peritoneal fluid from patients with ovarian cancer, bronchial alveolar lavage fluid from patients with lung cancer, bone marrow aspirates, urine, prostatic fluid, cerebrospinal fluid, gastric/biliary juice, and stool sam-

Table 14.2 Methylation Profiling of Tumor Cells and Tumor DNA in Body Fluids from Patients with Cancer

Cancer	Body Fluid	Genes	Detection Rate (%)	Ref.
Breast cancer	Ductal lavage fluid	*RAR*β Twist Cyclin D2	85	66
Cervical cancer	Papanicolaou smear	p16 E-cadherin	27–64	85
Colorectal cancer	Stool	SFRP2	77–90	86
Lung cancer	Bronchoalveolar lavage	p16	63	87
	Sputum	p16 MGMT	100	67
Pancreatic cancer	Pancreatic juice	ppENK pi6	11–67	88
Prostate cancer	Ejaculate	GSTP1	44	89
	Urine	GSTP1	73	90
Kidney cancer	Urine	p16 ARF APC VHL Timp-3 RASSF1A	88	60
Bladder cancer	Urine	p16 *DAPK* *RAR*β E-cadherin	14–68	59

ples from patients with a variety of cancers (Table 14.2).[2] Epigenetic alterations occur early in primary neoplasia, and promoter hypermethylation is an early phenomenon in premalignant or morphologically benign lesions. Many research groups have analyzed epigenetic markers in body fluids of patients with cancer as compared with healthy individuals. Tumor-associated methylated DNA was detected in exfoliated tumor cells isolated from body fluids (Table 14.2).[62] Thus far, DNA hypermethylation has been found in body fluids from patients with different cancer types, including lung cancer, liver cancer, breast cancer, prostate cancer, head and neck cancer, colon cancer, and acute leukemia.[12,29–31,34,55,63,64]

As opposed to conventional immunocytological detection, molecular detection in body fluids offers higher sensitivity. Prostate specific antigen (PSA) has been applied as a protein tumor marker for the diagnosis of prostate cancer with low specificity. Compared to protein tumor markers in serum, DNA extracted from body fluids can be easily amplified by PCR for higher sensitivity. For instance, *GSTP1* promoter hypermethylation was frequently detected in plasma, serum, blood cells, ejaculates, and urine from patients with prostate cancer at early stages with much higher specificity and sensitivity than those for PSA (Table 14.1 and Table 14.2).[30]

Specific methylation profiles in cancer cells are typically maintained during cancer progression.[65] DNA methylation profiles are thus heritable and therefore useful for cancer screening. Obviously, tumor biomarkers would be most efficiently detectable in the body fluids most intimately associated with the organs or systems where the tumors are located. For example, breast cancer cells, colorectal cancer cells, lung cancer cells, and prostate cancer cells may be readily detectable in ductal lavage fluid, stool, sputum, urine, and ejaculate (Figure 14.1; Table 14.2).[66,67] Semen, saliva, and bronchial brushings could also serve as sample sources for noninvasive diagnosis of prostate cancer, head and neck cancer, and lung cancer, respectively.[64,68,69]

As discussed in Chapter 9, cancer cells have on many occasions been detected in nipple aspirate fluid (NAF) of patients with breast cancer. In one such study, hypermethylation of *GSTP1, RAR*β*2, p16^{INK4a}, ARF, RASSF1A,* and *DAPK* promoters could be specifically detected in NAF from 82% of patients with breast cancer, including stage I cancer and ductal carcinoma *in situ.*[70] Tumor DNA in body fluids may also possibly reflect the tumor burden and help predict or monitor patients' response to therapy. For the other systems where intimately associated body fluids are not easily obtained, potentially relevant biomarkers could still possibly be detectable in plasma or serum. However, it is uncertain whether tumor cells that harbor tumor-specific DNA methylation will always release DNA into bloodstream or lumens such as the breast ducts of some patients with cancer.

14.10 QUALITATIVE AND QUANTITATIVE ANALYSES OF ABERRANT METHYLATION CHANGES: SENSITIVITY AND SPECIFICITY

DNA methylation can be analyzed by qualitative or quantitative PCR-based methods, including MSP,[71] bisulfite sequencing,[72,73] methylation-sensitive restriction enzyme PCR, combined bisulfite restriction analysis (COBRA),[74] methylation-sensitive single nucleotide primer extension (Ms-SNuPE),[75] and quantitative real-time MSP.[22] MSP, which couples the bisulfite modification of DNA and PCR, is fast, highly sensitive, and widely applied for DNA methylation analyses. Bisulfite modification converts all unmethylated cytosines to uracils, whereas methylcytosines remain unmodified.[72,73] MSP requires specific primer sets, which are designed to distinguish between methylated and unmethylated DNA sequences. MSP offers high sensitivity for detecting small amounts of methylated alleles in clinical samples, such as plasma, serum, other body fluids, blood cells, lymph nodes, biopsies, and paraffin-embedded tissues.[12,29,33]

The relative sensitivities of MSP for various genes ranged from 10^{-5} to 10^{-2}(1 methylated DNA copy among 10^5 to 10^2 unmethylated DNA copies).[12,29,30,32,33] The lower detection rates of methylation changes may be related to the lower analytical sensitivity of the MSP assay. The detection rates of methylation abnormalities can potentially be enhanced by using a greater amount of input DNA. Highly specific and sensitive MSP should be applicable for methylation analysis among the low percentage of tumor cells or tumor DNA and particularly useful for detecting MRD, tumor recurrence, or metastasis formation.

Bisulfite sequencing is relatively time-consuming, since large-scale sequencing of multiple plasmid clones is required to obtain the overall methylation pattern.[72,73] Methylation-sensitive restriction enzyme PCR combines methylation-sensitive restriction enzyme digestion and PCR.[76] After enzyme digestion, PCR products are obtained if the enzyme cannot digest at the methylated CpG sites within the specified DNA region. COBRA, Ms-SNuPE, and quantitative real-time MSP allow the quantitative analyses of DNA methylation. For cancer screening, real-time quantitative MSP may prove valuable for analyzing the fractional concentration of tumor DNA in plasma and serum from patients with cancer.[22,77] Methylation markers can be both tumor specific and tissue specific and thus useful for differential cancer diagnosis.[46]

14.11 HIGH-THROUGHPUT METHODS FOR METHYLATION PROFILING IN CANCER CELLS AND THE SELECTION OF TARGET GENES AS EPIGENETIC MARKERS IN BLOOD AND BODY FLUIDS

The combination of MSP with real-time PCR technology allows quantitative analyses of DNA methylation.[22] Quantitative assessment by enzymatic regional methylation assay (ERMA)[78] and microarray technologies, such as methylation-specific oligonucleotide microarray (MSO)[79] and expressed CpG island sequence tag (ECIST) microarray,[80] allow high-throughput analyses of the methylation profiles in different tumor types.

ERMA is a novel method for quantifying the methylation density of CpG sites within a particular DNA region, which may represent cellular or allelic methylation patterns in a biological sample.[78] After bisulfite modification of genomic DNA, the region of interest is PCR-amplified with primers containing two dam sites (GATC). PCR products are incubated with ^{14}C-labeled *S*-adenosyl-L-methionine (SAM) and dam methyltransferase for standardizing the DNA quantity as an internal control. Then, ^{3}H-labeled SAM and SssI methyltransferase are added for measuring the methylation density within the target DNA region. With the use of standard mixtures of cell line DNA with known methylation density in every assay, the methylation density of the region can be determined according to the ratio of ^{3}H to ^{14}C signal intensities.

MSO or differential methylation hybridization allows the global genomic analysis of DNA methylation by mapping methylation changes in multiple CpG islands and thus generating epigenetic profiles in cancer cells.[79] Genomic DNA is first modified with sodium bisulfite, which converts unmethylated cytosines into uracils but leaves methylated cytosines unmodified. After PCR amplification with the incorporation of the Cy5 fluorescent label, a pool of PCR products with differential methylation patterns as the targets are hybridized to an array of oligonucleotides that can discriminate between unmethylated and methylated cytosines at specific nucleotide positions. Quantitative differences in hybridization are then determined by fluorescence analysis.

MSO is suitable for examining a panel of genes among clinical samples.[81] This approach can generate a robust data set for discovering methylation profiles in cancer cells. Candidate epigenetic markers with diagnostic and prognostic implications can be identified. For example, after the profiling of methylation alterations of CpG islands in ovarian tumors, the duration of progression-free survival after chemotherapy was found to be significantly shorter for patients with Stage III/IV ovarian carcinoma with higher levels of concurrent methylation as compared with those possessing lower methylation levels.[82] A higher degree of CpG island methylation is associated with early tumor recurrence after chemotherapy. A selected group of

CpG island loci are potentially useful as epigenetic markers in plasma, serum, or other body fluids for predicting treatment outcome in patients with ovarian cancer.

ECIST microarray is a new method for dual screening of DNA hypermethylation and gene silencing in cancer cells.[80] ECISTs are DNA fragments typically located in the promoters and exon 1 regions of genes, with GC-rich sequences that can be screened for aberrantly methylated CpG sites in cancer cells. In addition, the exon-containing portions can be used to measure mRNA levels. Using an ECIST panel, both locus hypermethylation and gene silencing in cancer cells can be studied simultaneously. Therefore, ECISTs serve as effective markers for identifying novel genes with the expression silenced by CpG island hypermethylation. In a previous study, a total of 1162 loci met the criteria of ECISTs from an initial screening of 7776 CpG island tags.[80] Microarray profile analysis identified 30 methylation-silenced genes, which could be transcriptionally reactivated following demethylation.

The biological implications of methylation alterations in association with the biological behavior of tumor cells can be further clarified by using real-time quantitative MSP, which proves the biological significance of the methylation index.[22] The usefulness of fluorescence-based real-time MSP has also been validated by other groups who quantified *p16* and *hMLH1* methylation changes.[38,83] Quantitative assessment of methylation changes at specific CpG sites can be performed by using Ms-SNuPE,[75] and COBRA can measure the relative amounts of digested PCR products with particular methylated CpG sites among the total PCR products.[74] In contrast to real-time quantitative MSP, the latter two methods require gel electrophoresis, radioisotope incorporation, or restriction enzyme digestion. For molecular assessment and cancer monitoring, more robust real-time quantitative MSP may prove valuable for analyzing DNA methylation profiles in plasma, serum, and other body fluids from patients with cancer during clinical courses.

Application of methylation markers for cancer detection and monitoring has advantages over the detection of LOH markers since the sample source may consist of normal cells and normal DNA, which could lead to the underscoring of LOH results. Analysis of methylation markers also has an advantage over mRNA profiling since RNA is not as stable as DNA for the molecular assays.[84] Also, DNA can be easily isolated even from archived blood samples, urine samples, and body fluids. On the basis of encouraging results to date, DNA methylation-based screening assays in surrogate tissues may be employed for detecting many human malignancies at early and curable stages in the near future.

REFERENCES

1. Herman, J.G. and Baylin, S.B., Gene silencing in cancer in association with promoter hypermethylation, *N. Engl. J. Med.*, 349, 2042, 2003.
2. Jones, P.A. and Baylin, S.B., The fundamental role of epigenetic events in cancer, *Nat. Rev. Genet.*, 3, 415, 2002.
3. Baylin, S.B. and Herman, J.G., DNA hypermethylation in tumorigenesis: epigenetics joins genetics, *Trends Genet.*, 16, 168, 2000.

4. Costello, J.F. et al., Aberrant CpG-island methylation has non-random and tumour-type-specific patterns, *Nat. Genet.,* 24, 132, 2000.
5. Jones, P.A. and Laird, P.W., Cancer epigenetics comes of age, *Nat. Genet.,* 21, 163, 1999.
6. Herman, J.G. et al., Hypermethylation-associated inactivation indicates a tumor suppressor role for p15INK4B, *Cancer Res.*, 56, 722, 1996.
7. Takeuchi, S. et al., Analysis of a family of cyclin-dependent kinase inhibitors: p15/MTS2/INK4B, p16/MTS1/INK4A, and p18 genes in acute lymphoblastic leukemia of childhood, *Blood,* 86, 755, 1995.
8. Ogawa, S. et al., Loss of the cyclin-dependent kinase 4-inhibitor (p16; MTS1) gene is frequent in and highly specific to lymphoid tumors in primary human hematopoietic malignancies, *Blood,* 86, 1548, 1995.
9. Wong, I.H. et al., Transcriptional silencing of the p16 gene in human myeloma-derived cell lines by hypermethylation, *Br. J. Haematol.,* 103, 168, 1998.
10. Merlo, A. et al., 5′ CpG island methylation is associated with transcriptional silencing of the tumour suppressor p16/CDKN2/MTS1 in human cancers, *Nat. Med.,* 1, 686, 1995.
11. Kamb, A. et al., A cell cycle regulator potentially involved in genesis of many tumor types, *Science,* 264, 436, 1994.
12. Wong, I.H. et al., Aberrant p15 promoter methylation in adult and childhood acute leukemias of nearly all morphologic subtypes: potential prognostic implications, *Blood,* 95, 1942, 2000.
13. Stone, S. et al., Genomic structure, expression and mutational analysis of the P15 (MTS2) gene, *Oncogene,* 11, 987, 1995.
14. Hannon, G.J. and Beach, D., p15INK4B is a potential effector of TGF-beta-induced cell cycle arrest, *Nature,* 371, 257, 1994.
15. Bedossa, P. et al., Transforming growth factor-beta 1 (TGF-beta 1) and TGF-beta 1 receptors in normal, cirrhotic, and neoplastic human livers, *Hepatology,* 21, 760, 1995.
16. Esteller, M. et al., Inactivation of the DNA-repair gene MGMT and the clinical response of gliomas to alkylating agents, *N. Engl. J. Med.,* 343, 1350, 2000.
17. Wong, I.H. et al., Ubiquitous aberrant RASSF1A promoter methylation in childhood neoplasia, *Clin. Cancer Res.,* 10, 994, 2004.
18. Tang, X. et al., Hypermethylation of the death-associated protein (DAP) kinase promoter and aggressiveness in stage I non-small-cell lung cancer, *J. Natl. Cancer Inst.,* 92, 1511, 2000.
19. Toyota, M. et al., CpG island methylator phenotype in colorectal cancer, *Proc. Natl. Acad. Sci. U.S.A.,* 96, 8681, 1999.
20. Jones, P.A., DNA methylation errors and cancer, *Cancer Res.,* 56, 2463, 1996.
21. Hoon, D.S. et al., Profiling epigenetic inactivation of tumor suppressor genes in tumors and plasma from cutaneous melanoma patients, *Oncogene,* 23, 4014, 2004.
22. Lo, Y.M. et al., Quantitative analysis of aberrant p16 methylation using real-time quantitative methylation-specific polymerase chain reaction, *Cancer Res.,* 59, 3899, 1999.
23. Wong, I.H. et al., Hematogenous dissemination of hepatocytes and tumor cells after surgical resection of hepatocellular carcinoma: a quantitative analysis, *Clin. Cancer Res.,* 5, 4021, 1999.
24. Wong, I.H. et al., Circulating tumor cell mRNAs in peripheral blood from hepatocellular carcinoma patients under radiotherapy, surgical resection or chemotherapy: a quantitative evaluation, *Cancer Lett.,* 167, 183, 2001.

25. Wong, I.H. et al., Quantitative relationship of the circulating tumor burden assessed by reverse transcription-polymerase chain reaction for cytokeratin 19 mRNA in peripheral blood of colorectal cancer patients with Dukes' stage, serum carcinoembryonic antigen level and tumor progression, *Cancer Lett.,* 162, 65, 2001.
26. Wong, I.H. et al., Quantitative correlation of cytokeratin 19 mRNA level in peripheral blood with disease stage and metastasis in breast cancer patients: potential prognostic implications, *Int. J. Oncol.,* 18, 633, 2001.
27. Wong, I.H. et al., Quantitative analysis of circulating tumor cells in peripheral blood of osteosarcoma patients using osteoblast-specific messenger RNA markers: a pilot study, *Clin. Cancer Res.,* 6, 2183, 2000.
28. Adorjan, P. et al., Tumour class prediction and discovery by microarray-based DNA methylation analysis, *Nucleic Acids Res.,* 30, e21, 2002.
29. Wong, I.H. et al., Detection of aberrant p16 methylation in the plasma and serum of liver cancer patients, *Cancer Res.,* 59, 71, 1999.
30. Goessl, C. et al., Fluorescent methylation-specific polymerase chain reaction for DNA-based detection of prostate cancer in bodily fluids, *Cancer Res.*, 60, 5941, 2000.
31. Esteller, M. et al., Detection of aberrant promoter hypermethylation of tumor suppressor genes in serum DNA from non-small cell lung cancer patients, *Cancer Res.,* 59, 67, 1999.
32. Wong, I.H. et al., Tumor-derived epigenetic changes in the plasma and serum of liver cancer patients. Implications for cancer detection and monitoring, *Ann. N.Y. Acad. Sci.,* 906, 102, 2000.
33. Wong, I.H. et al., Frequent p15 promoter methylation in tumor and peripheral blood from hepatocellular carcinoma patients, *Clin. Cancer Res.,* 6, 3516, 2000.
34. Grady, W.M. et al., Detection of aberrantly methylated hMLH1 promoter DNA in the serum of patients with microsatellite unstable colon cancer, *Cancer Res.,* 61, 900, 2001.
35. Sanchez-Cespedes, M. et al., Gene promoter hypermethylation in tumors and serum of head and neck cancer patients, *Cancer Res.,* 60, 892, 2000.
36. Shapiro, B. et al., Determination of circulating DNA levels in patients with benign or malignant gastrointestinal disease, *Cancer,* 51, 2116, 1983.
37. Chen, X.Q. et al., Microsatellite alterations in plasma DNA of small cell lung cancer patients. *Nat. Med.,* 2, 1033, 1996.
38. Jahr, S. et al., DNA fragments in the blood plasma of cancer patients: quantitations and evidence for their origin from apoptotic and necrotic cells, *Cancer Res.,* 61, 1659, 2001.
39. Leon, S.A. et al., Free DNA in the serum of cancer patients and the effect of therapy, *Cancer Res.,* 37, 646, 1977.
40. Usadel, H. et al., Quantitative adenomatous polyposis coli promoter methylation analysis in tumor tissue, serum, and plasma DNA of patients with lung cancer, *Cancer Res.,* 62, 371, 2002.
41. Zou, H.Z. et al., Detection of aberrant p16 methylation in the serum of colorectal cancer patients, *Clin. Cancer Res.,* 8, 188, 2002.
42. Kawakami, K. et al., Hypermethylated APC DNA in plasma and prognosis of patients with esophageal adenocarcinoma, *J. Natl. Cancer Inst.*, 92, 1805, 2000.
43. von Knobloch, R. et al., Serum DNA and urine DNA alterations of urinary transitional cell bladder carcinoma detected by fluorescent microsatellite analysis, *Int. J. Cancer,* 94, 67, 2001.
44. Goessl, C. et al., DNA alterations in body fluids as molecular tumor markers for urological malignancies, *Eur. Urol.,* 41, 668, 2002.

45. Holdenrieder, S. et al., Circulating nucleosomes predict the response to chemotherapy in patients with advanced non-small cell lung cancer, *Clin. Cancer Res.,* 10, 5981, 2004.
46. Wong, I.H. et al., Epigenetic tumor markers in plasma and serum: biology and applications to molecular diagnosis and disease monitoring, *Ann. N.Y. Acad. Sci.,* 945, 36, 2001.
47. Sorenson, G.D., Detection of mutated KRAS2 sequences as tumor markers in plasma/serum of patients with gastrointestinal cancer, *Clin. Cancer Res.,* 6, 2129, 2000.
48. Lo, Y.M. et al., Rapid clearance of fetal DNA from maternal plasma, *Am. J. Hum. Genet.,* 64, 218, 1999.
49. Chen, D.S. et al., Serum alpha-fetoprotein in the early stage of human hepatocellular carcinoma, *Gastroenterology,* 86, 1404, 1984.
50. Kubo, Y. et al., Detection of hepatocellular carcinoma during a clinical follow-up of chronic liver disease: observations in 31 patients, *Gastroenterology,* 74, 578, 1978.
51. Wong, I.H. et al., Relationship of p16 methylation status and serum alpha-fetoprotein concentration in hepatocellular carcinoma patients, *Clin. Chem.,* 46, 1420, 2000.
52. Quesnel, B. et al., Methylation of the p15(INK4b) gene in myelodysplastic syndromes is frequent and acquired during disease progression, *Blood,* 91, 2985, 1998.
53. Muller, H.M. et al., DNA methylation in serum of breast cancer patients: an independent prognostic marker, *Cancer Res.,* 63, 7641, 2003.
54. Barrett, M.T. et al., Evolution of neoplastic cell lineages in Barrett oesophagus, *Nat. Genet.,* 22, 106, 1999.
55. Silva, J.M. et al., Presence of tumor DNA in plasma of breast cancer patients: clinicopathological correlations, *Cancer Res.,* 59, 3251, 1999.
56. Wong, I.H., Methylation profiling of human cancers in blood: molecular monitoring and prognostication (review), *Int. J. Oncol.,* 19, 1319, 2001.
57. Steiner, G. et al., Detection of bladder cancer recurrence by microsatellite analysis of urine, *Nat. Med.,* 3, 621, 1997.
58. van Rhijn, B.W. et al., Microsatellite analysis — DNA test in urine competes with cystoscopy in follow-up of superficial bladder carcinoma: a phase II trial, *Cancer,* 92, 768, 2001.
59. Chan, M.W. et al., Hypermethylation of multiple genes in tumor tissues and voided urine in urinary bladder cancer patients, *Clin. Cancer Res.,* 8, 464, 2002.
60. Cairns, P., Detection of promoter hypermethylation of tumor suppressor genes in urine from kidney cancer patients, *Ann. N.Y. Acad. Sci.,* 1022, 40, 2004.
61. Botezatu, I. et al., Genetic analysis of DNA excreted in urine: a new approach for detecting specific genomic DNA sequences from cells dying in an organism, *Clin. Chem.,* 46, 1078, 2000.
62. Sidransky, D., Emerging molecular markers of cancer, *Nat. Rev. Cancer,* 2, 210, 2002.
63. Belinsky, S.A. et al., Aberrant methylation of p16(INK4a) is an early event in lung cancer and a potential biomarker for early diagnosis, *Proc. Natl. Acad. Sci. U.S.A.,* 95, 11891, 1998.
64. Rosas, S.L. et al., Promoter hypermethylation patterns of p16, O6-methylguanine-DNA-methyltransferase, and death-associated protein kinase in tumors and saliva of head and neck cancer patients, *Cancer Res.,* 61, 939, 2001.
65. Markl, I.D. et al., Global and gene-specific epigenetic patterns in human bladder cancer genomes are relatively stable *in vivo* and *in vitro* over time, *Cancer Res.,* 61, 5875, 2001.

66. Evron, E. et al., Detection of breast cancer cells in ductal lavage fluid by methylation-specific PCR, *Lancet,* 357, 1335, 2001.
67. Palmisano, W.A. et al., Predicting lung cancer by detecting aberrant promoter methylation in sputum, *Cancer Res.,* 60, 5954, 2000.
68. Soria, J.C. et al., Aberrant promoter methylation of multiple genes in bronchial brush samples from former cigarette smokers, *Cancer Res.,* 62, 351, 2002.
69. Goessl, C. et al., DNA-based detection of prostate cancer in blood, urine, and ejaculates, *Ann. N.Y. Acad. Sci.,* 945, 51, 2001.
70. Krassenstein, R. et al., Detection of breast cancer in nipple aspirate fluid by CpG island hypermethylation, *Clin. Cancer Res.,* 10, 28, 2004.
71. Herman, J.G. et al., Methylation-specific PCR: a novel PCR assay for methylation status of CpG islands, *Proc. Natl. Acad. Sci. U S A.,* 93, 9821, 1996.
72. Grigg, G. and Clark, S., Sequencing 5-methylcytosine residues in genomic DNA, *Bioessays*, 16, 431, 1994.
73. Frommer, M. et al., A genomic sequencing protocol that yields a positive display of 5-methylcytosine residues in individual DNA strands, *Proc. Natl. Acad. Sci. U.S.A.,* 89, 1827, 1992.
74. Xiong, Z. and Laird, P.W., COBRA: a sensitive and quantitative DNA methylation assay, *Nucleic Acids Res.,* 25, 2532, 1997.
75. Gonzalgo, M.L. and Jones, P.A., Rapid quantitation of methylation differences at specific sites using methylation-sensitive single nucleotide primer extension (Ms-SNuPE), *Nucleic Acids Res.,* 25, 2529, 1997.
76. Yamada, Y. et al., A comprehensive analysis of allelic methylation status of CpG islands on human chromosome 21q, *Genome Res.,* 14, 247, 2004.
77. Wong, I.H. et al., Quantitative analysis of tumor-derived methylated p16INK4a sequences in plasma, serum, and blood cells of hepatocellular carcinoma patients, *Clin. Cancer Res.,* 9, 1047, 2003.
78. Galm, O. et al., Enzymatic regional methylation assay: a novel method to quantify regional CpG methylation density, *Genome Res.,* 12, 153, 2002.
79. Gitan, R.S. et al., Methylation-specific oligonucleotide microarray: a new potential for high-throughput methylation analysis, *Genome Res.,* 12, 158, 2002.
80. Shi, H. et al., Expressed CpG island sequence tag microarray for dual screening of DNA hypermethylation and gene silencing in cancer cells, *Cancer Res.,* 62, 3214, 2002.
81. Yan, P.S. et al., Differential methylation hybridization using CpG island arrays. *Methods Mol. Biol.,* 200, 87–100, 2002.
82. Wei, S.H. et al., Methylation microarray analysis of late-stage ovarian carcinomas distinguishes progression-free survival in patients and identifies candidate epigenetic markers, *Clin. Cancer Res.,* 8, 2246, 2002.
83. Eads, C.A. et al., MethyLight: a high-throughput assay to measure DNA methylation, *Nucleic Acids Res.,* 28, E32, 2000.
84. Wong, I.H. and Lo, Y.M., New markers for cancer detection, *Curr. Oncol. Rep.,* 4, 471, 2002.
85. Lin, W.M. et al., Molecular Papanicolaou tests in the twenty-first century: molecular analyses with fluid-based Papanicolaou technology, *Am. J. Obstet. Gynecol.,* 183, 39, 2000.
86. Muller, H.M. et al., Methylation changes in faecal DNA: a marker for colorectal cancer screening? *Lancet,* 363, 1283, 2004.
87. Ahrendt, S.A. et al., Molecular detection of tumor cells in bronchoalveolar lavage fluid from patients with early stage lung cancer, *J. Natl. Cancer Inst.,* 91, 332, 1999.

88. Fukushima, N. et al., Diagnosing pancreatic cancer using methylation specific PCR analysis of pancreatic juice, *Cancer Biol. Ther.,* 2, 78, 2003.
89. Suh, C.I. et al., Comparison of telomerase activity and GSTP1 promoter methylation in ejaculate as potential screening tests for prostate cancer, *Mol. Cell Probes,* 14, 211, 2000.
90. Goessl, C. et al., DNA-based detection of prostate cancer in urine after prostatic massage, *Urology,* 58, 335, 2001.

Section V

Future Considerations for Surrogate Tissue Profiling

CHAPTER 15

Regulatory and Technical Challenges in Incorporating Surrogate Tissue Profiling Strategies into Clinical Development Programs

Judith L. Oestreicher, Monica J. Cahilly, Deborah P. Mounts, Maryann Z. Whitley, Lisa A. Speicher, William L. Trepicchio, and Michael E. Burczynski

CONTENTS

15.1 INTRODUCTION

If transcriptional signatures in surrogate tissues have any hope of impacting clinical drug development or regulatory decision making,[1] transcriptional profiling strategies in surrogate tissues must first be formally incorporated into clinical trial designs when appropriate. Despite the growing body of information in the literature, the potential value of transcriptional profiling has not yet been fully appreciated by some traditional drug development groups, and the value of determining transcriptional signatures in surrogate tissues is certainly unproven to date. Nonetheless, powerful drivers — most importantly, the clinical accessibility of surrogate tissues and their potential relevance to a variety of disease indications — are enabling an era of exploration in this field.

A number of additional obstacles to the incorporation of pharmacogenomic strategies in real-time clinical trials exist. Transcriptional profiling methodology has not yet been "validated" to contribute to the drug development process, and the technology is considered expensive in an already cost-prohibitive environment. Additionally, stringent timelines to speed products through development render additional genomic sampling an often unwelcome component in what are already complex clinical trials.

In one paradigm, the complexity of incorporating genomic technologies in traditional drug development paradigms has led to the creation of smaller, focused groups in the pharmaceutical industry that carry much of the logistical burden for incorporating genomic technologies into the larger clinical research and development (CR&D) trials, thereby alleviating CR&D of the added responsibility for the genomic samples. In some scenarios such departments may also run independent, exploratory trials with pharmacogenomics as a primary objective of the trial. Exploratory pharmacogenomic (PG) trials can be designed to address biological questions that help guide dose selection for future trials and may provide early identification of expression patterns that are predictive of response to (or adverse effects of) treatment. These types of exploratory PG trials provide ideal conditions for the sampling of surrogate tissues for the purposes of biomarker identification in accessible tissues. One of the primary mandates of these groups is to deal with the regulatory and technical challenges associated with surrogate tissue profiling, and expression profiling in general, in clinical PG studies.

15.2 THE CLINICAL QUESTION OF INTEREST

With any large-scale expression profiling study conducted in human tissues, whether based on archived tissue or samples collected in real-time clinical trials, there are a number of practical issues that must be carefully addressed. The first and foremost is the identification of the clinical question of interest. What is the ultimate goal of the study? The chapter on surrogate tissue profiling in oncology (Chapter 4) reviews instances of investigators who explored the association of gene expression signatures with histological grades of tumors, molecular defects in tumors, and clinical outcomes in patients. Although clinical PG studies benefit greatly from a prospective clarification of the clinical question at the outset of a clinical trial, sometimes studies will be conducted in purely exploratory fashion in an attempt to discover hitherto unknown correlations between gene expression and any number of factors including ultimate clinical responses. This is often the case in surrogate tissue-based PG studies, where there are no precedents supporting the hypothesis that the transcriptome of the surrogate tissue profiled will bear any information relevant to the clinical study. Nonetheless, the more defined the clinical question (or constellation of possible clinical questions) at the outset, the easier the analysis and interpretation of the PG data at study conclusion.

If unsupervised approaches are to be used to address a clinical question of interest, critical decisions must be made concerning the relevance of sample, clinical, and demographic parameters that should be assessed between discovered subgroups. If supervised approaches are implemented, clinically guided decisions are required concerning how best to stratify patients in the training set on the basis of known clinical parameters to discover the most meaningful gene classifiers (clinically defined response categories, percent blast remission, time to progression, overall survival, etc). All of these critical steps are required to define a clinical study plan, even in exploratory studies, that will facilitate a successful expression profiling study.

15.3 PHARMACOGENOMIC STUDY LOGISTICS AND CLINICAL TRIAL DESIGN

15.3.1 Challenges Facing Incorporation of PG Sampling into Clinical Trials

Many of the challenges in the design and implementation of trials with PG end points mirror those of other types of tumor marker studies (for a review, see Reference 2). Pharmacogenomic studies also face additional unique challenges. Pharmacogenomic sampling is typically embedded in clinical studies designed to test the safety and activity of a new therapy. Because PGs are not the primary end point of these trials, it is often more difficult to ensure compliance from both the patients and the sites in PG sample acquisition. It is possible to mandate PG sampling in a trial, but this may have a negative effect on patient accrual and is not always accepted by institutional review boards (IRBs) or ethics committees. Careful site selection, staff training, and close interactions between the sponsor and the site can help optimize patient compliance with voluntary PG sampling. It is imperative that patients are adequately consented for these analyses, as outlined in the section below. With respect to surrogate tissue analysis, PG studies employing surrogate tissue profiling strategies are easier to implement, given the greater accessibility of surrogate tissues like blood and plasma or serum.

15.3.2 Informed Consent in Pharmacogenomic Studies

The informed consent process remains paramount to the protection of human subject participants in clinical trials. With the advent of genomic technologies, the process of informed consent has become increasingly complex. The reader is referred to a recent paper published on behalf of the Pharmacogenetics Working Group (PGW), which describes the elements of informed consent for PG research, much of which is applicable to PG sampling.[3] In this article, the PGW discusses the special considerations and disclosures in the informed consent process for pharmacogenetic research. It is important that clinical research subjects are encouraged to ask questions in light of the inherent complexity of both the technology and the terminology used to describe pharmacogenomics. A complete discussion should ensue on what transcriptional profiling does, and does not, constitute. If limited to transcriptional profiling, the investigator should distinguish transcriptional profiling from other genomic technologies where DNA samples are being collected. The collection of mRNA samples rather than DNA samples avoids some of the ethical issues surrounding privacy and informational risks associated with potential inadvertent or intentional disclosure of genetic information. Finally, the research subject should understand the nature of the type of testing planned.

Ethically, clinical research should produce benefits while minimizing or preventing potential risks.[4] As such, assessment of the risk-to-benefit ratio has become a standard component of protocol review by IRBs. Pharmacogenomic risks and benefits must be clearly outlined in the informed consent document, and the subject must have the right to refuse participation in or withdraw from the PG component of the trial. In conclusion, subjects who participate in clinical research deserve the utmost respect and clearest communication possible, for without them voluntary clinical research would not be possible.

15.3.3 Acquisition of Samples for Pharmacogenomic Analyses in Real-Time Clinical Trials

When incorporating PG sampling into clinical studies, careful consideration should be given to the PG objectives and end points including sample type (surrogate vs. target), sample collection time points, and alignment with other pharmacodynamic and/or clinical safety/activity end points. Clinical PG objectives typically fall into the identification of biomarkers in one of three categories: markers of disease, markers of drug exposure, and markers predictive or indicative of drug efficacy/safety. To meet PG objectives while not impeding the primary goals of the study, it is recommended that PG sample collection be coordinated, to the extent possible, with other scheduled clinical laboratory tests throughout the study. Surrogate tissue analysis (of blood, serum, or plasma) is particularly suited to this strategy, since the PG samplings can often be coordinated with scheduled visits for blood chemistry or pharmacokinetic sampling.

Coordination of PG sampling with other scheduled tests both minimizes the burden on the patient, (i.e., eliminates the need for a patient to return to the site for a specific PG sample), and simplifies the conduct of the study for the site personnel. When possible, PG kits should be supplied to the study sites to facilitate PG sample procurement. These kits should include all necessary material for sample collection, labeling, storage, and shipment. Appropriate user-friendly laboratory manuals and paperwork should accompany these kits to ensure samples can be properly tracked and located.

When available, a single or multiple baseline samples should be collected prior to the initiation of experimental therapy in the common adjuvant setting. This pretreatment sample can then be interrogated by transcriptional profiling for either prognostic or predictive markers of response. Harvesting of both pre- and post-drug treatment samples provide an opportunity to explore drug effects *in situ*, and possibly identify active drug-resistance profiles in addition to the baseline predictive profiles that can be obtained from a pretreatment sampling-only approach.

Real-time clinical PG studies require strong collaborative interactions among multiple disciplines within the hospital/clinical site, including medical oncology, surgery, and pathology, to obtain the harvested tissue in a timely manner. An experienced centralized coordinator of these site interactions is a key element for these studies to succeed. In addition, investigative sites should be encouraged to assign a well-trained staff member to remain with the sample specimen from the time of procurement through the final processing and storage to ensure adequate preservation of mRNA.

15.4 TRANSITIONING THE CLINICAL PHARMACOGENOMIC LABORATORY INTO A REGULATORY COMPLIANT ENVIRONMENT

Transcriptional profiling results derived from analysis of clinical samples harvested from clinical trials may ultimately support the development of clinically relevant diagnostics. Depending on the nature of the transcriptional profiling data (exploratory data vs. validation data), results from certain early clinical trials may constitute regulatory submissible data in support of a prospectively defined clinical assay, regardless of the source of the tissue. Clinical PG laboratories conducting transcriptional profiling as part of clinical studies should therefore strive to implement and maintain a documented quality control system that is specifically tailored to addressing key risks and objectives associated with these types of studies, including:

- Ensuring that the legal integrity of informed consents and clinical subject *(patient) confidentiality* are maintained throughout the conduct of clinical PG studies.
- Ensuring that *chain-of-custody* and *specimen traceability* are maintained throughout the conduct of clinical PG studies.
- Ensuring the *integrity (i.e., accuracy, completeness, and reliability) of data* reported from clinical PG studies.

The Food and Drug Administration (FDA) Good Laboratory Practice (GLP) regulation (21 Code of Federal Regulations [CFR] Part 58) is one example of an internationally accepted quality standard for laboratories that may be applied — at least in part — to PG laboratories to meet the above objectives. GLP regulations specifically apply to safety studies conducted in support of a research or marketing permit, such as toxicology studies conducted to evaluate the safety of a new drug or biologic. As such, there are several requirements of GLPs (such as requirements for maintaining animal test systems) that would not be suitable for exploratory clinical PG studies. However, firms may find it suitable to adopt a strategy for implementing a *risk-based* quality standard in their clinical PG laboratory that is consistent with "the spirit of GLPs." Moreover, within the pharmaceutical industry, there is some precedence to apply GLPs more loosely as a general quality standard to work for which FDA has not yet issued clear guidance. For example, quality programs "in the spirit of GLPs" have been implemented at many firms for clinical bioanalytical studies, such as bioequivalence or bioavailability studies.

Some example GLP requirements that may be applied to the conduct of transcriptional profiling studies in clinical PG laboratories include (1) the preparation and implementation of standard operating procedures (SOPs) for laboratory activities; (2) qualification, calibration, and maintenance of relevant equipment according to written schedules; (3) validation of computerized systems for their intended use; (4) establishment of the reliability, accuracy, and precision of analytical methods employed; (5) labeling, storage, and appropriate use of reagents solutions, and materials; (6) development of systems for training personnel in required tasks; (7) secure labeling of PG samples throughout the test process (from receipt to reporting); (8) contemporaneous and original recording of laboratory data in notebooks; and (9) second-party review of all transcribed data entries and supervisory-level review of all reported data at study conclusion. The implementation of SOPs in particular can greatly enhance the confidence associated with PG data, and provide a systematic framework for laboratory processes, data analysis, and data reporting. Table 15.1

Table 15.1 A List of Potential Standard Operating Procedures in the Clinical Pharmacogenomic Laboratory

	Title
1	Developing Standard Operating Procedures for Clinical Pharmacogenomics
2	Job Function Training
3	Personnel Roles and Responsibilities
4	Document Management
5	Maintaining Laboratory Notebooks
6	Data Entry into Laboratory Notebooks
7	End-User Administration and Control of Key Software Systems
8	Use of Key Software Systems in CP Studies
9	Software Validation
10	Documentation of PG Study Plans
11	Development and Documentation of PG Specimen Processing Methods
12	Documentation, Handling, and Trending Deviations and Quality Issues
13	Receipt, Labeling, Storage, Handling and Disposal of Reagents, Solutions, and Materials
14	Qualification and Monitoring of Specialty Labs
15	Equipment Qualification, Calibration and Maintenance
16	Retention and Disposal of Clinical Pharmacogenomic Specimens
17	Receipt, Labeling, and Storage of Clinical Pharmacogenomic Specimens
18	Processing of Clinical Pharmacogenomic Specimens
19	Verification of Clinical Pharmacogenomics Studies
20	QC Review of Clinical Pharmacogenomics Studies
21	Reporting of Clinical Pharmacogenomics Studies

Reprinted with permission. From Burczynski et al., *Curr. Mol. Med.*, 5, 83–102, 2005.

lists a set of standard SOPs that are being developed in the senior author's laboratory. While the development, implementation, and maintenance of compliance with these SOPs require significant effort, the benefit afforded by assurance of data quality during the discovery and development of clinically relevant assays is substantial. To utilize R&D resources most effectively, firms should specifically tailor the extent to which these and other requirements will be applied to clinical PG studies based on the relative risks posed to patient confidentiality, specimen traceability, and data integrity. In the next section we review general electronic system requirements recommended for ensuring data integrity in transcriptional profiling studies in clinical trials.

15.5 ASSURANCE OF DATA INTEGRITY GENERATED IN CLINICAL PHARMACOGENOMIC STUDIES: ESTABLISHING VALIDATED DATABASES AND DATA TRANSFERS

The voluntary guidance released by the FDA on PG studies[1] indicates that only studies reaching the lofty goal of determining criteria for clinical decision-making will fall under GLP and CFR 21 part 11 requirements. While many of the current studies may not yet meet this decision-making criterion, it is nonetheless advisable to implement or maintain processes that will ultimately enable conversion to full compliance with GLP and CFR 21 part 11 requirements. The complexity of the electronic systems needed to support expression profiling data for clinical studies translates to a very lengthy validation process and suggests that compliance efforts should begin well in advance of starting key pivotal trials. Starting with a minimal, risk-based validation approach can be quite effective and set the stage for full validation at a later point. The concept of a risk-based approach is to identify the components of the overall process that are most critical to conclusions and that are most prone to human error, and to prioritize validation efforts on those components.

As described earlier, the PG process begins with sample collection and chain of custody. In many cases samples are partially processed at one site and then transferred to another site for the expression profiling work, resulting in processing data tracked in several geographically distinct Laboratory Information Management Systems (LIMS). This is especially important in surrogate tissue–based PG studies conducted in the senior author's laboratory, where peripheral blood is collected into cell purification tubes (CPTs) at the clinical site, shipped at ambient temperature in temperature-controlled packaging to a central processing lab for PBMC isolation and RNA purification, and the RNA is finally transferred to the author's laboratory for gene chip analysis. To associate cell counts and other important parameters (for blood samples) with the ultimately generated profile, information on pre- and post-CPT purification cell counts must be electronically transferred from the central provider lab into the laboratory's LIMS system. Furthermore, the actual expression profiling data resides in the PG systems, while the patient information resides in a clinical data management system. The informatics staff is thus challenged with bringing all the pertinent information together (sample characteristics, expression data, and patient characteristics) in a validated environment, while assuring data integrity and security.

It is tempting in small initial studies to replicate the clinical data necessary for expression profiling analysis, either by data exports from clinical databases or by entering the data directly from clinical forms into the PG databases. However, this method is not sustainable as it is subject to manual error and therefore requires a large verification effort. In addition, any updates made to the clinical database during routine clinical data management validation are not represented in the replicated data, requiring constant discrepancy resolution on the PG databases. These considerations make it quite clear that the effort of maintaining the same data in multiple systems is not a useful exercise and is far greater than the effort of simply integrating or linking the necessary systems at the outset. Before embarking on an integration plan, however, it is important to recognize which components of each system are required for integration to occur, which are required in the final analysis data set, and which can be left behind, to be referenced as necessary. For instance, it may not be necessary to link expression profiles and clinical data with certain types of technical LIMS data. While day-to-day technical variation can lead to numerous false conclusions in expression profiling experiments, this type of bias can and should be evaluated and minimized prior to accessing clinical attributes of the sample profiles. Conversely, certain patient and sample identifiers must exist in each system to allow successful integration of the data. Depending on study design, these may include a patient number, study site, visit number, sample barcode, and sample collection dates.

Actual linking of the various data sources for analysis can occur in any number of ways, from simple database queries to elaborate graphical user interfaces. Whichever method is chosen must be flexible enough to account for different study designs. For instance, some studies will collect only a single PG sample per patient while others will follow the patients through several predetermined visits, and some studies may have sufficient controls within the study while others may need to leverage samples collected at other times. Studies for different indications, even within a single field such as oncology, will use different clinical end points and progression measures, and each analysis data set must include the appropriate values.

Finally, once the clinical data are locked and the study is ready to be unblinded, it is necessary to lock the PG data as well. Again, mechanisms for achieving this are varied in both cost and effort. From the risk-based approach, an appropriate process may be to archive one or more data files from which the data analysis could be reproduced. If for any reason the data needs to be restored in the analysis environment after the study is locked, it would be necessary to start with the archived data rather than the original mechanism that was used to extract the data from the databases.

Figure 15.1 illustrates one example of a system integration architecture that strives to ensure chain of custody and data integrity. All patient and sample information resides in a single Studies Management repository. This reduces the data integrity issues incurred when this information is duplicated in multiple systems. Only those patient and sample attributes necessary to accurately identify samples and link to the clinical database are stored in the repository. All other patient

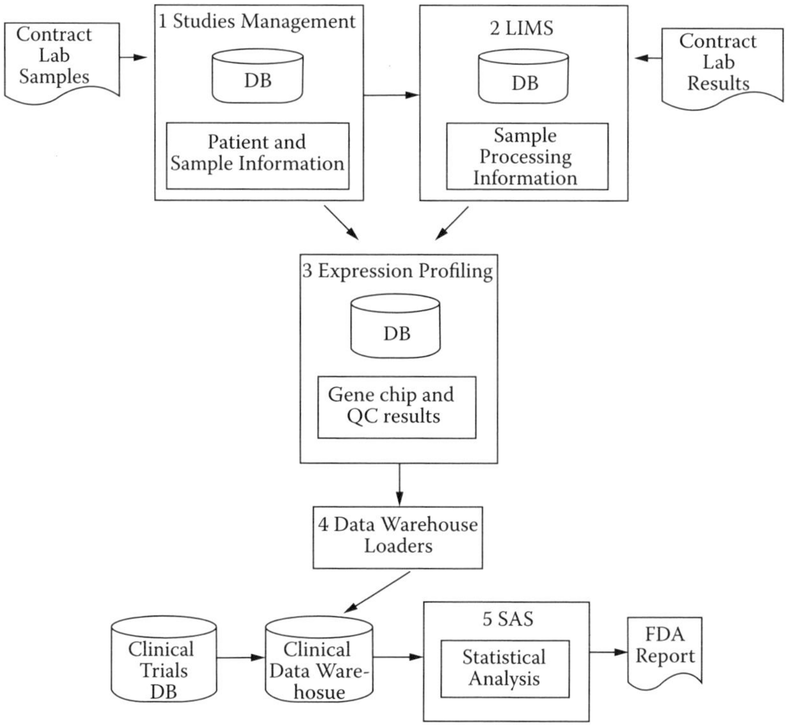

Figure 15.1 Data integrity and electronic data transfers in clinical pharmacogenomic studies. Depicted is one paradigm for linking various databases/warehouses of electronic information related to pharmacogenomic samples, clinical patient data, and expression profiles generated during clinical pharmacogenomic studies. Reprinted with permission. From Burczynski et al., *Curr. Mol. Med.*, 5, 83–102, 2005.

information resides solely in the clinical database. Information from contract laboratories (for surrogate tissues, this includes whole blood and purified PBMC cell count data along with a host of other relevant parameters) is programmatically loaded into the repository to eliminate manual data entry issues. Samples designated for gene expression in the Studies Management system are automatically logged into the LIMS and the Expression Profiling system, enforcing data integrity and ensuring chain of custody. As samples are tracked from RNA through to expression profiles in the LIMS, information necessary to process the gene chips is automatically transferred to the Expression Profiling system. Once the expression profiling experiments are complete, the gene expression results are programmatically transferred to the clinical data warehouse where they can be analyzed along with the clinical end point data. To begin this process, patient and sample information from the repository (e.g., patient ID, visit number) is compared with information from the clinical data warehouse to ensure proper linking of expression results. Once in the clinical data warehouse, gene expression results can be selected, normalized, and extracted along with clinical end point data for analysis using validated Statistical Analysis Software (SAS) modules. Once the final transfer of gene expression results to the clinical data warehouse is performed, the transfer file is permanently stored and permissions on the Studies Management, LIMS, and Expression Profiling systems are modified to ensure that information for the transferred protocol can no longer be modified.

While the example above represents one solution to the complex task of moving a clinical PG laboratory to a regulatory environment, other pathways certainly exist. It is clear that the task of providing reliable PG data with validated links to clinical information across multiple trials requires a large investment of time, resources, and strategic planning, and requires flexibility to adapt to alterations in the existing infrastructures and sample processes that will likely be encountered in the future.

15.6 REGULATORY CONSIDERATIONS AND TRIAL DESIGN ISSUES DURING PHARMACOGENOMIC MARKER DEVELOPMENT

One of the first concepts to be clarified during initial discussions of the voluntary genomic data submission proposal[1] was the idea that the initial discovery of a gene classifier in a clinical trial was not sufficient to immediately propose that classifier as a diagnostic, no matter how strong the apparent correlation between the transcriptional signature and the desired clinical parameter. This concept was echoed in a recent follow-up meeting between representatives from industry and the FDA.[5] Pharmacogenomic marker development therefore is anticipated to follow three stages during the clinical development process, with specific milestones as depicted in Figure 15.2.

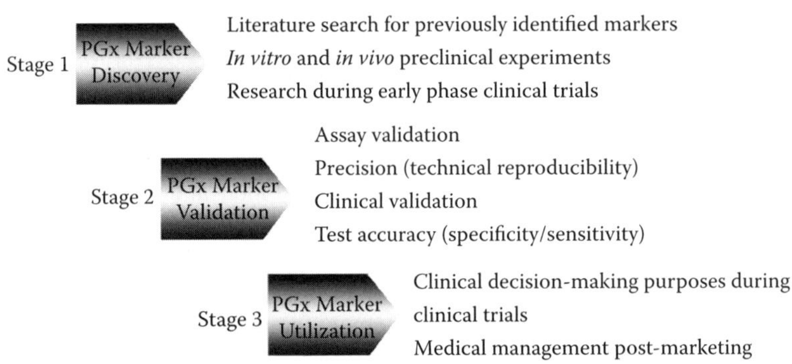

Figure 15.2 Stages of pharmacogenomic marker discovery, validation, and implementation. In the first stage pharmacogenomic markers are postulated or discovered in descriptive pharmacogenomic studies in early-phase clinical trials (Phase 1, Phase 2a). In the second-stage pharmacogenomic markers can be prospectively validated in subsequent clinical trials (Phase 2a, Phase 2b) where the apparent accuracy of the pharmacogenomic assay is established and assay characteristics (sensitivity and specificity) are determined. In the final stage, pharmacogenomic markers that have been prospectively validated can be utilized in later phases of drug development (Phase 3) for patient stratification and, if approved as a co-developed diagnostic assay, could be employed in therapeutic decision making in the post-marketing phase. Reprinted with permission. From Burczynski et al., *Curr. Mol. Med.*, 5, 83–102, 2005.

In the first phase, PG markers are discovered by profiling samples from a clinical trial and discovering a correlation between expression signatures and the desired clinical outcome. With the exception of inflammatory diseases, there are few if any precedents in the literature supporting the concept that transcriptional signatures in peripheral blood will be correlated with clinical outcomes. Thus, in this first phase the discovered correlations between surrogate tissue transcriptional profiles and clinical outcomes are almost certain to represent descriptive PG results. In the second phase, descriptive PG markers discovered in a previous trial must be validated. The question of how to validate transcriptional patterns appropriately is an important one; the requirements for PG validation of a transcriptional signature-based diagnostic are one of the main subjects of current dialogue between the pharmaceutical industry and the FDA.[5] In the second stage of PG marker development, parameters for the technical conduct of the assay will need to be established (precision of the assay) and parameters describing the overall performance of the clinical aspects of the assay should be defined (accuracy, in terms of both specificity and sensitivity with respect to correct assignment of clinical outcomes). In the final phase of PG marker development, if the assay's characteristics appear sufficiently robust, the PG marker can be utilized either for the purposes of patient enrichment in clinical trials or for guiding clinical decision making in the post-marketing phase of the drug's life cycle.

To reach these milestones in a timeframe that is consistent with the clinical development program, prospective analysis plans are critical, from a regulatory perspective, so that any discoveries are well positioned to be considered "validated" in subsequent independent trial(s). It therefore behooves the pharmaceutical industry to implement PG profiling studies as early as possible in those scenarios when there is a perceived benefit of doing so (i.e., when preselection of candidates is not possible because translational biomarkers predictive of efficacy are not available for the therapy or are less than optimal). Discovery of potentially important PG profiles in early Phase 1 and Phase 2a studies means that gene classifiers or predictive models can be assessed independently in later phase studies. In this way, if the PG markers enhance the safety or efficacy of the drug in a given subpopulation of oncology patients, they can be co-developed with the therapeutic before the oncology drug candidate undergoes approval. Pharmacogenomic co-development was the subject of a recent Drug Information Association meeting between industry and FDA representatives, highlighting the importance of this emerging concept.

One of the difficulties experienced by clinical PG laboratories to date is the uncertainty concerning regulatory requirements for the translation of transcriptional profiling discoveries into validated assays. Obstacles, both theoretical and practical, abound at multiple levels. For the purposes of this chapter we focus on practical considerations that are applicable to both surrogate and target tissue profiling studies, but it should be noted prior to this discussion that there are fundamental difficulties associated with assay development based on expression profiling studies. Most importantly, there is the risk of overfitting the data generated in clinical PG analyses. The numbers of covariates measured by microarray technology are in vast excess of the number of samples analyzed in any clinical trial. This raises a fundamental statistical difficulty (the so-called $p >> n$ problem) that increases the uncertainty that gene classifiers discovered in a set of samples are actually linked to the clinical stratification of interest. For this reason alone it is critical that PG strategies be implemented at the earliest possible stages of drug development to (1) provide the greatest likelihood that discovered classifiers can be independently validated one or more times in independent test sets of samples; and (2) gather evidence and elucidate mechanistic links to the discovered classifiers in a manner that further supports their status as emergent biomarkers. This latter goal is both the most difficult and potentially rewarding task facing surrogate tissue profiling strategies, since it is unclear in most cases how the transcriptional signatures in a surrogate tissue are linked to the disease and/or clinical outcome of interest in indications outside the area of inflammation. Nonetheless, if descriptive PG correlations in surrogate tissues are upheld and validated in subsequent studies, these results will engender entirely new avenues of research into the role of surrogate tissues, like the circulating cells of peripheral blood, in unanticipated diseases.

A case in point is the author's laboratory's identification of disease-associated signatures in peripheral blood of patients with RCC (see Chapter 4). Although unanticipated, many of the disease-associated genes have been recapitulated in an independent study of patients with RCC (data not shown) and have led to an entirely new avenue of inquiry as to the relevance of PBMC transcriptional profiles in the context of RCC. Further research will reveal whether these differences reflect active or passive physiological responses to the presence of this type of solid tumor, and whether certain of the disease-associated transcripts (and/or the cells from which they are derived) in PBMCs play important roles in the immune surveillance of these tumors.

The identification of PG markers depends on the synthesis and analysis of data from a variety of sources, as summarized in the previous sections. The complexities involved in this process have already been covered in detail; however, the subtleties of the practical caveats associated with this process and their implications for PG assay development are not so obvious. One of the most important considerations in drug development is that the phases of clinical trials do not necessarily occur in an orderly, linear fashion. For instance, Phase 3 protocols can be prepared and submitted for regulatory approval on the basis of preliminary encouraging Phase 2 data. Thus, when a relevant PG marker or pattern discovered in a Phase 2 trial would require validation in a Phase 3 clinical trial, the constellation of clinical end points observed in the Phase 2 study must be sufficiently "mature" prior to finalization of the Phase 3 protocol to enable supervised identification of transcriptional correlates in clinically relevant subgroups of patients (responders and nonresponders, short-term survivors vs. long-term survivors, etc.). If the transcriptional correlation is identified prior to submission of the Phase 3 clinical protocol, it is possible to define the signature and describe its prospective validation plan in the upcoming Phase 3 study. Validation of markers in Phase 3 is outside the scope of this review, but single-arm and multiple-arm strategies in which the therapeutic of interest and a standard of care are compared in patient subpopulations bearing, or lacking, the transcriptional signature predictive of favorable outcome are possible. These strategies should provide an opportunity to determine whether transcriptional signatures observed in Phase 2 single-arm studies are generally prognostic of outcome regardless of therapy, or are actually "theranostic" and specifically predict outcome in the context of the therapy in question.

Finally, it should be noted that clinical databases are seldom locked/cleaned for a Phase 2 study prior to initiation of a Phase 3 trial. It may therefore be necessary to formally propose in the Phase 3 protocol a prospective validation of an "apparent" theranostic/predictive gene classifier observed in a Phase 2 study, pending the fidelity of the clinical data collected in real time during the clinical trial prior to the final database lock. These and other subtleties will need to be addressed as all interested parties strive to incorporate, and regulate, transcriptional profiling and other global expression profiling strategies in surrogate tissues during the drug decision-making process.

15.7 SUMMARY

Surrogate tissue profiling activities conducted in the context of clinical trials for the purposes of drug development will have to comply with the same requirements as target tissue-based clinical pharmacogenomic analyses. Transcriptional profiling approaches can be critical for therapeutics with underdeveloped translational biomarker strategies, but transcriptional profiling can also enhance the safety and efficacy of therapeutics accompanied by robust predictive biomarkers as well. There are many considerations and complexities involved in the real-time implementation of PG sampling in ongoing clinical studies, including but not limited to cost, logistical difficulties, and regulatory uncertainties. The promise afforded by genome-wide transcriptional profiling technology in surrogate tissues is great and will be poised for realistic evaluation in upcoming years as pharmaceutical companies continue to employ surrogate tissue profiling strategies in decision making during the drug development process.

REFERENCES

1. Food and Drug Administration Center for Drug Evaluation and Research. 2003. Draft Guidance for Industry: Pharmacogenomic Data Submissions. http://www.fda.gov/cder/guidance/5900dft/pdf. Accessed May 29, 2004.
2. Sargent, D. and Allegra, C. *Semin. Oncol.*, 29, 222–230, 2002.
3. Anderson, D.C., Gomez-Mancilla, B., Spear, B.B. et al. *Pharmacogenomics J.,* 2, 284–292, 2002.
4. Englelhardt, H.T. In *The Foundations of Bioethics*, 2nd ed. Oxford University Press, London, 1996.
5. Trepicchio, W.L., Williams, G.A., Essayan, D., Hall, S.T., Harty, L.C., Shaw, P.M., Spear, B., Wang, S.J., and Watson, M.L. *Pharmacogenomics*, 5, 519–524, 2004.
6. Rockett, J.C., Burczynski, M.E., Fornace, A.J., Jr., Hermann, P.C., Krawetz, S.A., and Dix, D.J. *Tox. Appl. Pharmacol.*, 194, 189–199, 2004.

CHAPTER 16

Considerations in the Economic Assessment of the Value of Molecular Profiling

Sarah C. Stallings, Anthony J. Sinskey, and Stan N. Finkelstein

CONTENTS

16.1 INTRODUCTION

Economic evaluations can be implemented to inform decisions about new technologies by balancing the outcomes and the costs of the new technology and are particularly applicable for examining the trade-offs inherent in implementing new technologies in the pharmaceutical industry. As innovation in science and engineering progresses, new technologies emerge, and their value influences in what manner and how quickly they are adopted. Economic analysis provides estimates of that value to support or challenge the integration of the new technology. Economic evaluation of the incorporation of pharmacogenomic strategies can provide incentive and guidance for the following:

- Integrating molecular profiling technologies into drug discovery and development processes
- Aligning diagnostic co-development with market and pipeline realities
- Implementing the necessary changes in our health care, drug development, medical practice, regulatory, and social systems for adopting molecular marker profiling technologies and products

This chapter focuses on the evaluation of the economic incentives for incorporating general pharmacogenomic strategies into drug development. However, the principles are equally applicable to the more specialized application of surrogate tissue profiling, as appropriate, in the drug development process. In this chapter, we first review pharmacoeonomics and its role in evaluating choices. We describe pharmacoeconomic evaluations of pharmacogenomic strategies in clinical practice as an illustration of that role. Finally, we construct a case for the economic advantages molecular profiling could bring to drug development.

16.2 HEALTH ECONOMICS, PHARMACOECONOMICS, AND OVERCOMING INERTIA IN THE ADOPTION OF PHARMACOGENOMIC STRATEGIES

Health care is both a scarce resource and a unique commodity. It is apportioned according to some combination of patient demand, health care payer organization regulation, and health care provider supply. Its product — health — is unlike any other; setting a health standard that can be used to measure improvements in health care outcomes is increasingly difficult as we bump into our society's resource limits.

Health economics research involves *describing* the health care situation at a point in time, *explaining* how it might change with time, and *evaluating* health care practices for their efficient use of the available resources. Describing health care includes collecting statistics on morbidity and mortality, describing the health care supply, and determining a definition of and a value for "health." Explaining health care lies in finding models to determine how the health care situation has changed over time and what we might expect in the future. Evaluating health care consists of judging both the macroeconomic (health care policy, reimbursement policy, insurance) and microeconomic (individual health care interventions) aspects of health care for their performance in providing equitable health care within resource limitations (Jacobs and Rapoport, 2002). In this context, the branch of evaluative economics can provide a measure of the value of pharmacogenomics to health care and pharmaceutical development.

Pharmacoeconomics is a form of evaluative economics that ranks alternative pharmaceutical goods and services according to their relative costs and outcomes. The four main tools of pharmacoeconomics are cost-effectiveness analysis, cost–benefit analysis, cost-minimization analysis, and cost-utility analysis. Readers interested in in-depth treatment of pharmacoeconomics methods can consult excellent texts dedicated to this subject (Drummond et al., 1987; Gold et al., 1996;

Table 16.1 Different Pharmacoeconomic Methods

Analytical Method	Question Addressed
Cost-effectiveness	Comparison of interventions with different costs and different effectiveness; results give the cost per unit of effectiveness of an intervention
Cost–benefit	Comparison of interventions with different costs and different outcomes where all outcomes and costs are measured in monetary terms; used for resource allocation
Cost utility	Comparison of interventions with different costs and different outcomes expressed as a quality of life or societal preference
Cost minimization	Comparison of equally effective interventions to determine which costs less

Pettiti, 2000). Costs and outcomes are defined and enumerated slightly differently in the four methods (Table 16.1). Cost-effective analyses commonly return an incremental ratio (Equation 16.1) that estimates the cost per unit of effectiveness of one treatment alternative over another (Phillips et al., 2003):

$$\text{Incremental Cost Effectiveness} = \frac{\text{Cost of intervention A} - \text{Cost of intervention B}}{\text{Effectiveness of A} - \text{Effectiveness of B}} \tag{16.1}$$

An average cost-effectiveness ratio is measured for an intervention without regard to any alternative interventions and is only equivalent to the incremental ratio in the special case where the alternative has no cost and is ineffective.

In addition to the four primary methods of pharmacoeconomics are simple cost and cost offset analyses. Simple cost analyses of different treatment regimens for an indication can be used in cost-offset studies, where the use of one aspect of health care, such as pharmaceutical interventions, may reduce utilization of other, possibly more expensive, aspects of health care (emergency department visits, surgery, etc.). Finally, cost of illness studies, where total costs to society for care of people with a particular illness are compared to costs for people without the illness, provide burden of illness statistics that can be used in more formal cost/outcome evaluations.

As demands for health care value increase, measures of the value of health care interventions proliferate. The results of pharmacoeconomic analyses give one view of the trade-offs in choosing one drug therapy over another. As such, they are important input for health care resource allocation, and they influence formulary development, health care payer policies, and clinical practice guidelines. Increasingly, pharmacoeconomic data are used to influence pharmaceutical industry initiatives for certain drugs, as an additional measure for managing pharmaceutical candidate portfolios, and to rally support from the government, regulatory agencies, and society for certain drug development efforts geared toward unmet medical needs.

A similar focus on value is directing other aspects of the pharmaceutical industry. Evaluative economic methods similar to pharmacoeconomics can measure not only the value of therapeutic intervention on health, but also the relative values of implementing alternatives in the research and development process itself in relation to costs. In decisions of whether to continue using known technologies or to invest money and opportunity to fully adopt and integrate a new technology, cost–benefit type analyses can provide valuable support or challenge for change.

In the case of assessing the value of profiling technologies in drug development or medical practice, there are legitimate concerns about validation of methods and about the longitudinal meaning of the data collected. In addition to these technological concerns, there is inertia against their adoption that derives from the far-reaching implications of their data. The significance and future health implications of the profiling data are not necessarily unequivocal at the time of their collection. This uncertainty generates fear of information mishandling and fear of litigation based on a future understanding of a profile's meaning to a person's health. It generates questions about the regulatory evaluation of the data and what regulatory agencies will ask from sponsors based on the data. This uncertainty also makes obtaining truly informed consent difficult. If those difficulties are not daunting enough, the macromolecular profile inherent in the technology is a highly personal and politically charged set of information — especially if it is a genetic profile. This topic has surfaced during discussions between industry representatives and scientists at the U.S. Food and Drug Administration (FDA) concerning genomic data submission that ultimately led to the voluntary genomic data submission proposal (Lesko and Atkinson, 2001).

It is becoming increasingly evident that profiling technologies could be central to a future health care that is focused on predicting and preventing disease-causing cellular pathways rather than observing and ameliorating late-stage disease symptoms. Profiling technologies geared toward the identification of informative biomarkers could also become central to a more efficient drug development routine that evaluates drugs based on their ability to access their target, modulate their target, and affect the causative pathogenic pathway, rather than on symptom-based metrics such as their ability to prolong life or increase symptom-free days. Implemented under optimal conditions, profiling strategies can, at least in theory, reduce the time, cost, and imprecision of clinical trials.

Despite the promise of these technologies, a key issue faces the pharmaceutical industry: how to ensure the availability and efficient allocation of resources to invest in the technological advances that promise to transform the drug development process and eventually health care in general. If profiling technologies are to be fully adopted by all stakeholders in the health care system — patients, providers, payers, pharmaceutical developers, regulators, and health policy makers — their costs and consequences need to be explicitly appraised. That is a job for evaluative economics.

16.3 ECONOMIC EVALUATIONS OF MOLECULAR PROFILING IN CLINICAL PRACTICE

Molecular profiling is not unfamiliar in the clinic. A routine physical today would not be complete without the blood work, where the concentrations of different macromolecules and biochemical intermediates indicate the patient's general health status. Some molecular markers have even risen to the level of surrogate markers for disease progression that can be used for treatment decisions and in clinical trials. For example, since the development of the statin family of drugs — HMG-CoA reductase inhibitors, such as lovostatin and simvastatin — cholesterol levels have been part of the prescribing recommendations for those drugs. As investigators have studied the role of statins in lowering cholesterol levels and reducing coronary events requiring hospitalization and surgery, hypercholesterolemia has emerged as an indication in itself that can be alleviated with statins (American Heart Association, 2002). Through their use as diagnostic tools, as guides for drug development, and as clinically significant measures of disease progression and drug effectiveness, molecular markers have been increasingly important to clinical practice for some time. With continued improvements in the ability to detect the markers and relate them to potential disease, and with continued health care emphasis on detecting diseases earlier in hopes of precluding more expensive health care interventions through prevention and prophylactic drug treatment, molecular markers promise to exert an ever-growing influence on clinical practice.

Molecular profiling in clinical practice would entail the use of clinically relevant and validated diagnostic tests meant to screen patients for disease propensity, for likely drug response prior to treatment, or for early indicators of successful therapy during treatment. However, not all profiles would make useful diagnostics. Diagnostic tests are characterized several ways. The sensitivity and specificity are technical measures of the test's false negative and false positive rates, respectfully. In other words, how many samples that are truly positive are determined as positive with the test and how many samples determined positive with the test are truly positive? A descriptive test's positive predictive value is the likelihood that a positive test will give the predicted outcome, and is given by the number of true positives over the sum of true positives and test positives. The attributable risk of a marker describes what proportion of all people with the outcome also has the marker (Holtzman and Marteau, 2000; Higashi and Veenstra, 2003). In truth, many interesting markers may make useless diagnostics, either because the predictive value of the marker is low, the attributable risk represents a very small portion of the at-risk population as a whole, or simply because they predict an outcome for which there is no intervention, leaving patients with positive test results with little productive recourse.

Using marker diagnostics sounds appealing — but the value is not assured. Pharmacogenomics, a subset of molecular profiling in which genomic markers are used to predict drug response, provides an example of the promise of profiling bumping into the burden of integrating a new technology. Because of its importance to the future of health care, the question of incorporating pharmacogenomics into drug development and clinical practice has shifted from If? to How? and How soon? An informal poll taken by the FDA showed that the use of pharmacogenomic data in INDs and NDAs is increasing rapidly, and the agency issued Guidelines for Industry for Pharmacogenomic Data Submission in November 2003 (Lesko et al., 2003). Pharmaceutical companies claim that the risks of including pharmacogenomic data in their FDA submissions overshadow the potential of pharmacogenomics for expediting new drug development. Health care payers, providers, and patients face issues of reimbursement and interpretation of genomic data in clinical decisions, as well as overarching concerns about privacy and liability. Resolving these issues will demand reliable assessments of the potential value for pharmacogenomics to each of these stakeholder groups.

To date, there have been but a few empirical studies evaluating pharmacogenomics in clinical practice (Higashi and Veenstra, 2003; Phillips et al., 2003). These studies have looked at data from targeted patient populations using available pharmacogenomic-based test/treatment combinations and used modeling to derive cost-effectiveness measures of alternative treatment decision paths. The results of these evaluations vary depending on the indication, the cost of screening, the cost of treatment, and the prevalence of the pharmacogenomic variant. In some cases the benefit of pharmacogenomic screening outweighs the costs, but not in all cases. These results suggest that the economic viability of pharmacogenomics will depend on specific circumstances of its use. However, a method for predicting the circumstances of economic viability before investing resources into pharmacogenomic marker discovery and diagnostic test development would be of great use.

Cost-effectiveness analyses conducted prospectively in controlled clinical trials have become a commonplace and useful application of pharmacoeconomics. Yet, clinical trials are done on a short timeframe with relatively few patients while cost-effectiveness analyses with a societal perspective typically look to understand the value for a large population over many years. Economic analyses of chronic diseases often employ modeling to extend the data collected during the clinical trial over a longer time period or to a larger population (Drummond et al., 1987; Gold et al., 1996; Pettiti, 2000). In an example of using modeling to extend the time of the study, Weinstein et al. (2001) modeled a lifetime of HIV infection in a million simulated patients and found that using genotypic resistance testing to guide therapy in HIV disease was more cost-effective when the prevalence of resistance was higher or when it was used following initial treatment failure.

A commonly used cost-effectiveness protocol in pharmacogenomics is to use decision analysis to map the potential clinical decisions that would be affected by the use of a pharmacogenomic-based screening test and a model of the course of the disease, typically a Markov-type or state-transition model, to predict the population-level health outcomes over a long period of time. Maitland-van der Zee et al. (2004) used this approach to determine the cost-effectiveness of genotyping the angiotensin-converting enzyme (ACE) of male patients with hypercholesterolemia prior to prescribing statins when a prospective clinical trial had identified a difference in statin effectiveness due to ACE genotype. This represents a prototypical situation that pharmacogenomics is expected to alleviate — one in which a genetic test could screen for drug response, preventing both the costs of unnecessary medication and the associated nonresponse. The cost-effectiveness analysis found that ACE genotype screening saved money but did not affect life expectancy in the model.

Higashi et al. (2002) took a similar approach to determine the clinical situations in which screening for genetic susceptibility for periodontal disease might be cost-effective. The modeling results highlighted three clinical variables with strong influence over the potential value of the genetic screening. They were (1) compliance with maintenance therapy, (2) the effectiveness of nonsurgical treatments for periodontal disease, and (3) the relative risk of disease progression for test positive patients. The modeling was made more difficult by the unusually broad modeling assumptions necessitated by an incomplete knowledge of the periodontal disease and its treatments and of the positive predictive value and attributable risk of the genetic screening test. That the model could be helpful, however, indicates the value of these types of analyses to those making clinical and reimbursement decisions.

Stallings et al. (2005, submitted) took a wholly different approach designed to estimate the potential economic value of pharmacogenomics in the absence of a specific pharmacogenomic-based diagnostic. They used a stochastic model with asthma patients' data from a retrospective health claims database to investigate the cost offset realized using a hypothetical pharmacogenomic test to determine a preferred initial therapy. They compared the annual costs distributions under two clinical strategies: testing all patients for a nonresponse genotype prior to treating and testing none. Were it possible through a diagnostic test to determine who would not respond to a given therapy, the costs of nonresponse could be eliminated — a cost offset realized. Because the framework specified neither the genetic marker tested by the diagnostic nor the drugs used in treatment, it is very general. Other indications in which a population can be stratified by response could be analyzed similarly. They found that the cost of testing in advance is highly likely to be offset by avoiding costs associated with nonresponse. The results indicated that genetic variant prevalence, test cost, nonresponder rate, and the cost of choosing the wrong treatment are key parameters in the economic viability of pharmacogenomics in clinical practice. This prospective analysis of parameters influencing the economic viability of using molecular profiling in the clinic is an important new tool for evaluating as-yet-undeveloped diagnostic tests based on advances in pharmacogenomics.

Measured economically, then, genomic marker-based diagnostics are and will be valuable in clinical practice, in specific and definable circumstances. What remain between the potential value of interesting markers and their promise as therapeutic guides are standards and validation for everything from assay platforms to data reporting. In looking at a range of genetic association studies in breast cancer, scientists at Celera Diagnostics found little agreement in experimental protocols (for example, different end points and different genes in studies of the same indication), irreproducibility of results, and lack of standards for reporting and evaluating results. Without uniformity in how data are collected and presented, the evaluation, validation, and regulatory acceptance of markers becomes even more difficult (Colburn and Lee, 2003). These results magnify the current uncertainty regarding molecular profiling strategies that appears to outweigh the potential economic value afforded by inclusion of these technologies in the long run.

16.4 MOLECULAR PROFILING IN DRUG DEVELOPMENT

The cost-effectiveness predicted for pharmacogenomic-based diagnostics in clinical practice increases the value of diagnostic biomarkers with clinical relevance to the pharmaceutical industry. However, molecular profiling may also be valuable as a drug development tool. A growing ability to measure pathogenic biological events and the response to drugs has already resulted in therapies with greater efficacy and fewer side effects, and the trend toward understanding increasingly complex biological systems continues. Molecular profiling yields biomarkers that can be used to more accurately measure biological events. Better measurements generate information that makes the process of developing drugs more efficient by increasing confidence in early drug development decisions and thereby optimizing R&D resource allocation. Cost and impact comparisons could determine the potential value of new drug action measurement technologies like molecular profiling for improving the efficiency of the drug development process.

It is generally accepted that drug development is more efficient when ultimately unsuccessful candidates are culled earlier in the process. During the course of drug development, groups within companies make go/no-go decisions that determine a drug candidate's fate. Because the bulk of the financial burden of drug development occurs during clinical trials, the success rate of drug candidates entering this complex phase of the development process is central to the cost of drug development. Historically, the rate-limiting step of drug development was discovering molecules with therapeutic potential. In that circumstance, a failed candidate during clinical testing represented a risk predominantly to the money spent during that candidate's development. With new technologies like high-throughput screening (HTS) and genomics-based target discovery efforts, drug discovery has become increasingly systematic, providing more molecules for more pharmaceutical targets. As a consequence, the rate-limiting step in drug development has shifted from finding drug candidate molecules to selecting optimal drug candidate molecules with the greatest potential from a large series of candidates. With more candidates vying for the same development resources, losing a candidate during development means losing the opportunity for other potential candidates left behind and losing the misallocated resources, both financial and human.

Countering this economic pressure to abandon projects earlier is the fact that the certainty surrounding decisions to advance or abandon a specific candidate increases as drug development progresses. The most reliable way to know if a drug is safe, effective, marketable, and clinically important is to dispense it to many people and assess the effect. Unfortunately, that is also the most expensive and ethically questionable method to collect this information. For this reason pharmaceutical companies evaluate a molecule's potential at sequential stages using any and all available information about its efficacy, its toxicity, its tolerability, its pharmacokinetics and pharmacodynamics, its demand, and its manufacturability — information assembled from a wide variety of tests calculating those characteristics.

Pharmaceutical developers, then, have an interest in allocating development resources toward candidates that are cost-effective, and pharmacoeconomics is used increasingly by the pharmaceutical industry to evaluate prospective leads and drug candidates for their cost-effectiveness. Industry pharmacoeconomics groups have become involved earlier in drug development in response to the pressure to reduce late-stage failures, with cost-effectiveness data figuring into candidate advance or abandon decisions (Data et al., 1995; Grabowski, 1997; DiMasi et al., 2001). In a similar manner, it is likely that evaluative economics will be used to consider balancing the costs of adding tissue profiling technologies with the effectiveness of these technologies in providing markers for more efficient drug development, as measured by the proper allocation of development resources toward successful drug candidates (DiMasi et al., 2001).

New scientific and technological advances like molecular profiling could be valuable for several reasons. Molecular profiling represents the growing ability to objectively measure increasingly complex biological systems such as drug intervention in pathogenesis. Better measurement technologies are the foundation of a streamlined drug development pipeline broadened not by backup candidates for important indications, but by candidates for an increasing number of more specific disease states than currently known (Stallings et al., 2001). The potential value to the industry is increased efficiency and unprecedented innovation. Further, since biomarkers resulting from molecular profiling can be used to classify patients, to determine if a drug hits its target and does its intended job, and to detect off-target and potential adverse effects, their use can improve attrition efficiency, reduce clinical trial populations while maintaining statistical power, and align industry efforts in co-diagnostic development. Finally, since drugs fail on the basis of their relative efficacy in the general population, biomarkers could, in effect, rescue failed drugs by identifying a subpopulation for which the therapeutic index is optimal.

Overlaying the potential impact of pharmacogenomics in decision making onto the actual transformation that occurred when pharmacokinetics became recognized as an important decision-making tool during drug development illustrates how valuable profiling technologies can be. The practice of using pharmacokinetic data in decisions earlier in drug development has reduced considerably late-stage drug candidate attrition rates attributable to metabolic failure (Figure 16.1). This has changed remarkably from the early 1990s, when pharmacokinetic (PK) or bioavailability issues represented the majority of late-stage failures. By 2000, PK/bioavailability accounted for less than 10% of clinical development failures, down from nearly 40% in 1991 (Frank and Hargreaves, 2003). The reduction in clinical failures from unacceptable PK/bioavailability is traced to making the relevant information available for earlier decisions (Eddershaw et al., 2000). An increased knowledge base in physicochemical and pharmacokinetic properties of molecules allowed more prospective use of metabolic information during discovery for guiding lead design, optimization, and selection. Improved analytical technologies for collecting metabolism data made it feasible to incorporate pharmacology data into earlier stages of

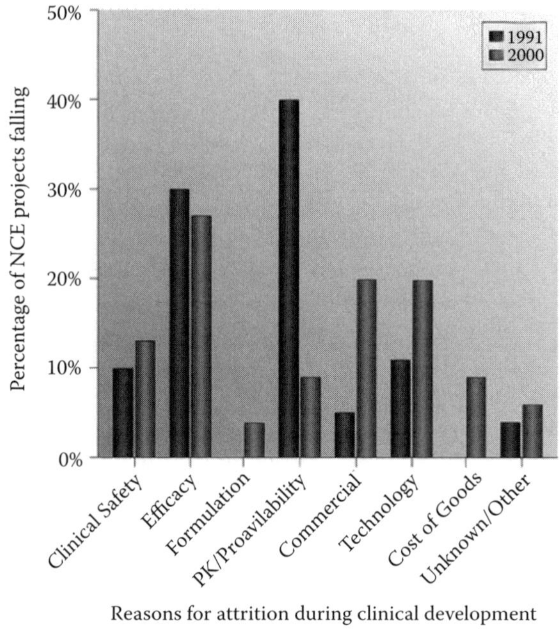

Figure 16.1 Comparing reasons for attrition, expressed as a percentage of all projects abandoned during clinical development, between 1991 and 2000. From Frank, R. and Hargreaves, R. (2003). *Nat. Rev. Drug Discov.* 2(7), 566–580. With permission.

discovery than it had previously been (Humphrey, 1996; Watt et al., 2000; White, 2000). Similarly, gathering and applying relevant biomarker data earlier could address the current burden of late stage failures from efficacy and toxicology issues, providing measurable economic benefit from the integration of these technologies (Frank and Hargreaves, 2003).

16.5 CONCLUSION

Innovative methods for analyzing tissues — both target and surrogate — in search of disease-related profiles are emerging. Searching for markers broadly makes sense, given the complexity of biological systems and of the pathogenic mechanisms that cause disease. For the common complex diseases that are taxing health and health care resources — diabetes, obesity, and cardiovascular disease, among others — single markers may not provide the predictive power necessary for clinically powerful diagnostic tests. However, molecular profiling may identify patterns of molecular signals for disease likelihood or for drug response with potential uses in improving therapeutic and drug development efficiencies. In addition, the markers discovered along the way to a clinical diagnostic may prove useful for improving the drug development process itself. The definitive proof will await retrospective analysis of changes in clinical practice and drug development in the wake of tissue profiling. In the meantime, prospectively evaluating the economic benefits of these innovative technologies as discussed in this chapter, can provide needed incentive to drive greater adoption until their actual value is adequately demonstrated in multiple cases.

ACKNOWLEDGMENT

This work is based on research conducted within the MIT Program on the Pharmaceutical Industry with support from the Alfred P. Sloan Foundation, the Merck Foundation, and Millenium Pharmaceuticals, Inc.

REFERENCES

American Heart Association. (2002). *Cholesterol Lowering Drugs*, American Heart Association. www.americanheart.org/presenter.jhtml?identifier=4510.

Colburn, W.A. and Lee, J.W. (2003). Biomarkers, validation and pharmacokinetic-pharmacodynamic modelling. *Clin. Pharmacokinet.* 42(12), 997–1022.

Data, J.L., Willke, R.J. et al. (1995). Re-engineering drug development: integrating pharmacoeconomic research into the drug development process. *Psychopharmacol. Bull.* 31(1), 67–73.

DiMasi, J.A., Caglarcan, E. et al. (2001). Emerging role of pharmacoeconomics in the research and development decision-making process. *Pharmacoeconomics* 19(7), 753–766.

Drummond, M.F., Stoddart, G.L. et al., Eds. (1987). *Methods for the Economic Evaluation of Health Care Programmes*. Oxford University Press, New York.

Eddershaw, P.J., Beresford, A.P. et al. (2000). ADME/PK as part of a rational approach to drug discovery. *Drug Discov. Today* 5(9), 409–414.

FDA (2003). Pharmacogenomic Data Submissions. 2004.

Frank, R. and Hargreaves, R. (2003). Clinical biomarkers in drug discovery and development. *Nat. Rev. Drug Discov.* 2(7), 566–580.

Gold, M.R., Siegel, J.E. et al., Eds. (1996). *Cost-Effectiveness in Health and Medicine*. Oxford University Press, New York.

Grabowski, H. (1997). The effect of pharmacoeconomics on company research and development decisions. *Pharmacoeconomics* 11(5), 389–397.

Higashi, M.K., Veenstra, D.L. et al. (2002). The cost-effectiveness of interleukin-1 genetic testing for periodontal disease. *J. Periodontol.* 73(12), 1474–1484.

Higashi, M.K. and Veenstra, D.L. (2003). Managed care in the genomics era: assessing the cost effectiveness of genetic tests. *Am. J. Manag. Care* 9(7), 493–500.

Holtzman, N.A. and Marteau, T.M. (2000). Will genetics revolutionize medicine? *N. Engl. J. Med.* 343(2), 141–144.

Humphrey, M.J. (1996). Application of metabolism and pharmacokinetic studies to the drug discovery process. *Drug Metab. Rev.* 28(3), 473–489.

Jacobs, P. and Rapoport, J. (2002). *The Economics of Health and Medical Care*. Aspen Publishers, Gaithersburg, MD.

Lesko, L.J. and Atkinson, A.J., Jr. (2001). Use of biomarkers and surrogate endpoints in drug development and regulatory decision making: criteria, validation, strategies. *Annu. Rev. Pharmacol. Toxicol.* 41, 347–366.

Lesko, L.J., Salerno, R.A. et al. (2003). Pharmacogenetics and Pharmacogenomics in Drug Development and Regulatory Decision Making: Report of the First FDA-PWG-PhRMA-DruSafe Workshop. *J. Clin. Pharmacol.* 43(4), 342–358.

Maitland-van der Zee, A.H., Klungel, O.H. et al. (2004). Pharmacoeconomic evaluation of testing for angiotensin-converting enzyme genotype before starting beta-hydroxy-beta-methylglutaryl coenzyme A reductase inhibitor therapy in men. *Pharmacogenetics* 14(1), 53–60.

Pettiti, D.B. (2000). *Meta-Analysis, Decision Analysis, and Cost-Effectiveness Analysis: Methods for Quantitative Synthesis in Medicine.* Oxford University Press, New York.

Phillips, K.A., Veenstra, D. et al. (2003). An introduction to cost-effectiveness and cost-benefit analysis of pharmacogenomics. *Pharmacogenomics* 4(3), 231–239.

Stallings, S.C., Rubin, R.H., et al. (2001–2002). Technological innovation in pharmaceuticals. *Pharmaceutical Discovery and Development.* PharmaVentures, Ltd., Oxford, U.K.

Stallings, S.C., Witt, W.P. et al. (2003). POPI Working Paper #64-03: An Economic Framework for Evaluating Personalized Medicine. Program on the Pharmaceutical Industry (Massachusetts Institute of Technology), Cambridge, MA.

Stallings, S.C., Huse, D., et al. (2005) A Framework to Evaluate the Impact of Pharmacogenics, manuscript submitted.

Watt, A.P., Morrison, I.I. et al. (2000). Approaches to higher-throughput pharmacokinetics (HTPK) in drug discovery. *Drug Discov. Today* 5(1), 17–24.

Weinstein, M.C., Goldie, S.J. et al. (2001). Use of genotypic resistance testing to guide HIV therapy: clinical impact and cost-effectiveness. *Ann. Intern. Med.* 134(6), 440–450.

White, R.E. (2000). High-throughput screening in drug metabolism and pharmacokinetic support of drug discovery. *Annu. Rev. Pharmacol. Toxicol.* 40, 133–157.

CHAPTER 17

The Impact and Challenges of Pan-Omic Approaches in Pharmaceutical Discovery and Development

William D. Pennie, Jennifer L. Colangelo, and Michael P. Lawton

CONTENTS

17.1 INTRODUCTION

The challenges facing the pharmaceutical industry at the beginning of the current millennium are manifold. The economic realities of drug discovery and development are forcing both a reconsideration of R&D priorities and the implementation of innovative solutions to combat compound attrition.[1] Despite a significant increase in research and development spending (estimated at approaching threefold increase over the course of 1990 to 2000), the number of new drugs being approved for public use has remained relatively constant. Although heralded with much promise, the "genomics revolution" has not yet had a demonstrable impact on new drug survival during the costly development process; this too has remained fairly constant at an approximately 90% failure rate. Improving survivability of compounds in the development phase is therefore a clear goal of the industry. This will require technical and scientific innovation, certainly, but also possibly an even more aggressive application of genomics sciences (or "omics," as they have become collectively known) to improve the quality of candidate molecules as early in the discovery phase as practical.

Progress toward improving the quality of new pharmaceuticals with genomic tools and genomics-derived knowledge is clearly being made, however. Genomic sciences are helping drug discovery scientists to identify new targets for drugs and to screen for compounds that interact with them. While discovery scientists focus attention on single biomolecules (or pathways) as drug targets, pharmaceutical toxicity testing attempts to predict or determine a novel compound's effect on a very wide range of biological end points. As the pace of new drug discovery increases, traditional toxicology is challenged to deliver quality candidate safety information without becoming the rate-limiting step in compound advancement. The application of genomics sciences to the discipline of toxicology in an industrial setting is therefore a key issue and will be illustrated throughout this chapter. In the clinical setting the challenges are also considerable. Understanding the impact of genomic differences in responsiveness to therapy (or susceptibility to adverse effects) is the cornerstone of "individualized medicine," a concept that has received much attention in the scientific community but that faces significant scientific, economic, regulatory, and legal challenges before it evolves to common practice.

In this chapter we offer a brief overview of the genomics technologies most relevant to pharmaceutical discovery and development, with a particular emphasis on applications to enhance our predictive capability and mechanistic understanding of drug safety. Reducing the failure of compounds in clinical development for safety reasons continues to be a huge opportunity for the industry from a cost perspective and, more significantly, to improve the safety of medicines in both the clinical trials phases and for the broader patient population.

Finally, while this chapter focuses broadly on the application of these maturing technologies in a drug discovery and development setting, their importance to surrogate tissue analysis is clear. Surrogate tissue markers of the type discussed throughout this volume are very likely to be discovered in the future by molecular profiling technologies such as proteomics, metabonomics, transcript profiling, and genetics. Also, while the development and application of emerging technologies is often fostered in a preclinical setting, an understanding of the extrapolation (or linkage) of preclinical markers of efficacy or safety to *humans* is of critical importance in advancing the best candidate molecules. Inappropriate preclinical models will give less than desirable predictive capability, resulting in less than desirable compound failure in the clinical setting. To break this cycle, predictive markers, used in a preclinical setting, need to be validated for their power to predict clinical outcome.

For many end points, building this concordance (or validation) data will require non-invasive sampling during clinical trials and thus markers derived from, or measured in, surrogate tissue or fluid samples are likely to be most amenable to this approach. An excellent illustration of the challenges of this approach is attempting to predict and characterize chemically induced vascular damage, and this is discussed later in the chapter.

17.2 THE GENOMICS SCIENCES: PREDICTIVE AND INVESTIGATIVE OPPORTUNITIES

There are multiple disciplines falling under the umbrella of genomics sciences; the major ones of practical utility to drug discovery and development are *genetics, genomics, proteomics, metabonomics, chemogenomics,* and *informatics.* These approaches can be applied broadly to discover the molecular basis for disease and to help find new molecular targets, to improve efficiency at screening for molecular interactions with "druggable" targets, to increase mechanistic understanding of toxicity, to better extrapolate preclinical toxicology findings to human risk, and to understand the basis for individual responsiveness to therapy or idiosyncratic adverse drug reactions.

17.2.1 Genetics

Over the last several decades, many investigators have been successful in identifying genes responsible for, or at least associated with, specific diseases. In most of these cases there is a single gene or a few causal genes involved, and the consequences of genetic differences are easy to diagnose. The candidate-gene approach to determining individual responsiveness to drugs does not appear to have had the anticipated impact. Family-based linkage studies may be more valuable in mapping genes associated with therapy response in common, but genetically complex, diseases such as asthma.[2] Mapping the susceptibility genes for complex human traits is inevitably a huge challenge and novel molecular and statistical approaches are needed to reveal the molecular basis for variations in responsiveness to therapy or susceptibility to potential adverse effects.

The human genome project (HGP) has made an important contribution to our knowledge base around these issues. Recent estimates suggest that the human genome consists of approximately 20,000 to 25,000 genes,[3] Sequencing efforts have resulted in multiple technical innovations facilitating the identification of literally millions of DNA sequence variants and the development of genotyping tools to map them. Large-scale consortium efforts have been instrumental in developing well-characterized sets of DNA sequence polymorphisms; for example, single nucleotide polymorphisms (SNPs) have been identified by the HapMap Project and the SNP Consortium (TSC). These efforts have provided positional information and allele frequencies of these polymorphisms as well as developed specific assays for genotyping them. The identification of polymorphisms that characterize disease and treatment response represents merely the first step in leveraging genetic technologies in drug discovery and development, however. The next step, genetic analysis of an individual's genome, has traditionally been an expensive and labor-intensive task, but recent technical developments such as the Affymetrix DNA chip and the Illumina BeadArrays allow tens of thousands of markers to be genotyped simultaneously. This rate of throughput will be essential in validating polymorphic differences across the large study designs that will be necessary to characterize the significance of these variations.

17.2.2 Genomics

With development of microarray technologies, the expression level of practically the entire mammalian genome can be measured. Expression changes in transcripts may serve as biomarkers for exposure and also serve to aid in understanding the mechanism of action of the stimuli as well as the cellular pathways involved in response. Within the pharmaceutical discovery and development process, an illustrative application of transcript profiling is the use of microarrays (employing cDNA or oligonucleotide probes) to predict or investigate toxicity, an approach that has become known as toxicogenomics. Several recent publications on toxicogenomics have been published and can be reviewed for a more detailed discussion of its general principles.[4–6] Treatment of test systems with known reference toxicants (with similar toxic end point, mechanism, chemical structure, target organ, etc.) permits the identification of diagnostic gene expression patterns for particular toxic outcomes. The ability of transcript profiling to distinguish between distinct classes of compounds has now been demonstrated by many laboratories,[7] and the development and application of predictive toxicology models based on gene transcript changes has been effectively commercialized by a number of biotechnology companies. Such pattern recognition facilitates the discovery, and subsequent validation, of biomarkers useful for application in higher-throughput approaches to help "de-risk" novel chemical series in the discovery process. One illustration of this approach, by Burczynski and colleagues,[8] involved the analysis of a prototypic single representative from two compound classes to look for consistent diagnostic expression changes while avoiding background noise. Computer-based prediction tools were employed to expand the list of consistent gene expression events to those genes that could distinguish accurately between the chemical classes (DNA damaging agents and the anti-inflammatory drugs) in a 100-compound learning set. More recently, Thomas and colleagues[9] have taken a broadly comparable approach to distinguishing five hepatotoxicant classes, based on a learning set assembled from 24 reference compounds, profiled using a custom 1200-gene microarray. Perhaps surprisingly, these results suggest that the gene expression fingerprint that allows such classification can consist of merely dozens of genes, again raising the possibility that the approach could be modified to a high-throughput format. Correlating gene expression changes with clinical chemistry and pathology findings should put gene expression data in context with more established endpoints, as demonstrated recently by Waring and colleagues.[10]

A more fundamental application of these technologies is the investigation of the regulation events that underpin the development of an adverse biological response, rather than those that allow prediction of the outcome. This approach should allow a more mechanism-based assessment of risk, particularly where applied to characterize a finding found in a regulatory study performed in a preclinical test species (e.g., a rodent or canine study). With appropriate experimental design this approach can generate "candidate" gene lists that can be used to formulate hypotheses as to the mechanism by which a compound gives rise to a toxicity finding. Proving the causative involvement of any of these candidates requires detailed follow-up work, most probably employing more traditional functional genomics and biochemical approaches. Transcript profiling is also being employed to understand the relationship between *in vivo* and *in vitro* models. For example, the de-differentiation of hepatocytes following explant has been characterized over time at the transcriptional level,[11] demonstrating that the isolation of hepatocytes can have marked effects on pathways known to be involved in toxicant response.

17.2.3 Proteomics

Proteins are integral components of biochemical pathways; in essence they represent the functional manifestation of genetic information. Characterizing the protein components of a biological system and understanding their functions are key factors in understanding changes in physiology that are causative for the disease state or a consequence of compound administration. Proteomic technologies, such as two-dimentional gel electrophoresis and mass spectrometry, provide avenues for measuring the changing expression levels of proteins and providing further characterization, specifically protein modifications, function, and activity.[12,13] There are multiple technologies of potential utility in identifying protein biomarkers of efficacy and toxicity, each with strengths and weaknesses.[14] New technologies for proteome analysis continue to emerge, expanding the capabilities for these types of analyses.

The potential impact of proteomics approaches to the drug discovery and development process is broad.[15] A major advantage of proteomics profiling is the opportunity to sample body fluids — serum, urine, cerebrospinal fluid (CSF), synovial fluid — for surrogate protein markers. This capability allows both surrogate tissue analysis (e.g., through profiling of lymphocytes) and the measurement of alterations in secreted protein profiles or of proteins released as a consequence of tissue damage.[16] As observed for genomics, many investigations of the proteome have been conducted to address issues in toxicology. The most common form of analysis is differential expression profiling, which provides the expression levels of proteins within a system relative to other proteins in that system.

In the toxicology sciences, for example, dose–response proteome "fingerprint changes" have been described for a number of drugs.[17–20] The technologies have also been applied by toxicologists in the identification of potential biomarkers for target tissue damage[21,22] and to give better mechanistic insight of toxicology findings.[23] As with genomics, proteomics may also assist in better species comparison experiments by increasing our understanding of the functional differences and responsiveness of preclinical test species and humans. An understanding of the functional proteome of specific organ systems in specific species will offer insight into mechanisms of action and the biochemical processes behind induced toxicities.[24,25] Other proteomic analyses include profiling protein isoforms and modifications, investigations of protein–protein interactions, and characterization of protein binding sites that may be related to toxic events.[26,27]

Major challenges in proteomics include determining the best technology platform on which to perform the analysis, to process and interpret the experimental data, and to place the findings in the correct biological context. New platforms for differential expression analysis continue to rapidly emerge, with many in the validation phase.[28] Difficulty arises not only when trying to compare data sets that have been acquired on different platforms, but when comparing those taken at different time periods and within different laboratories. These variations can produce data sets that might not lead to the same conclusion.[29] Integrating other types of experimental data, such as genomics data sets, provide additional value and aid in interpretation.[30,31]

Characterizing various proteomes and then applying those findings to recognizing and understanding toxicological events is an enormous undertaking. The Human Proteome Organization (HUPO) was formed in 2001 and consists of members from various government, industry, and academic organizations.[32] One of their goals is to compare the various technology platforms that can be used to profile proteomes. It also plans to develop a comprehensive characterization of the proteins found in human serum and plasma, evaluate differences within the human population, and create a global knowledge base and data repository. Concerted efforts such as this will aid in expediting the task of understanding the proteome, and similar efforts will be needed to address proteomic factors in disease and disease treatment.

17.2.4 Metabonomics

Metabonomics is defined as the study of metabolic responses to drugs, environmental changes, and diseases. In essence, the approach involves the quantitative measurement of changes in multiparametric metabolic response of living systems to internal or external stimuli, or as a consequence of genetic change.[33] The term is often used in an interchangeable fashion with metabolomics, which more specifically relates to the analysis of *all* metabolites in a biological sample. Clearly, the emerging field of metabonomics is a logical extension to the more established fields of genetics, genomics, and proteomics and, increasingly, is being used as a valuable research tool in characterizing chemically induced changes in physiological processes. The technique normally involves the processing of biofluid samples (e.g., urine, plasma, CSF) or other tissue preparations followed by analyzing high-resolution nuclear magnetic resonance (NMR) spectra to identify the metabolites present. As with genomics and proteomics, data mining and *in silico* biochemical pathway analyses are critical to characterizing the resultant data.[34] This is particularly important when data from multiple omics sources are used to attempt to give a more holistic understanding of mechanistic toxicology. For example, mechanistic understanding of even relatively well characterized agents can be increased by such a combinatorial approach, as recently demonstrated in studies on acetaminophen, which have been characterized by both genomics and metabonomics end points.[35]

Many researchers have described the potential utility of this approach in the pharmaceutical industry by better characterizing potential adverse drug effects[33,36,37] and as a complementary approach to other omics technologies in toxicology research.[37] In this regard, the pharmaceutical sector has visibly partnered with academia in the COMET consortium (Consortium for Metabonomic Toxicology) to define appropriate methodologies and to generate metabolic "fingerprints" of potential utility in preclinical screening of candidate drugs.[38] Metabonomics analyses in a clinical trial setting can clearly be complicated by a multitude of factors, including a trial participant's other medications, variations in diet, etc., and therefore careful experimental design and rigorous statistical analysis are essential.

17.2.5 Chemogenomics

Another emerging discipline with applications to drug development and discovery is chemogenomics,[39] where computational chemistry and genomics are modeled together to give better rationalized drug design and mechanistic tools to understand the downstream consequences of drug-target interactions.[40–42] Chemogenomics approaches should be enhanced by improvements in structure prediction and homology modeling of three-dimensional protein structures,[43] and in simulations of molecular docking between drug and target.[44]

17.2.6 Informatics and Systems Biology

As a discipline, bioinformatics is evolving from a genome information annotation, comparison, and analytical tool to having a significant role in understanding the fundamental biology behind disease processes and mechanisms to identify and test new therapeutic strategies in the pharmaceutical industry.[45] The genomics sciences, by definition, generate large-volume data sets, and therefore offer the opportunity to characterize biological processes in terms of patterns or of changes rather than the traditional biomarker approaches that have tended to concentrate on changes in a single (or discrete number of) molecules as the measured end point. This potential has challenged our existing definition of biomarkers; we now recognize that patterns or fingerprints of individual changes (themselves composed of potentially dozens of individual markers) may be the biomarkers of the future.[46]

Creating a more holistic view of biological processes, including an understanding of the regulation of, and interactions between, regulatory pathways is an emerging discipline often described under the broad term "systems biology." As a consequence of the availability of higher-volume omics data, these approaches appear to be maturing at a rapid pace and are beginning to have demonstrable impact in the interpretation of large-volume data sets generated in the course of drug discovery and development.[47] Clearly, the sharing of nonproprietary genomics data among academia, industry, and regulators will be critical in the development of this field. Public software and databases are being developed at a number of institutions such as the National Centre for Toxicogenomics Research with its ArrayTrack software for toxicogenomics data management and analysis[48] and the European Bioinformatics Institute's ArrayExpress database.[49] Consortia efforts among academia, industry, and regulatory scientists are also generating data to "seed" these public domain databases. A notable example is the International Life Sciences Institute Genomics consortium, which has worked with some 30 member companies to develop a toxicogenomics data set and release it into the public domain through collaboration with the European Bioinformatics Institute.[50] Integrating genomics data with more traditional end points will hopefully help with a holistic understanding of the genotype–phenotype relationship and has been included in a number of efforts to build analytical tools, databases, and data exchange standards.[48–50] This issue is particularly important if genomics data are going to be compared or extrapolated across species.[51]

While the development of robust bioinformatics analysis tools continues apace, there have been several examples of using pattern-matching tools from other disciplines to characterize omics data sets. For example, voice-speech pattern algorithms have been used in concert with transcript profiling experiments to classify and predict the therapeutic response of patients with ovarian cancer.[52]

One step in the maturation of systems biology may be more complete and better descriptions of functional units in biology (i.e., a characterization of the major pathways and physiological processes into a discrete number of units). This is an important extension of ongoing efforts to take biochemical pathway information (such as that found in the KEGG database) to a more highly annotated and linked relational database.[53] Biological processes can then be defined in terms of interactions between major pathways rather than individual genes or proteins.[54,55] Such initiatives are complemented by protein informatics, which aims to predict the putative function of uncharacterized (or hypothetical) proteins based on structural features and pathway mapping.[56,57]

17.3 ONCOLOGY AND DRUG-INDUCED VASCULITIS: EXAMPLES OF PROGRESS AND PRACTICAL CONSIDERATIONS IN APPLYING GENOMICS TECHNIQUES

The potential significance of the genomics sciences to drug discovery and development can be demonstrated through examples that are representative of current interest and activity. In this regard we briefly consider the therapeutic area of oncology and a safety assessment issue of drug-induced vascular injury (vasculitis).

17.3.1 Oncology

The completion of the human genome project was heralded as a major facilitating factor in advancing detection, treatment, and monitoring of cancer.[58] Genomics sciences allow a complex disease such as cancer to be characterized in more depth and allow the consideration of alternate classes of "druggable" targets discovered through large-scale analysis techniques. These targets could conceivably cover many aspects of tumor progression and metastasis, facilitating the discovery of new therapies to modulate differentiation, drug uptake, or metabolism, or cell–cell interactions.[59] In addition to target discovery, genomics approaches should help in identifying appropriate surrogate markers for proliferation and differentiation, genetic damage, growth regulation, and alterations in cell physiology.[60] The application of these markers may help in characterizing state of progression, or benign tumors from malignancies as established in proteomics-based approaches[61] to distinguishing colorectal cancer from colorectal adenoma or normal tissues.[61] They may also help to link proteomics expression maps to the morphology and tumorigenicity of cells in culture.[62] Specific tumors may have a genetic background that is more or less responsive to a particular drug therapy, and treatment for diseases such as non-Hodgkin's lymphoma might be more tailored based on the genetic profiling of a patient's tumor.[63] A tumor's genetic profile could conceivably alter a number of important pharmacokinetic/pharmacodynamic parameters such as genes involved in drug catabolism, drug transport, apoptosis, and the drug target itself (structure, distribution, and function) in that individual.[64]

Application of genomics sciences in a clinical setting has also been used to detect and analyze putative direct or surrogate markers of therapeutic effect. Proteomics characterization has been used in a number of studies to characterize tumor metastatic potential and drug resistance, including studies using clinical samples.[65,66] Dowlati and colleagues[67] demonstrated that obtaining sequential tumor biopsies in an early-phase clinical trial setting can be achieved with appropriate skill and protocol design. Markers of pharmacodynamic effect can thus conceivably be monitored over the course of compound administration and linked, ultimately, to therapeutic outcome. It should be noted that surrogate marker analysis of human clinical trial material is confounded by a number of factors (including amount of material available, timing window for sampling, potential high miss rate for localized phenotypic changes) that may not be evident in developing surrogate marker strategies in preclinical models.[68]

17.3.2 Drug-Induced Vasculitis

Vasculitis is a lesion characterized by infiltration of inflammatory cells and necrosis of blood vessel walls. In preclinical toxicology testing, drug-induced vasculitis has been observed with a number of structurally and pharmacologically diverse compounds. Although several mechanisms of vascular toxicity have been proposed, the exact mechanism(s) by these drugs damage blood vessels is not known. In addition, there are no specific biomarkers that can be used preclinically or clinically to either predict or diagnose vasculitis; histopathology is currently the only method to detect vasculitis in preclinical animal toxicity studies. Moreover, the clinical relevance of drug-induced vasculitis observed in animal models is unclear. To further our understanding of vasculitis, multiple omics technologies can help contribute to mechanistic insight into the lesion, molecular basis of species differences, development of gene-based screens to improve selection of compounds for drug development, and identification of more sensitive and specific biomarkers.

Toxicogenomics is one of several experimental approaches that can facilitate the identification of vasculitis biomarkers. By monitoring genes that are differentially expressed in blood vessels isolated from animals treated with compounds that induce vasculitis, it might be possible to narrow the search for candidate biomarkers by specifically focusing on genes that encode cell-surface or secreted proteins. These genes encode potential circulating biomarkers that could be further characterized directly in plasma or serum using immunoassays or other diagnostic methods. Profiling circulating leukocytes, which can be easily obtained from whole blood, might also identify surrogate biomarkers. Alcorta et al.[69] have reported that gene expression changes in circulating leukocytes from patients with a variety of renal diseases, including small vessel vasculitis (ANCA disease), can be clustered according to disease type. The use of whole blood for gene profiling should be particularly valuable for characterizing vasculitis in humans. For genomics, there are a number of technical issues that must be considered when working with blood vessels. Certainly one of the most significant is that most blood vessels are small, making it difficult to obtain sufficient quantities of RNA for most profiling methods. However, the availability of increasingly sensitive amplification techniques means that starting RNA amounts will not be a limitation in the future. Moreover, like most tissues, blood vessels are complex tissues comprising multiple cell types and are found in close association with surrounding tissues such as fat, pancreas, or lymph nodes. When interpreting expression data generated from vascular tissue, this complexity can make it difficult to distinguish the contributions of endothelial cells from vascular smooth muscle cells (VSMCs), infiltrating leukocytes, and any surrounding tissues that may been removed along with the blood vessels during the dissection procedure. Performing *in situ* hybridization or immunohistochemistry to localize the cellular source of specific transcripts or gene products of interest, assuming the necessary antibody and cDNA reagents are available or can be generated, is therefore recommended for further characterization of candidate markers. Laser capture microdissection (LCM) can also be used to isolate enriched populations of endothelial cells, VSMCs, or other targeted cell types.[70]

While genomic profiling of tissues can be used to identify candidate biomarkers of vasculitis, the ability to detect vasculitis signals in urine, plasma, or serum might also lead to novel biomarkers. Two techniques that have been used for this are proteomics and metabonomics. In both cases statistical methods, such as dimension reduction (e.g., principal components analysis, multidimensional scaling) or hierarchical clustering, are used to compare treatment groups and identify individual molecules or groups of them. Structural determination of candidate biomarkers, whether they are spots on a two-dimensional gel or peaks in a NMR spectrum, generally involves a subsequent mass spectrometry step. A number of recent publications have demonstrated the value of metabonomics for studying vasculitis.[71,72]

As with all large-scale expression profiling experiments, data analysis and interpretation remain a key challenge. Lists of genes, proteins, or metabolites that are up- and downregulated in tissues or fluids from animals with vasculitis will easily be generated, but how can we separate cause from effect? This task will be facilitated by solid experimental designs that include time courses (i.e., take samples at early time points to capture potential initiating events), dose–response (i.e., include a low dose that does not cause toxicity to help separate pharmacologically mediated changes in gene expression from toxicological ones), and careful choice of positive and negative controls (i.e., generating expression data from animals treated with compounds that cause inflammation but not vasculitis will facilitate the search for more specific biomarkers). Furthermore, bioinformatic approaches to link differentially expressed genes to altered metabolic and signaling pathways and well-designed and focused follow-up studies are critical to confirm new hypothesis that global gene expression approaches might generate.

Ultimately, the success of identifying biomarkers for vasculitis and understanding mechanisms of vasculitis will likely require application of multiple technologies, including genomics, proteomics, metabonomics, flow cytometry, imaging, etc. This creates the additional challenge of combining disparate data from these various approaches and integrating them to allow cross-platform querying and extraction of biological knowledge to gain a more holistic understanding of vasculitis.

17.4 MOVING FORWARD

In the context of mechanism-based research, it is probably best to regard results obtained using omics technologies as the springboard to more detailed and focused investigations that would confirm or refute the significance of the observed changes. Concern has been voiced regarding possible misinterpretation, or over-interpretation, of such high-volume data analyses, particularly in the context of safety assessment. It must be recognized that the interaction of any chemical with a biological system will without fail result in changes measurable by these sensitive techniques. It is therefore important that omics observations are followed through with traditional approaches and analyzed fully to establish if the measured changes are background noise, adaptive, beneficial, or potentially harmful. Where there are no physiological or pathological indicators of harmful effect it is clearly important to not over-interpret genomic, proteomics, or metabonomics data. These are points on which it is critically important to foster the development of consensus and common understanding among industry, academia, and regulatory bodies.

From a regulatory perspective, opportunities to integrate genomics sciences into clinical practice have been recognized by the U.S. Food and Drug Administration (FDA),[73] as has the role of these technologies in discovering and developing new molecular diagnostics for use in clinical monitoring and the need for engagement of the entire scientific community if this potential is going to be realized.[74] In regard to this latter point, in November 2003 the FDA released draft guidelines on pharmacogenomics data submission (http://fda.gov/cder/guidance/index.htm) and, following an open period for consideration of public comments, the final guidelines were released in 2005. The draft evolved through a very open consultation between industry and the FDA, most notably through the participation of trade groups and consortia such as Drusafe, PhRMA and ILSI/HESI. In their current form, the guidelines clearly recognize that discovery applications (where the technology is being used to streamline candidate selection in the pharmaceutical industry) are considered research applications and as such submission of this data is not required, with the exception of circumstances where "known" or "probable" valid biomarker signatures are flagged by a compound treatment. In the absence of these signature patterns, data are only required for submission with an investigational or new drug application if it is being used to support a safety argument (such as species relevance), clinical trial design (for example, patient stratification or the monitoring of a pharmacogenomic marker in a dose escalation), or in support of a labeling issue. Although the draft clarifies out-of-scope applications through worked examples, there remains some lack of clarity around biomarker validation and data submission and communication processes. These guidelines, when finalized, should help remove much of the ambiguity regarding the reportability of the data and allow decisions on the appropriate application of these technologies to be driven by sound science, public safety, and appropriate business drivers rather than an unfounded fear of regulatory repercussions.

Looking ahead, the application of omics to develop and apply diagnostic biomarkers of efficacy, responsiveness, and safety in individualized medicine may well be approaching,[75] but how close we are to practical application is debated by many.[76,77] Technical developments, such as the use of protein array chips, may be required to enable broader usage.[78] Beyond the practicality of using relatively expensive and technically specialized tools in clinical practice (or the physician's office), ethical and legal issues need to be considered. In particular, informed consent, the disclosure of genetic information, and financial compensation to those whose genetic information is used in the development of these tests need to be further resolved if maximal utility is to be achieved.[79] Ultimately, economic factors may limit broad usage of even robust diagnostic tools.[80]

In conclusion, maximal utility of these approaches is likely to require their aggressive integration into drug discovery and development processes and will rely on the further development of reference data and analytical tools. There are inevitably many "cultural" factors in the pharmaceutical industry with regard to application of evolving methodologies, particularly their relationship with well-established regulatory processes such as safety assessment. Ultimately the opportunities afforded by genomics sciences are unlikely to be limited by technologies themselves, but rather by their rate of application to pharmaceutical discovery and development portfolios.

REFERENCES

1. Frank, R.G. New estimates of drug development costs. *J. Health Econ.* 22(2), 325, 2003.
2. Halapi, E., Stefansson, K., and Hakonarson, H. Population genomics of drug response. *Am. J. Pharmacogenomics* 4, 73, 2004.
3. International Human Genome Sequencing Consortium. Finishing the euchromatic sequence of the human genome. *Nature* 431(7011), 931, 2004.
4. Ulrich, R. and Friend, S.H. Toxicogenomics and drug discovery: will new technologies help us produce better drugs? *Nat. Rev. Drug Discov.* 1(1), 84, 2004.
5. Pennie, W.D., Woodyatt, N.J., Aldridge, T.C., and Orphanides, G. Application of genomics to the definition of the molecular basis for toxicity. *Toxicol. Lett.* 120, 353, 2001.
6. Castle, A.L., Carver, M.P., and Mendrick, D.L. Toxicogenomics: a new revolution in drug safety. *Drug Discov. Today* 7, 728, 2002.
7. Goodsaid, F.M. Genomic biomarkers of toxicity. *Curr. Opin. Drug Discov. Dev.* 6(1), 41, 2003.

8. Burczynski, M.E., McMillian, M., Ciervo, J., Li, L., Parker, J.B., Dunn, R.T., II, Hicken, S., Farr, S., and Johnson, M.D. Toxicogenomics-based discrimination of toxic mechanism in HepG2 human hepatoma cells. *Toxicol. Sci.* 58, 399–415, 2000.
9. Thomas, R.S., Rank, D.R., Penn, S.G., Zastrow, G.M., Hayes, K.R., Pande, K., Glover, E., Silander, T., Craven, M.W., Reddy, J.K., Jovanovich, S.B., and Bradfield, C.A. Identification of toxicologically predictive gene sets using cDNA microarrays. *Mol. Pharmacol.* 60(6), 1189, 2001.
10. Waring, J.F., Jolly, R.A., Ciurlionis, R., Lum, P.Y., Praestgaard, J.T., Morfitt, D.C., Buratto, B., Roberts, C., Schadt, E., and Ulrich, R.G. Clustering of hepatotoxins based on mechanism of toxicity using gene expression profiles. *Toxicol. Appl. Pharmacol.* 175(1), 28, 2001.
11. Baker, T.K., Carfagna, M.A., Gao, H., Dow, E.R., Li, Q., Searfoss, G.H., and Ryan, T.P. Temporal gene expression analysis of monolayer cultured rat hepatocytes. *Chem. Res. Toxicol.* 14, 1218, 2001.
12. Bandara, L. and Kennedy, S. Toxicoprotomics — a new preclinical tool. *DDT* 7, 411, 2002.
13. Kennedy, S. The role of proteomics in toxicology: identification of biomarkers of toxicity by protein expression analysis. *Biomarkers* 7, 269, 2002.
14. Hale, J.E., Gelfanova, V., Ludwig, J.R., and Knierman, M.D. Application of proteomics for discovery of protein biomarkers. *Brief Funct. Genomic Proteomic* 2, 185, 2003.
15. Walgren, J.L. and Thompson, D.C. Application of proteomic technologies in the drug development process. *Toxicol. Lett.* 149(1–3), 377, 2004.
16. Kennedy, S. Proteomic profiling from human samples: the body fluid alternative. *Toxicol. Lett.* 120(1–3), 379, 2001.
17. Fountoulakis, M. and Suter, L. Proteomic analysis of the rat liver. *J. Chromatogr. B* 782, 197, 2002.
18. Chaurand, P., DaGue, B., Pearsall, R., Threadgill, D., and Caprioli, R. Profiling proteins from azoxymethane-induced colon tumors at the molecular level by matrix-assisted laser desorption/ionization mass spectrometry. *Proteomics* 1, 1320, 2001.
19. Petricoin, E., Rajapaske, V., Herman, E., Arekani, A., Ross, S., Johann, D., Knapton, A., Zhang, J., Hitt, B., Conrads, T., Veenstra, T., Liotta, L., and Sistare, F. Toxicoproteomics: serum proteomic pattern diagnostics for early detection of dug induced cardiac toxicities and cardioprotection. *Toxicol. Pathol.* 32(Suppl. 1), 122, 2004.
20. Meneses-Lorente, G., Guest, P., Lawrence, J., Muniappa, N., Knowles, M., Skynner, H., Salim, K., Cristea, I., Mortishire-Smith, R., Gaskell, S., and Watt, A. A proteomic investigation of drug-induced steatosis in rat liver. *Chem. Res. Toxicol.* 17, 605, 2004.
21. Gao, J., Garulacan, L., Storm, S., Hefta, S., Opiteck, G., Lin, J., Moulin, F., and Dambach, D. Identification of *in vitro* protein biomarkers of idiosyncratic liver toxicity. *Toxicol. Vitro* 18, 533, 2004.
22. Dare, T., Davies, H., Turton, J., Lomas, L., Williams, T., and York, M. Application of surface-enhanced laser desorption/ionization technology to the detection and identification of urinar parvalbumin: a biomarker of compound-induced skeletal muscle toxicity in the rat. *Electrophoresis* 23, 3241, 2002.
23. Jones, J., Kaphalia, L., Treinen-Moslen, M., and Leibler, D. Proteomic characterization of metabolites, protein adducts, and biliary proteins in rats exposed to 1,1-dichloroethylene or diclofenac. *Chem. Res. Toxicol.* 16, 1306, 2003.
24. Fountoulakis, M., Berndt, P., Boelsterli, U., Crameri, F., Winter, M., Albertini, S., and Suter, L. Two-dimensional database of mouse liver proteins: changes in hepatic protein levels following treatment with acetaminophen or its nontoxic regioisomer 3-acetamidophenol. *Electrophoresis* 21, 2148, 2000.

25. Da Cruz, S., Xenarios, I., Langridge, J., Vilbois, F., Parone, P., and Martinou, J. Proteomic analysis of the mouse liver mitochondrial inner membrane. *J. Biol. Chem.* 278, 41566, 2003.
26. Leonoudakis, D., Conti, L., Anderson, S., Radeke, C., McGuire, L., Adams, M., Froehner, S., Yates, J., and Vandenberg, C. Protein trafficking and anchoring complexes revealed by proteomic analysis of inward rectifier potassium channel (kir2.x)-associated proteins. *J. Biol. Chem.* 279, 22331, 2004.
27. Nisar, S., Lane, C., Wilderspin, A., Welham, K., Griffiths, W., and Patterson, L. A proteomic approach to the identification of cytochrome P450 isoforms in male and female rat liver by nanoscale liquid chromatography-electrospray ionization-tandem mass spectrometry. *Drug Metab. Disp.* 32, 382, 2004.
28. Zhu, H., Bilgin, M., and Snyder, M. Proteomics. *Annu. Rev. Biochem.* 72, 783, 2003.
29. Baggerly, K., Morris, J., and Coombes, K. Reproducibility of SELDI-TOF protein patterns in serum: comparing datasets from different experiments. *Bioinformatics* 20, 777, 2004.
30. Heijne, W., Stierum, R., Slijper, M., van Bladeren, P., and van Ommen, B. Toxicogenomics of bromobenzene hepatotoxicity: a combined transcriptomics and proteomics approach. *Biochem. Pharmacol.* 65, 857, 2003.
31. Ruepp, S., Tonge, R., Shaw, J., Wallis, N., and Pognam, F. Genomics and proteomics analysis of acetaminophen toxicity in mouse liver. *Toxicol. Sci.* 65, 135, 2002.
32. Omenn, G. The Human Proteome Organization Plasma Proteome Project pilot phase: reference specimens, technology platform comparisons, and standardized data submissions and analyses. *Proteomics* 4, 1235, 2004.
33. Nicholson, J.K.., Connelly, J. et al. (2002). Metabonomics: a platform for studying drug toxicity and gene function. *Nat. Rev. Drug Discov.* 1(2), 153, 2002.
34. Forster, J., Gombert, A.K. et al. A functional genomics approach using metabolomics and in silico pathway analysis. *Biotechnol. Bioeng.* 79, 703, 2002.
35. Coen, M., Ruepp, S.U. et al. Integrated application of transcriptomics and metabonomics yields new insight into the toxicity due to paracetamol in the mouse. *J. Pharm. Biomed. Anal.* 35(1), 93, 2004.
36. Robosky, L.C., Robertson, D.G. et al. *In vivo* toxicity screening programs using metabonomics. Comb. Chem. High Throughput Screen 5, 651, 2002.
37. Reo, N.V. NMR-based metabolomics. *Drug Chem. Toxicol.* 25(4), 375, 2002.
38. Lindon, J.C., Holmes, E. et al. Metabonomics technologies and their applications in physiological monitoring, drug safety assessment and disease diagnosis. *Biomarkers* 9, 1, 2004.
39. Bredel, M. and Jacoby, E. Chemogenomics: an emerging strategy for rapid target and drug discovery. *Nat. Rev. Genet.* 5(4), 262–275, 2004.
40. Engelberg, A. Iconix Pharmaceuticals, Inc. — removing barriers to efficient drug discovery through chemogenomics. *Pharmacogenomics* 5(6), 741, 2004.
41. Mills, J.S. and Showell, G.A. Exploitation of silicon medicinal chemistry in drug discovery. *Expert Opin. Invest. Drugs* 13(9), 1149, 2004.
42. Mestres, J. Computational chemogenomics approaches to systematic knowledge-based drug discovery. *Curr. Opin. Drug Discov. Dev.* 7(3), 304, 2004.
43. Takeda-Shitaka, M., Takaya, D., Chiba, C., Tanaka, H., and Umeyama, H. Protein structure prediction in structure based drug design. *Curr. Med. Chem.* 11(5), 551, 2004.
44. Parsons, L. and Orban, J. Structural genomics and the metabolome: combining computational and NMR methods to identify target ligands. *Curr. Opin. Drug Discov. Dev.* 7(1), 62, 2004.

45. Whittaker, P.A. What is the relevance of bioinformatics to pharmacology? *Trends Pharmacol. Sci.* 24(8), 434, 2003.
46. Bailey, W.J. and Ulrich, R. Molecular profiling approaches for identifying novel biomarkers. *Expert Opin. Drug Saf.* 3(2), 137, 2004.
47. Butcher, E.C., Berg, E.L., and Kunkel, E.J. Systems biology in drug discovery. *Nat. Biotechnol.* 22(10), 1253, 2004.
48. Tong, W., Harris, S., Cao, X., Fang, H., Shi, L., Sun, H., Fuscoe, J., Harris, A., Hong, H., Xie, Q., Perkins, R., and Casciano, D. Development of public toxicogenomics software for microarray data management and analysis. *Mutat. Res.* 549(1–2), 241, 2004.
49. Rocca-Serra, P., Brazma, A., Parkinson, H., Sarkans, U., Shojatalab, M., Contrino, S., Vilo, J., Abeygunawardena, N., Mukherjee, G., Holloway, E., Kapushesky, M., Kemmeren, P., Lara, G.G., Oezcimen, A., and Sansone, S.A. ArrayExpress: a public database of gene expression data at EBI. *C. R. Biol.* 326(10–11), 1075, 2003.
50. Pennie, W., Pettit, S.D., and Lord, P.G. Toxicogenomics in risk assessment: an overview of an HESI collaborative research program. *Environ. Health Perspect.* 112(4), 417, 2004.
51. Twigger, S.N., Nie, J., Ruotti, V., Yu, J., Chen, D., Li, D., Mathis, J., Narayanasamy, V., Gopinath, G.R., Pasko, D., Shimoyama, M., De La Cruz, N., Bromberg, S., Kwitek, A.E., Jacob, H.J., and Tonellato, P.J. Integrative genomics: in silico coupling of rat physiology and complex traits with mouse and human data. *Genome Res.* 14(4), 651, 2004.
52. Selvanayagam, Z.E., Cheung, T.H., Wei, N., Vittal, R., Kit Lo, K.W., Yeo, W., Kita, T., Ravatn, R., Hung Chung, T.K., Wong, Y.F., and Chin, K.V. Prediction of chemotherapeutic response in ovarian cancer with DNA microarray expression profiling. *Cancer Genet. Cytogenet.* 154(1), 63, 2004.
53. Kanehisa, M., Goto, S., Kawashima, S., Okuno, Y., and Hattori, M. The KEGG resource for deciphering the genome. *Nucleic Acids Res.* 32(Database issue:D277), 2004.
54. Ge, H., Walhout, A.J., and Vidal, M. Integrating "omic" information: a bridge between genomics and systems biology. *Trends Genet.* 19(10), 551, 2003.
55. Ozsoyoglu, Z.M., Nadeau, J.H., and Ozsoyoglu, G. Pathways database system. *OMICS* 7(1), 123, 2003.
56. Kinoshita, K. and Nakamura, H. Protein informatics towards function identification. *Curr. Opin. Struct. Biol.* 13(3), 396–400, 2003.
57. Wu, C.H., Huang, H., Yeh, L.S., and Barker, W.C. Protein family classification and functional annotation. *Comput. Biol. Chem.* 27(1), 37, 2003.
58. Zoon, K.C. Future directions in cancer research: impact of the completion of the human genome. *Toxicol. Path.* 32 Suppl 1:1–12, 2004.
59. Sausville, E.A. Optimizing target selection and development strategy in cancer treatment: the next wave. *Curr. Med. Chem. Anti-Cancer Agents* 4(5), 445, 2004.
60. Kelloff, G.J., Sigman, C.C., Johnson, K.M., Boone, C.W., Greenwald, P., Crowell, J.A., Hawk, E.T., and Doody, L.A. Perspectives on surrogate end points in the development of drugs that reduce the risk of cancer. *Cancer Epidemiol. Biomarkers Prev.* 9(2), 127, 2000.
61. Yu, J.K., Chen, Y.D., and Zheng, S. An integrated approach to the detection of colorectal cancer utilizing proteomics and bioinformatics. *World J. Gastroenterol.* 10(21), 3127, 2004.

62. Harris, R.A., Yang, A., Stein, R.C., Lucy, K., Brusten, L., Herath, A., Parekh, R., Waterfield, M.D., O'Hare, M.J., Neville, M.A., Page, M.J., and Zvelebil, M.J. Cluster analysis of an extensive human breast cancer cell line protein expression map database. *Proteomics* 2(2), 212, 2002.
63. Loni, L., De Braud, F., Zinzani, P.L., and Danesi, R. Pharmacogenetics and proteomics of anticancer drugs in non-Hodgkin's lymphoma. *Leuk Lymphoma* 44(Suppl. 3), 15, 2003.
64. Di Paolo, A., Danesi, R., and Del Tacca, M. Pharmacogenetics of neoplastic diseases: new trends. *Pharmacol. Res.* 49, 331–342, 2004.
65. Hathout, Y., Gehrmann, M.L., Chertov, A., and Fenselau, C. Proteomic phenotyping: metastatic and invasive breast cancer. *Cancer Lett.* 210(2), 245, 2004.
66. Celis, J.E., Gromova, I., Moreira, J.M., Cabezon, T., and Gromov, P. Impact of proteomics on bladder cancer research. *Pharmacogenomics* 5(4), 381, 2004.
67. Dowlati, A., Haaga, J., Remick, S.C., Spiro, T.P., Gerson, S.L., Liu, L., Berger, S.J., Berger, N.A., and Willson, J.K. Sequential tumor biopsies in early phase clinical trials of anticancer agents for pharmacodynamic evaluation. *Clin. Cancer Res.* 7(10), 2971–2976, 2001.
68. Lipkin, M., Bhandari, M., Hakissian, M., Croll, W., and Wong, G. Surrogate endpoint biomarker assays in phase II chemoprevention clinical trials. *J. Cell Biochem. Suppl.* 19, 47, 1994.
69. Alcorta, D., Preston, G., Munger, W., Sullivan, P., Yang, J.J., Waga, I., Jennette, J.C., and Falk, R. Microarray studies of gene expression in circulating leukocytes in kidney diseases. *Exp. Nephrol.* 10(2), 139, 2002.
70. Stagliano, N.E., Carpino, A.J., Ross, J.S., and Donovan, M. Vascular gene discovery using laser capture microdissection of human blood vessels and quantitative PCR. *Ann. N.Y. Acad. Sci.* 947, 344, 2001.
71. Robertson, D.G., Reily, M.D., Albassam, M., and Dethloff, L.A. Metabonomic assessment of vasculitis in rats. *Cardiovasc. Toxicol.* 1(1), 7, 2001.
72. Slim, R.M., Robertson, D.G., Albassam, M., Reily, M.D., Robosky, L., and Dethloff, L.A. Effect of dexamethasone on the metabonomics profile associated with phosphodiesterase inhibitor induced vascular lesions in rats. *Toxicol. Appl. Pharmacol.* 183(2), 108, 2002.
73. Lesko, L.J. and Woodcock, J. Translation of pharmacogenomics and pharmacogenetics: a regulatory perspective. *Nat. Rev. Drug Discov.* 3(9), 763, 2004.
74. Ardekani, A.M., Petricoin, E.F., III, and Hackett, J.L. Molecular diagnostics: an FDA perspective. *Expert Rev. Mol. Diagn.* 3(2), 129, 2003.
75. Evans, W.E. and Relling, M.V. Moving towards individualized medicine with pharmacogenomics. *Nature* 429(6990), 464, 2004.
76. Nebert, D.W. and Vesell, E.S. Advances in pharmacogenomics and individualized drug therapy: exciting challenges that lie ahead. *Eur. J. Pharmacol.* 500(1–3), 267, 2004.
77. Weinshilboum, R. and Wang, L. Pharmacogenomics: bench to bedside. *Nat. Rev. Drug Discov.* 3(9), 739, 2004.
78. Jain, K.K. Role of pharmacoproteomics in the development of personalized medicine. *Pharmacogenomics* 5(3), 331, 2004.
79. Bear, J.C. "What's my DNA worth, anyway?": a response to the commercialization of individuals' DNA information. *Perspect. Biol. Med.* 47(2), 273, 2004.
80. Phillips, K.A., Veenstra, D.L., Ramsey, S.D., Van Bebber, S.L., and Sakowski, J. Genetic testing and pharmacogenomics: issues for determining the impact to healthcare delivery and costs. *Am. J. Manag. Care* 10(7 Pt. 1), 425, 2004.

CHAPTER 18

Current and Future Aspects of Surrogate Tissue Analysis

Michael E. Burczynski

CONTENTS

18.1 INTRODUCTION

The chapters of this textbook have been assembled to provide an overview of some of the most exciting areas of recent research employing novel methodologies in the field of surrogate tissue analysis. Nonetheless, it is readily appreciated that the actual topic of this textbook is neither novel nor comprehensive in its scope: for instance, there are many examples of analytes in surrogate tissues that have already become accepted, if not validated, biomarkers of disease (prostate specific antigen, serum cholesterol, etc.), and the work presented in this book highlights a fraction of the expression analysis approaches currently being conducted to identify additional biomarkers in surrogate tissues. The recent "omic" explosion (certainly not yet fully erupted) has enabled exploration of surrogate tissues to an extent never before possible.[1] Undoubtedly, the coming years of scientific investigation will bring a large number of novel biomarkers in clinically accessible tissues to light, with the hope that these biomarkers will influence human health and biomedical knowledge to an extent never before imagined.

While transcriptional profiling has surged ahead of other massively parallel expression profiling platforms due to the amenable nature of Watson–Crick base pairing, it is exceedingly likely that it will only be a matter of time before proteomic and metabolomic platforms are equally global in nature. Already proteomic investigations can encompass several thousand proteins, and metabolomic technologies can load and analyze samples at a rate of about one every 2 minutes.

The important questions for surrogate tissue analysis and data mining in the near future may not necessarily relate to understanding exactly which downstream platforms will be developed for the detection of these analytes, but rather focus on a number of other issues, both technological and theoretical, which are discussed briefly in the sections below.

18.2 TRANSLATIONAL MEDICINE, BIOMARKERS, AND SURROGATE TISSUES

One of the most important applications of surrogate tissue profiling will be in the context of the translational medicine initiatives that have been undertaken in recent years within pharmaceutical companies. The value of biomarkers during clinical drug development is now well understood — they can serve as indicators of pharmacodynamic effect in "first in man" studies and help guide dose selection in subsequent clinical trials.[2,3] Additional types of novel biomarkers such as transcriptional signatures may ultimately identify efficacious (or toxic) therapeutic regimens or even predict patient responses.[4]

One of the most challenging aspects of translational medicine, as the name implies, is the absolute requirement to "translate" biomarker assays that indicate drug effect in preclinical models into biomarker assays that can also indicate drug effect in human beings. As mentioned in the preface, a biochemical phosphorylation event in the CA3 region of the hippocampus may be a perfect indicator of drug effect in a mouse model during lead compound optimization and preclinical drug development, but this certainly will not be a suitable assay for use in the clinic. For these types of preclinical biomarker assays, which work in target tissues that are not feasible in clinical settings, there is an obvious need for translational activities.

One of the recent successful paradigms in translational research, therefore, has been the screening of surrogate tissues for alternative indicators of drug effect. Biochemical events (transcription, translation, post-translational modification, or metabolic evidence of enzymatic/nonenzymatic activity) in surrogate tissues that can be used for the translational goal of indicating drug effect appear to fall into three main categories: (1) events occurring at the site of the target tissue that are subsequently released into (and detected within) the surrogate tissue; (2) events that occur in the surrogate tissue itself via a *direct effect* of the drug on the surrogate tissue; or (3) events that occur in the surrogate tissue *in response* to a direct effect of the drug on the target tissue. Examples of all three types were introduced in the first chapter and presented in detail in various chapters throughout this textbook and are briefly recapitulated below.

18.2.1 Biochemical Events in Target Tissues Subsequently Detected in Surrogate Tissues

These markers involve analytes that are formed and secreted, or lost, by the primary tissue of interest through a physiological, pathophysiological, toxicological, or pharmacological process. The chapter by Petricoin et al. (Chapter 7) demonstrates how a portion of the low-molecular-weight circulatory proteome may consist of aberrantly processed protein fragments produced in the tumor microenvironment and intimates that a suitably sensitive method may have diagnostic implications for early cancer detection and monitoring disease progression. Similar applications can be found for these types of biomarkers in other chapters. The chapter by Wong (Chapter 14) demonstrates how methylation profiling can examine methylation patterns of DNA "lost" from a primary tumor and detected in the circulation, while the chapter by Ghossein et al. (Chapter 13) demonstrates a series of examples how sensitive RT-PCR methodologies can be used to detect circulating tumor cells in blood. Sauter (Chapter 9) demonstrates how a noncirculatory surrogate tissue (nipple aspirate fluid) can be interrogated to determine the absence or presence of breast cancer disease.

Metabolomic-based interrogations of surrogate tissues also fall into this category. Several descriptions of metabolomic fingerprints in a variety of experimental settings, as mentioned in both the chapter by Griffin and Waters (Chapter 10) and the chapter by Ritchie (Chapter 11), indicate that metabolomes in surrogate tissues may be dynamically affected by metabolic events in target tissues. The chapter by Clish and Serhan (Chapter 12) reflects the same theme. In several of their studies these authors have even evaluated the relationship between transcript levels and metabolites of mechanistic relevance. The source of these types of biomarkers is easy to understand, since the markers are indicators that originate in an inaccessible primary tissue and are liberated (or are freely diffusible) and hence detectable in an accessible tissue.

18.2.2 Biochemical Events as a Result of Direct Drug/Toxicant Effects in Surrogate Tissues

A second type of marker is a biochemical event that occurs due to a direct effect of a drug or toxicant on the surrogate tissue itself. In the case of therapeutics, these types of markers provide an excellent opportunity to use allometric-scaling-type approaches to extrapolate the relationship between dose–response effects in a target tissue and simultaneous measurements of the same or similar pathway in a surrogate tissue. In the case of toxicants, these types of markers also provide an excellent opportunity to use similar approaches to estimate toxicant exposures. The chapter by Rockett (Chapter 5) demonstrates the overlap in gene expression effects of 17-beta estradiol in both the target (placenta) and surrogate (peripheral blood) tissues. Since 17-beta estradiol can have either therapeutic or toxic effects depending on the dose (and subject), the biomarkers discovered in these studies may actually be useful for indicating efficacy or toxicity. The chapter by Ostermeier and Krawetz (Chapter 6) demonstrates how spermatazoal RNA levels may provide good dosimeters of toxicant exposure in the testis, with implications for assessing toxicity in a specific organ.

18.2.3 Biochemical Events in Surrogate Tissues as Responses to Distal Effects in Target

A third type of marker encountered in surrogate tissues is one in which the surrogate tissue itself responds to the presence of disease or a physiological event or a pharmacological intervention. Several instances of these types of markers have been included in this text as well. With respect to transcriptional profiles in peripheral blood, it is apparent that the surrogate tissue likely represents at least a portion of "the target tissue" in inflammatory conditions like inflammatory bowel disease, psoriasis, rheumatoid arthritis, and lupus, where the pathophysiologically initiating event is due to aberrant responses in circulating cells of the immune system.[5] While inflammatory types of studies have not been reviewed in this textbook, non-inflammatory diseases may also give rise to relevant signatures in circulating PBMCs. The chapter by Tang et al. (Chapter 3) reviews transcriptional responses of peripheral blood mononuclear cells to neurologic diseases, while our own laboratory has begun to define transcriptional responses of PBMCs to the presence of solid tumors (Chapter 4). The chapter by Reddy et al. (Chapter 8) reviews data that suggest lymphocyte integrin expression is not only a response to a physiological event (embryo implantation) but in fact may actively modulate whether implantation will successfully occur. While still in relative infancy, analysis of surrogate tissues that reflect physiologic responses to events in distal tissues may provide a rich source of biomarkers in the future.

18.3 VARIABILITY, REFERENCE RANGES, AND REFERENCE STANDARDS IN SURROGATE TISSUE ANALYSIS

For any biomarker assay (irrespective of the number of analytes) it is ultimately necessary to understand the "reference range" associated with the level of the biomarker in the tissue of interest. Using transcriptional markers in peripheral blood as an example, Whitney et al. initially assessed variability of transcriptional markers in peripheral blood in a set of 75 disease-free subjects,[6] and identified transcripts that appeared variable and/or correlated with various parameters like cell composition or physical parameters of the blood samples.

A more comprehensive catalog of transcriptional variability in human peripheral blood will likely comprise a useful resource for biomedical researchers in many fields of inquiry. For drug developers this will inform researchers as to the likely suitability of transcriptional markers of drug effect. For instance, an *ex vivo* culture assay may suggest that a transcript in peripheral blood may be an excellent biomarker for evaluating drug effects *in vivo*. In this case an understanding of the candidate transcript's natural variability in disease-free subjects will be informative regarding its likely utility as a pharmacodynamic biomarker in a first-in-man study in healthy volunteers. Highly variable transcripts may be deemed unsuitable while normally stable transcripts may be viable candidates.

One of the most important issues facing the expression profiling of peripheral blood as a surrogate tissue (indeed, facing any type of massively parallel analysis of any type of surrogate tissue) will be whether a harmonization of sample processing method(s) will allow comparison of biomarker levels in surrogate tissue samples from different laboratories and ultimately an understanding of the true range of variability of these multitudinous analytes in given surrogate tissues. The same issues facing transcriptional profiling lie at the heart of proteomic and metabolomic-based interrogations of surrogate tissues as well.

In addition, reference standards are lacking for laboratories conducting microarray-based evaluations of surrogate tissues. Given the complexity (and fragility) of any cellular RNA source, it seems an almost insurmountable task to imagine the generation of a reference RNA sample that could be used to "qualify" a microarray processing center on the basis of achieving accurate determinations of the entire transcriptome in a reference sample. Interactions between stakeholders in industry, academia, and government — e.g., Food and Drug Administration (FDA), Environmental Protection Agency (EPA), and the National Institute of Standards and Technologies (NIST) — could help foster the development of RNA reference standards.

Our laboratory initially prepared large RNA samples from large volumes (500 ml) of leukophoresed blood, prepared several large aliquots of the homogeneous aqueous RNA mixture, aliquoted one of these into several hundred 2-μg aliquots in 10 μl of DEPC-treated water, and stored the remainder of the initial large aqueous aliquots as ethanol-precipitates. These 2-μg aliquot samples were then run as controls with every batch of peripheral blood samples processed on Affymetrix chips to serve as an external QC indicator of the overall gene chip process. These types of QC samples work well within a single laboratory over the course of several hundred experiments, but even this large reference is finite. In addition, informal stability assessments indicated that even in the absence of freeze–thaw cycles small amounts of RNA stored as aqueous samples at –80°C eventually become unsuitable for microarray analysis as evidenced by 3′ to 5′ ratios for beta-actin and GAPDH deteriorating in these aliquots over the course of more than a year.

One can imagine that future reference standards (if they are developed) will likely not be entire cellular transcriptomes, but rather multiplex standard mixtures of transcripts of defined quality that are present at known concentrations in the reference sample. Maintaining these reference standards over long periods will be a challenge. However, the type of analytical precision that will be afforded by these types of reference samples will go a long way toward improving the quality of microarray data as currently generated. Similar to the situation for a nonexistent "transcriptome reference standard," there are neither proteome nor metabolome reference standards. Similar strategies as those proposed above, in which a reference standard does not encompass the entire body of measurable analytes, but simply a cross section of analytes at different levels, may find suitability in these applications as well.

A quick and easy alternative to all of the above obstacles may be to leave whole transcriptome/proteome/metabolome profiling experiments as research-grade endeavors that only assess relative differences between samples within an experiment. This strategy then leaves the burden of assay validity and quantitative certainty on alternative lower throughput methods (quantitative multiplexed RT-PCR assays, multiplexed immunoassays, high-resolution mass spectrometry) that can be applied to the subset of analytes of interest discovered by exploratory omic analyses.

18.4 SURROGATE TISSUE PROFILING WILL ULTIMATELY FOSTER BASIC DISCOVERIES IN BIOLOGICAL RESEARCH

One of the most exciting aspects of surrogate tissue profiling in the years to come will be the elucidation of novel molecular entities that may be involved in disease processes, therapeutic activities, or toxic effects. Concomitant with that knowledge will come the burden of understanding how the observations "fit together." An example is the novel finding that transcriptional signatures in PBMCs of patients with solid tumor appear different from transcriptional signatures in PBMCs of healthy subjects.[7] While a portion of this signature appears to be due to differential cell compositions in healthy vs. diseased PBMCs, this parameter does not explain all observed variability. It is very likely that some of the distinct differences between transcript levels in PBMCs of healthy individuals and patients with solid tumors are due to altered transcriptional responses of circulating PBMCs to the presence of these tumors. Additional experimental data generated in clinical studies conducted with whole-blood stabilization methodologies and analysis of transcriptional differences in isolated cell types will shed much needed light on these initial compelling findings.

It is highly likely that at least a portion of the transcriptional differences in PBMCs of patients with RCC (relative to healthy controls) reflect the differential transcriptional response of circulating PBMCs to the presence of RCC tumors. Little statistical evidence was uncovered indicating that PBMC profiles were dependent on the type of renal tumor, but this could have been due to a lack of statistical power since the majority of patients in this study possessed clear cell carcinomas. Understanding why (and by what mechanism) peripheral blood responds to the presence of renal and other tumors may provide insight into understanding how tumors ultimately evade immune system surveillance once proliferation outpaces cell death, and tumor progression and ultimately metastasis occur. Other insights are to be gleaned from exploring the "why" involved with the apparent responses of peripheral blood to the presence of neurological disease states, or lymphocyte integrin expression during implantation, or any other number of intriguing observations that are beginning to come to light in this age of postgenomic analysis of surrogate tissues. Valuable insights will likely be gained by detailed analysis of any disease setting, pharmacological treatment, or toxicant exposure where biomarkers measured in surrogate tissues are found to constitute responses of the surrogate tissue to occurrences in distal targets.

While the discovery of novel diagnostics and prognostic indicators may very well be an important outcome of current research efforts in the area of surrogate tissue analysis, it is also clear that the new omic technologies listed within the pages of this textbook (and those waiting to be developed and employed) will generate exciting new hypotheses and sustain biomedical research in multiple fields for many years to come.

REFERENCES

1. Rockett, J.C., Burczynski, M.E., Fornace, A.J., Jr., Hermann, P.C., Krawetz, S.A., and Dix, D.J. Surrogate tissue analysis: monitoring toxicant exposure and health status of inaccessible tissues through the analysis of accessible tissues and cells. *Tox. Appl. Pharmacol.*, 194, 189–199, 2004.
2. Swanson, B.N. Delivery of high-quality biomarker assays. *Dis. Markers,* 18, 47–56, 2002.
3. Park, J.W., Kerbel, R.S., Kelloff, G.J., Barrett, J.C., Chabner, B.A., Parkinson, D.R., Peck, J., Ruddon, R.W., Sigman, C.C., and Slamon, D.J. Rationale for biomarkers and surrogate endpoints in mechanism-driven oncology drug development. *Clin. Cancer Res.,* 10, 3885–3896, 2004.
4. Burczynski, M.E., Oestreicher, J.L., Cahilly, M.J., Mounts, D.P., Whitley, M.Z., Speicher, L.A., and Trepicchio, W.L. Clinical pharmacogenomics and transcriptional profiling in early phase oncology clinical trials. *Curr. Mol. Med.*, 5, 83–102, 2005.
5. Maas, K., Chan, S., Parker, J., Slater, A., Moore, J., Olsen, N., and Aune, T.M. Cutting edge: molecular portrait of human autoimmune disease. *J. Immunol.*, 169, 5–9, 2002.
6. Whitney, A.R., Diehn, M., Popper, S.J., Alizadeh, A.A., Boldrick, J.C., Relman, D.A., and Brown, P.O. Individuality and variation in gene expression patterns in human blood. *Proc. Natl. Acad. Sci. U.S.A.*, 100, 1896–1901, 2003.
7. Twine, N.C, Stover, J.A., Marshall, B., Dukart, G., Hidalgo, M., Stadler, W., Logan, T., Dutcher, J., Hudes, G., Dorner, A.J., Slonim, D.K., Trepicchio, W.L., and Burczynski, M.E. Disease-associated expression profiles in peripheral blood mononuclear cells from patients with advanced renal cell carcinoma. *Cancer Res.,* 63, 6069–6075, 2003.

Index

A

B

C

D

E

F

G

H

M

N

O

P

Q

R

S

T

U

V

W